Set Theory: A First Course

Set theory is a rich and beautiful subject whose fundamental concepts permeate virtually every branch of mathematics. One could say that set theory is a unifying theory for mathematics, since nearly all mathematical concepts and results can be formalized within set theory.

This textbook is meant for an upper undergraduate course in set theory. In this text, the fundamentals of abstract sets, including relations, functions, the natural numbers, order, cardinality, transfinite recursion, the axiom of choice, ordinal numbers, and cardinal numbers, are developed within the framework of axiomatic set theory.

The reader will need to be comfortable reading and writing mathematical proofs. The proofs in this textbook are rigorous, clear, and complete, while remaining accessible to undergraduates who are new to upper-level mathematics. Exercises are included at the end of each section in a chapter, with useful suggestions for the more challenging exercises.

Daniel W. Cunningham is a Professor of Mathematics at SUNY Buffalo State, specializing in set theory and mathematical logic. He is a member of the Association for Symbolic Logic, the American Mathematical Society, and the Mathematical Association of America. Cunningham's previous work includes *A Logical Introduction to Proof*, which was published in 2013.

CAMBRIDGE MATHEMATICAL TEXTBOOKS

Cambridge Mathematical Textbooks is a program of undergraduate and beginning graduate level textbooks for core courses, new courses, and interdisciplinary courses in pure and applied mathematics. These texts provide motivation with plenty of exercises of varying difficulty, interesting examples, modern applications, and unique approaches to the material.

Set Theory

A FIRST COURSE

DANIEL W. CUNNINGHAM
SUNY Buffalo State, USA

CAMBRIDGE
UNIVERSITY PRESS

CAMBRIDGE
UNIVERSITY PRESS

One Liberty Plaza, 20th Floor, New York, NY 10006, USA

Cambridge University Press is part of the University of Cambridge.

It furthers the University's mission by disseminating knowledge in the pursuit of education, learning, and research at the highest international levels of excellence.

www.cambridge.org
Information on this title: www.cambridge.org/9781107120327

© Daniel W. Cunningham 2016

First published 2016

Printed in the United States of America by Sheridan Books, Inc.

A catalog record for this publication is available from the British Library.

Library of Congress Cataloging in Publication Data
Names: Cunningham, Daniel W.
Title: Set theory : a first course / Daniel W. Cunningham, Buffalo State, NY, USA.
Description: New York NY : Cambridge University Press, 2016. | Series: Cambridge mathematical textbooks | Includes bibliographical references and index.
Identifiers: LCCN 2015047021 | ISBN 9781107120327 (hardback : alk. paper)
Subjects: LCSH: Set theory – Textbooks.
Classification: LCC QA248 .C86 2016 | DDC 511.3/22–dc23
LC record available at http://lccn.loc.gov/2015047021

ISBN 978-1-107-12032-7 Hardback

CONTENTS

PREFACE

Set theory is a rich and beautiful subject whose fundamental concepts permeate virtually every branch of mathematics. Yet, most mathematics students receive only a cursory overview of the theory of sets in their lower division courses. Set theory is a subject that is sufficiently important and interesting to merit its own undergraduate course. The book is intended to offer the reader, or department, just such a course.

This introductory textbook delivers a one-semester course in undergraduate set theory at the upper-division level. Sets, of course, are the central objects that shall be investigated. Since essentially every significant concept in modern mathematics can be defined in terms of sets, students who possess a fairly deep understanding of set theory will find a first course in abstract algebra or real analysis much less formidable and, perhaps, much more accessible. My first undergraduate course in set theory, for example, gave me a definite advantage over other students in my ensuing mathematics courses.

In this book, the fundamental facts about abstract sets–relations, functions, natural numbers, order, cardinality, transfinite recursion, the axiom of choice, ordinal numbers, and cardinal numbers–are covered and developed within the framework of axiomatic set theory. Mathematicians have shown that virtually all mathematical concepts and results can be formalized within set theory. This has been recognized as one of the greatest achievements of modern mathematics, and, consequently, one can now say that "set theory is a unifying theory for mathematics."

The textbook is suitable for a broad range of readers, from undergraduate to graduate students, who desire a better understanding of the fundamental topics in set theory that may have been, or will be, overlooked in their other mathematics courses. I have made an effort to write clear and complete proofs throughout the text.

Many modern books in undergraduate set theory are written for a reader who is well versed in mathematical argument and proof. My primary goal was to produce a book that would be accessible to a relatively unsophisticated reader. Nevertheless, I have composed completely rigorous proofs. In addition, these proofs favor detail over brevity. Another goal was to write a book that is focused on those topics that a student is likely to see in advanced courses (including graduate courses). Thus, the book is comparatively concise and can be covered in one semester. Most other undergraduate set theory texts cannot possibly be covered in a semester.

On the Origins of Set Theory

In the nineteenth century, mathematicians were investigating problems that concerned functions in real analysis. Some of these problems involved the representation of a function by a trigonometric series and the convergence of such series. The young Georg Cantor then proved a uniqueness theorem for trigonometric series. Cantor, in his proof, introduced a process for constructing collections of real numbers that involved an infinite iteration of the limit operation. Cantor's novel proof led him to a deeper investigation of sets of real numbers and then to the concept of transfinite numbers.

It has been said that set theory, as a subject in mathematics, was created on the day that Georg Cantor proved that the set of real numbers is uncountable, that is, there is no one-to-one correspondence between the real numbers and the natural numbers. In the two decades following this proof, Cantor developed a fascinating theory of ordinal and cardinal numbers.

Cantor's revolutionary research raised many difficult issues and paradoxes. As the axiomatic method had assumed an important role in mathematics, Ernst Zermelo developed an axiomatic system for set theory and published the first axiomatization of set theory in 1908. Zermelo's axioms resolved the difficulties introduced by Cantor's development of set theory. After receiving some proposed revisions from Abraham Fraenkel, in 1930 Zermelo presented his final axiomatization of set theory, now known as the Zermelo–Fraenkel axioms.

Topics Covered

The book presents the axioms of Zermelo–Fraenkel set theory and then uses these axioms to derive a variety of theorems concerning functions, relations, ordering, the natural numbers, and other core topics in mathematics. These axioms shall also be used to investigate infinite sets and to generalize the concepts of induction and recursion.

The Zermelo–Fraenkel axioms are now generally accepted as the standard foundation for set theory and for mathematics. On the other hand, there are textbooks that introduce set theory using a naive point of view. As naive set theory[1] is known to be inconsistent, such a theory cannot seriously be offered as a foundation for mathematics or set theory.

The basics of logic and elementary set theory are first discussed in Chapter 1. Since students, typically, easily grasp the topics covered in this chapter, it can be covered at a fairly quick pace. The chapter ends with a brief overview of the Zermelo–Fraenkel axioms.

Chapter 2 examines the first six axioms of Zermelo–Fraenkel set theory and begins proving theorems, justified by the axioms, about sets. It is also shown that the set operations discussed in Chapter 1 can be derived from these six axioms.

A relation is defined as a set of ordered pairs in Chapter 3, where equivalence relations and induced partitions are also discussed. Then a precise set-theoretic definition of a function in terms of ordered pairs is presented. The chapter ends with a section on order relations and a section on the concepts of congruence and preorder.

In Chapter 4, after representing the natural numbers as sets, the fundamental principles of number theory (for example, proof by mathematical induction) are derived from a few very basic notions involving sets.

Cantor's early work on the "size of infinite sets" is the subject of Chapter 5. In particular, we explore the notion of a one-to-one correspondence between two sets and the concept of a countable set.

Well-ordered sets and transfinite recursion are examined in Chapter 6. In Chapter 7, it is shown that the axiom of choice implies Zorn's Lemma, the Ultrafilter Theorem, and the Well-Ordering Theorem. The theory of ordinals is carefully developed in Chapter 8, and the theory of cardinals is presented in Chapter 9. The last section of Chapter 9 discusses closed unbounded sets and stationary sets.

How to Use the Book

It is strongly recommended that the reader be acquainted with the basics of sets, functions, relations, logic, and mathematical induction. These topics are typically introduced in a "techniques of proof" course (for example, see [1]), and they will be more seriously discussed in this book. As the emphasis will be on theorems and their proofs, the reader should be comfortable reading and writing mathematical proofs. For example, one should know how to prove that a function is one-to-one and to prove that one set is a subset of another set.

In the book, naturally, there is a progressive increase in complexity. The first five chapters cover the important topics that every mathematics undergraduate should know about the theory of sets, including the Recursion Theorem and the Schröder–Bernstein Theorem. The last four chapters present more challenging material, beginning in Chapter 6 with the principle of transfinite recursion. When applying definition by transfinite recursion there are two functional forms that may be used: a set function or a class function. In both of these cases, we prove that one can construct a new function by recursion. The two proofs are similar, except that the class form requires the replacement axiom.

Moreover, the class version implies the set version. Students will likely find these technical proofs relatively difficult to follow or appreciate; however, they should be assured that such recursive definitions are valid and that one can read the rest of the book without possessing a deep understanding of these proofs.

If time is short, certain topics can be overlooked. For example, the following sections can be skipped without loss of continuity:

- 3.5 Congruence and Preorder
- 4.2.1 The Peano Postulates
- 7.2 Filters and Ultrafilters
- 8.4 The Cumulative Hierarchy
- 9.3 Closed Unbounded and Stationary Sets.

Furthermore, the proofs presented in Chapter 7 of Zorn's Lemma 7.1.1 and of the Well-Ordering Theorem 7.3.1 do not appeal to transfinite recursion on the ordinals; however, such proofs are carefully outlined in Exercises 14 and 15, respectively, beginning on page 193 of Chapter 8. Therefore, in Chapter 7, one could just highlight the proofs of 7.1.1 and 7.3.1 and then, after completing Section 8.2, assign these shorter "ordinal recursion" proofs as exercises.

The symbol Ⓢ marks the end of a solution, and the symbol □ identifies the end of a proof. Exercises are given at the end of each section in a chapter. An exercise marked with an asterisk ∗ is one that is cited, or referenced, elsewhere in the book. Suggestions are also provided for those exercises that a newcomer to upper-division mathematics may find more challenging.

Acknowledgments

This book began as a set of notes for an undergraduate course in set theory that I taught and designed at SUNY Buffalo State. I want to thank Michael Filipski, Anthony Laffrado, Hongmei Lin, and Joshua Terhaar for taking this set theory course and for their helpful suggestions. The anonymous reviewers offered many significant recommendations that greatly improved the book, and I thank each of them. David Anderson deserves to be recognized and thanked for his meticulous copy-editing. I am especially grateful to Kaitlin Leach, editor at Cambridge University Press, for her guidance and her enthusiasm. Thank you, Marianne Foley, for your support and our discussions on English composition. Finally, I must thank Springer Science and Business Media for granting me copyright permission to use in this book some of the language, examples, and figures that I composed and created in Chapters 1–2 and 5–7 of [1].

THE GREEK ALPHABET

A	α		alpha
B	β		beta
Γ	γ		gamma
Δ	δ		delta
E	ϵ	ε	epsilon
Z	ζ		zeta
H	η		eta
Θ	θ	ϑ	theta
I	ι		iota
K	κ		kappa
Λ	λ		lambda
M	μ		mu
N	ν		nu
O	o		omicron
Ξ	ξ		xi
Π	π		pi
P	ρ		rho
Σ	σ		sigma
T	τ		tau
Υ	υ		upsilon
Φ	ϕ	φ	phi
X	χ		chi
Ψ	ψ		psi
Ω	ω		omega

1 Introduction

Georg Cantor devoted much of his mathematical career to the development of a new branch of mathematics, namely, Set Theory. Little did he know that his pioneering work would eventually lead to a unifying theory for mathematics. In his earlier work, Cantor took a set of real numbers P and then formed the *derived set* P' of all limit points of P. After iterating this operation, Cantor obtained further derived sets P'', P''', ... These derived sets enabled him to prove an important theorem on trigonometric series. This work led Cantor to investigate sets in a more general setting and to develop an abstract theory of sets that would dramatically change the course of mathematics.

The basic concepts in set theory are now applied in virtually every branch of mathematics. Furthermore, set theory serves as the basis for the definition and explanation of the most fundamental mathematical concepts: functions, relations, algebraic structures, function spaces, etc. Thus, it is often said that set theory provides a foundation for mathematics.

1.1 Elementary Set Theory

The set concept is one that pervades all of mathematics. We shall not attempt to give a precise definition of a set; however, we will give an informal description of a set and identify some important properties of sets.

A set is a collection of objects. The objects in such a collection are called the *elements* of the set. We write $a \in A$ to assert that a is an *element*, or a *member*, of the set A. We write $a \notin A$ when a is *not* an element of the set A. A set is merely the result of collecting objects of interest, and it is usually identified by enclosing its elements with braces (curly brackets). For example, the collection $A = \{3, 7, 11, \pi\}$ is a set that contains the four elements $3, 7, 11, \pi$. So $7 \in A$, and $8 \notin A$.

Sets are exceedingly important in mathematics; in fact, most mathematical objects (e.g., numbers, functions) can be defined in terms of sets. When one first learns about sets, it appears that one can naively define a set to be *any* collection of objects. In Section 1.4, we will see that such a naive approach can create serious problems.

Certain sets routinely appear in mathematics. In particular, the sets of natural numbers, integers, rational, and real numbers are regularly discussed. These sets are usually denoted by:

1. $\mathbb{N} = \{0, 1, 2, 3, \ldots\}$ is the set of natural numbers.
2. $\mathbb{Z} = \{\ldots, -3, -2, -1, 0, 1, 2, 3, \ldots\}$ is the set of integers.
3. $\mathbb{Q}$ is the set of rational numbers. Thus, $\frac{3}{2} \in \mathbb{Q}$.
4. $\mathbb{R}$ is the set of real numbers, and so $\pi \in \mathbb{R}$.

Basic Definitions of Set Theory

In this section, we discuss the basic notation and concepts that are used in set theory. An object x may or may not belong to a given set A; that is, either $x \in A$ or $x \notin A$, but not both.

Definition 1.1.1. The following set terminology is used extensively throughout mathematics:

1. Let A and B be sets. We write $A = B$ when both sets have exactly the same elements.
2. For sets A and B we write $A \subseteq B$ to mean that the set A is a **subset** of the set B, that is, every element of A is also an element of B.
3. We say that the set A is a **proper** subset of the set B, denoted by $A \subset B$, when $A \subseteq B$ and $A \neq B$, that is, when every element of A is an element of B but there is at least one element in B that is not in A.
4. We write $\varnothing$ for the *empty set*, or the *null set*. The set $\varnothing$ has no elements.
5. Two sets A and B are **disjoint** if they have no elements in common.

It follows that $A \subseteq A$ and $\varnothing \subseteq A$, for any set A. To see why $\varnothing \subseteq A$, suppose that $\varnothing \nsubseteq A$. Then there exists an $x \in \varnothing$ such that $x \notin A$. As there is no x such that $x \in \varnothing$, we arrive at a contradiction. Therefore, we must have that $\varnothing \subseteq A$.

A *Venn diagram* is a configuration of geometric shapes, which is commonly used to depict a particular relationship that holds between two or more sets. In Figure 1.1(a), we present a Venn diagram that illustrates the subset relation. Figure 1.1(b) portrays two sets that are disjoint.

A *property* is a statement that asserts something about one or more variables (for more detail, see Section 1.3). For example, the two statements "x is a real number" and "$y \in \mathbb{R}$ and $y \notin \mathbb{N}$" are clearly properties that assert something, respectively, about x and y. One way to construct a subset is called the method

(a) Venn diagram of $A \subseteq B$ (b) Disjoint sets A and B

Figure 1.1. Two set relationships

of *separation*. Let A be a set. Given a property $P(x)$ about the variable x, one can construct the set of objects $x \in A$ that satisfy the property $P(x)$; namely, we can form the *truth set* $\{x \in A : P(x)\}$. Thus, we can separate the elements in A that satisfy the property from those elements that do not satisfy the property.

Problem 1. Evaluate each of the truth sets:

1. $A = \{x \in \mathbb{N} : 3 < x < 11\}$.
2. $B = \{y \in \mathbb{Z} : y^2 = 4\}$.
3. $C = \{z \in \mathbb{N} : z \text{ is a multiple of } 3\}$.

Solution. $A = \{4, 5, 6, 7, 8, 9, 10\}$, $B = \{2, -2\}$, and $C = \{0, 3, 6, 9, \dots\}$.
ⓢ

An *interval* is a set consisting of all the real numbers that lie between two given real numbers a and b, where $a < b$:

1. The open interval (a, b) is defined to be $(a, b) = \{x \in \mathbb{R} : a < x < b\}$.
2. The closed interval $[a, b]$ is defined to be $[a, b] = \{x \in \mathbb{R} : a \leq x \leq b\}$.
3. The left-closed interval $[a, b)$ is defined to be $[a, b) = \{x \in \mathbb{R} : a \leq x < b\}$.
4. The right-closed interval $(a, b]$ is defined to be $(a, b] = \{x \in \mathbb{R} : a < x \leq b\}$.

For each real number a, we can also define intervals called rays or half-lines:

1. The interval (a, ∞) is defined to be $(a, \infty) = \{x \in \mathbb{R} : a < x\}$.
2. The interval $[a, \infty)$ is defined to be $[a, \infty) = \{x \in \mathbb{R} : a \leq x\}$.
3. The interval $(-\infty, a)$ is defined to be $(-\infty, a) = \{x \in \mathbb{R} : x < a\}$.
4. The interval $(-\infty, a]$ is defined to be $(-\infty, a] = \{x \in \mathbb{R} : x \leq a\}$.

The symbol ∞ denotes "infinity" and this symbol does not represent a number. The notation ∞ is often used to represent an interval "without a right endpoint." Similarly, the mathematical notation $-\infty$ is used to denote an interval "having no left endpoint."

Definition 1.1.2. Let A be a set. The **power set** of A, denoted by $\mathcal{P}(A)$, is the set whose elements are all of the subsets of A. That is, $\mathcal{P}(A) = \{X : X \subseteq A\}$.

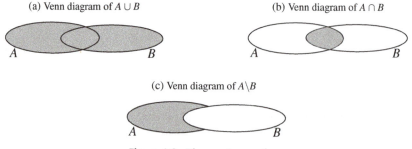

(a) Venn diagram of $A \cup B$

(b) Venn diagram of $A \cap B$

(c) Venn diagram of $A \backslash B$

Figure 1.2. Three set operations

Thus, $X \in \mathcal{P}(A)$ if and only if $X \subseteq A$. If A is a finite set with n elements, then one can show that the set $\mathcal{P}(A)$ has 2^n elements. The set $A = \{1, 2, 3\}$ has three elements, and so $\mathcal{P}(A)$ has eight elements, namely,

$$\mathcal{P}(A) = \{\varnothing, \{1\}, \{2\}, \{3\}, \{1, 2\}, \{1, 3\}, \{2, 3\}, \{1, 2, 3\}\}.$$

Set Operations

For a pair of sets A and B, there are three fundamental operations that we can perform on these sets. The *union* operation unites, into one set, the elements that belong either to A or to B. The *intersection* operation forms the set of elements that belong both to A and to B. The *difference* between A and B (in that order) is the set of all elements that are in A and not in B.

Definition 1.1.3. Given sets A and B, we can build new sets using the following **set operations**:

(a) $A \cup B = \{x : x \in A \text{ or } x \in B\}$ is the **union** of A and B.[1]
(b) $A \cap B = \{x : x \in A \text{ and } x \in B\}$ is the **intersection** of A and B.
(c) $A \setminus B = \{x : x \in A \text{ and } x \notin B\}$ is the **set difference** of A and B (also stated in English as A "minus" B).

The set operations in Definition 1.1.3 are illustrated in Figures 1.2(a), 1.2(b), and 1.2(c). Shading is used to identify the result of a particular set operation. For example, in Figure 1.2(c) the shaded area represents the set $A \setminus B$.

When the elements of sets A and B are clearly presented, then one can easily evaluate the operations of union, intersection, and difference.

Example 2. Let $A = \{1, 2, 3, 4, 5, 6\}$ and $B = \{2, 4, 6, 8, 10, 12\}$. Then

- $A \cup B = \{1, 2, 3, 4, 5, 6, 8, 10, 12\}$ and $A \cap B = \{2, 4, 6\}$.

- $A \setminus B = \{1, 3, 5\}$ and $B \setminus A = \{8, 10, 12\}$.
- $(A \setminus B) \cup (B \setminus A) = \{1, 3, 5, 8, 10, 12\}$.
- $(A \setminus B) \cap (B \setminus A) = \varnothing$.

Exercises 1.1

Let A, B, and C be sets.

1. If $a \notin A \setminus B$ and $a \in A$, show that $a \in B$.

2. Show that if $A \subseteq B$, then $C \setminus B \subseteq C \setminus A$.

3. Suppose $A \setminus B \subseteq C$. Show that $A \setminus C \subseteq B$.

4. Suppose $A \subseteq B$ and $A \subseteq C$. Show that $A \subseteq B \cap C$.

5. Suppose $A \subseteq B$ and $B \cap C = \varnothing$. Show that $A \subseteq B \setminus C$.

6. Show that $A \setminus (B \setminus C) \subseteq (A \setminus B) \cup C$.

7. Show that if $A \setminus B \subseteq C$ and $A \nsubseteq C$, then $A \cap B \neq \varnothing$.

8. Let $P(x)$ be the property $x > \frac{1}{x}$. Are the assertions $P(2)$, $P(-2)$, $P(\frac{1}{2})$, $P(-\frac{1}{2})$ true or false?

9. Show that each of the following sets can be expressed as an interval:
   (a) $(-3, 2) \cap (1, 3)$.
   (b) $(-3, 4) \cup (0, \infty)$.
   (c) $(-3, 2) \setminus [1, 3)$.

10. Express the following sets as truth sets:
   (a) $A = \{1, 4, 9, 16, 25, \dots\}$.
   (b) $B = \{\dots, -15, -10, -5, 0, 5, 10, 15, \dots\}$.

11. Show that each of the following sets can be expressed as an interval:
   (a) $\{x \in \mathbb{R} : x^2 - 1 \leq 3\}$.
   (b) $\{x \in \mathbb{R} : x > 0 \text{ and } (x - 1)^2 < 1\}$.

12. Evaluate the truth sets:
   (a) $A = \{x \in \mathbb{N} : 0 < x^2 < 24\}$.
   (b) $B = \{y \in \mathbb{Z} : y \text{ evenly divides } 12\}$.
   (c) $C = \{z \in \mathbb{N} : 4 \text{ evenly divides } z\}$.
   (d) $D = \{y \in \mathbb{R}^- : 1 \leq y^2 \leq 4\}$.

Exercise Notes: For Exercise 6, $x \notin B \setminus C$ means that $x \notin B$ or $x \in C$.

1.2 Logical Notation

Before introducing the fundamentals of set theory, it will be useful to identify some relevant aspects of language and logic. The importance of correct logical notation to set theory, and to mathematics, cannot be overstated. Formal logical notation has one important advantage: statements can be expressed much more concisely and much more precisely. Set theory often expresses many of its important concepts using logical notation. With this in mind, we now discuss the basics of logic.

Propositions and Logical Connectives

A *proposition* is a declarative sentence that is either true or false, but not both. When discussing the logic of propositional statements, we shall use symbols to represent these statements. Capital letters, for instance, P, Q, R, are used to symbolize propositional statements and are called *propositional components*. Using the five logical connectives $\wedge, \vee, \neg, \rightarrow, \leftrightarrow$ together with the components, we can form new logical sentences called *compound sentences*. For example,

1. $P \wedge Q$ (means "P and Q" and is called *conjunction*).
2. $P \vee Q$ (means "P or Q" and is called *disjunction*).
3. $\neg P$ (means "not P" and is called *negation*).
4. $P \rightarrow Q$ (means "if P, then Q" and is called a *conditional*).
5. $P \leftrightarrow Q$ (means "P if and only if Q" and is called a *biconditional*).

Using propositional components as building blocks and the logical connectives as mortar, we can construct more complex compound sentences, for example, $(P \wedge \neg Q) \vee (\neg S \rightarrow R)$. Parentheses ensure that our compound sentences will be clear and readable; however, we shall be using the following conventions:

1. The *outermost* parentheses need not be explicitly written; that is, one can write $A \wedge B$ to denote $(A \wedge B)$.
2. The negation symbol shall apply to as little as possible. We can therefore use $\neg A \wedge B$ to denote $(\neg A) \wedge B$.

Truth Tables

The truth value of a compound sentence in propositional logic can be evaluated from the truth values of its components. The logical connectives $\wedge, \vee, \neg, \rightarrow$, and $\leftrightarrow$ yield the natural truth values given in Table 1.1, where T means "true" and F means "false."

Table 1.1. Basic truth tables

P	Q	$P \wedge Q$
T	T	T
T	F	F
F	T	F
F	F	F

(1) Conjunction

P	Q	$P \vee Q$
T	T	T
T	F	T
F	T	T
F	F	F

(2) Disjunction

P	$\neg P$
T	F
F	T

(3) Negation

P	Q	$P \rightarrow Q$
T	T	T
T	F	F
F	T	T
F	F	T

(4) Conditional

P	Q	$P \leftrightarrow Q$
T	T	T
T	F	F
F	T	F
F	F	T

(5) Biconditional

Table 1.1(1) has four rows (not including the header). The columns beneath P and Q list all the possible pairs of truth values that can be assigned to the components P and Q. For each such pair, the corresponding truth value for $P \wedge Q$ appears to the right. For example, consider the third pair of truth values in this table, F T. Thus, if the propositional components P and Q are assigned the respective truth values F and T, we see that the truth value of $P \wedge Q$ is F.

Table 1.1(2) shows that if P and Q are assigned the respective truth values T and F, then the truth value of $P \vee Q$ is T. Moreover, when P and Q are assigned the truth values T and T, then the truth value of $P \vee Q$ is also T. In mathematics, the connective "or" has the same meaning as "and/or"; that is, $P \vee Q$ is true if and only if either P is true *or* Q is true, or both P *and* Q are true. Table 1.1(3) shows that the negation of a statement reverses the truth value of the statement.

Table 1.1(4) states that when P and Q are assigned the respective truth values T and F, then the truth value of $P \rightarrow Q$ is F; otherwise, it is T. In particular, when P is false, we shall say that $P \rightarrow Q$ is *vacuously* true. Table 1.1(5) shows that $P \leftrightarrow Q$ is true when P and Q are assigned the same truth value; when P and Q have different truth values, then the biconditional is false.

Using the truth tables for the sentences $P \wedge Q$, $P \vee Q$, $\neg P$, $P \rightarrow Q$, and $P \leftrightarrow Q$, we will now discuss how to build truth tables for more complicated compound sentences. Given a compound sentence, we identify the "outside" connective to be the "last connective that one needs to evaluate." Once the outside connective has been identified, one can break up the sentence into its "parts." For example, in the compound sentence $\neg P \vee (Q \wedge P)$ we see that $\vee$ is the outside connective with two parts $\neg P$ and $Q \wedge P$.

Problem 1. Construct a truth table for the sentence $\neg P \rightarrow (Q \wedge P)$.

Solution. The components P and Q will each need a column in our truth table. Since there are two components, there are four possible combinations of truth values for P and Q. We will enter these combinations in the two left most columns in the same order as that in Table 1.1(1). The outside connective of the propositional sentence $\neg P \rightarrow (Q \wedge P)$ is $\rightarrow$. We can break this sentence into the two parts $\neg P$ and $Q \wedge P$. So these parts will also need a column in our truth table. As we can break the sentences $\neg P$ and $Q \wedge P$ only into components (namely, P and Q), we obtain the following truth table:

P	Q	$\neg P$	$Q \wedge P$	$\neg P \rightarrow (Q \wedge P)$
T	T	F	T	T
T	F	F	F	T
F	T	T	F	F
F	F	T	F	F
STEP # 1	1	2	3	4

We will describe in steps how one obtains the truth values in the above table. STEP 1: Specify all of the truth values that can be assigned to the components. STEP 2: In each row, use the truth value assigned to the component P to obtain the corresponding truth value for $\neg P$, using Table 1.1(3). STEP 3: In each row, use the truth values assigned to Q and P to determine the corresponding truth value in the column under $Q \wedge P$ via Table 1.1(1). STEP 4: In each row, use the truth values in the columns under $\neg P$ and $Q \wedge P$ to evaluate the matching truth value for the final column under the sentence $\neg P \rightarrow (Q \wedge P)$, employing Table 1.1(4). Ⓢ

Tautologies and Contradictions

After constructing a truth table for a compound sentence, suppose that every entry in the final column is true. The sentence is thus true no matter what truth values are assigned to its components. Such a sentence is called a tautology.

Definition 1.2.1. A compound sentence is a **tautology** when its truth value is true regardless of the truth values of its components.

So a compound sentence is a tautology if it is always true. One can clearly see from the following truth table that the sentence $P \vee \neg P$ is a tautology:

P	$\neg P$	$P \vee \neg P$
T	F	T
F	T	T

Definition 1.2.2. A compound sentence is a **contradiction** when its truth value is false regardless of the truth values of its components.

Therefore, a compound sentence is a contradiction if it is always false. One can easily show that the sentence $P \wedge \neg P$ is a contradiction.

Logical Equivalence

A *propositional sentence* is either a compound sentence or just a component. The next definition describes when two propositional sentences are logically equivalent, that is, when they mean the same thing. Mathematicians frequently take advantage of logical equivalence to simplify their proofs, and we shall do the same in this book. In this section, we will use Greek letters (e.g., α, β, φ, and ψ; see page xiii) to represent propositional sentences.

Definition 1.2.3. Let ψ and φ be propositional sentences. We will say that ψ and φ are **logically equivalent**, denoted by $\psi \Leftrightarrow \varphi$, whenever the following holds: For every truth assignment applied to the components of ψ and φ, the resulting truth values of ψ and φ are identical.

Problem 2. Show that $\neg(P \vee Q) \Leftrightarrow \neg P \wedge \neg Q$.

Solution. After constructing truth tables for the two statements $\neg(P \vee Q)$ and $\neg P \wedge \neg Q$, we obtain the following:

P	Q	$P \vee Q$	$\neg(P \vee Q)$
T	T	T	F
T	F	T	F
F	T	T	F
F	F	F	T

P	Q	$\neg P$	$\neg Q$	$\neg P \wedge \neg Q$
T	T	F	F	F
T	F	F	T	F
F	T	T	F	F
F	F	T	T	T

As the final columns in the truth tables for $\neg(P \vee Q)$ and $\neg P \wedge \neg Q$ are identical, we can conclude from Definition 1.2.3 that they are logically equivalent.

Ⓢ

Whenever φ and ψ are logically equivalent, we shall say that $\psi \Leftrightarrow \varphi$ is a *logic law*. We will now present two important logic laws that are often used in mathematical proofs. These laws were first identified by Augustus De Morgan.

De Morgan's Laws (DML)

1. $\neg(P \vee Q) \Leftrightarrow \neg P \wedge \neg Q$.
2. $\neg(P \wedge Q) \Leftrightarrow \neg P \vee \neg Q$.

Let ψ and φ be propositional sentences. If one can apply a truth assignment to the components of ψ and φ such that the resulting truth values of ψ and φ disagree, then ψ and φ are *not logically equivalent*. We will use this fact in our next problem, which shows that the placement of parentheses in a compound sentence is very important. Note: A regrouping can change the meaning of the sentence.

Problem 3. Show that sentences $P \vee (Q \wedge \neg P)$ and $(P \vee Q) \wedge \neg P$ are not logically equivalent.

Solution. We shall use the truth table

P	Q	$P \vee (Q \wedge \neg P)$	$(P \vee Q) \wedge \neg P$
T	T	T	F
T	F	T	F
F	T	T	T
F	F	F	F

Since their final columns are not identical, we conclude that the propositional sentences $P \vee (Q \wedge \neg P)$ and $(P \vee Q) \wedge \neg P$ are not equivalent.   Ⓢ

Propositional Logic Laws

If a propositional component appears in a logic law and each occurrence of this component is replaced with a specific propositional sentence, then the result is also a logic law. Thus, in the above De Morgan's Law

$$\neg(P \vee Q) \Leftrightarrow \neg P \wedge \neg Q,$$

if we replace P and Q, respectively, with propositional sentences ψ and φ, then we obtain the logic law

$$\neg(\psi \vee \varphi) \Leftrightarrow \neg \psi \wedge \neg \varphi,$$

which is also referred to as De Morgan's Law.

Listed below are some important laws of logic, where ψ, φ, and χ represent any propositional sentences. These particular logic laws are frequently applied in mathematical proofs. They will also allow us to derive theorems concerning certain set operations.

De Morgan's Laws (DML)

1. $\neg(\psi \lor \varphi) \Leftrightarrow \neg\psi \land \neg\varphi$.
2. $\neg(\psi \land \varphi) \Leftrightarrow \neg\psi \lor \neg\varphi$.

Commutative Laws

1. $\psi \land \varphi \Leftrightarrow \varphi \land \psi$.
2. $\psi \lor \varphi \Leftrightarrow \varphi \lor \psi$.

Associative Laws

1. $\psi \lor (\varphi \lor \chi) \Leftrightarrow (\psi \lor \varphi) \lor \chi$.
2. $\psi \land (\varphi \land \chi) \Leftrightarrow (\psi \land \varphi) \land \chi$.

Idempotent Laws

1. $\psi \land \psi \Leftrightarrow \psi$.
2. $\psi \lor \psi \Leftrightarrow \psi$.

Distributive Laws

1. $\psi \land (\varphi \lor \chi) \Leftrightarrow (\psi \land \varphi) \lor (\psi \land \chi)$.
2. $\psi \lor (\varphi \land \chi) \Leftrightarrow (\psi \lor \varphi) \land (\psi \lor \chi)$.
3. $(\varphi \lor \chi) \land \psi \Leftrightarrow (\varphi \land \psi) \lor (\chi \land \psi)$.
4. $(\varphi \land \chi) \lor \psi \Leftrightarrow (\varphi \lor \psi) \land (\chi \lor \psi)$.

Double Negation Law (DNL)

1. $\neg\neg\psi \Leftrightarrow \psi$.

Tautology Law

1. $\psi \land$ (a tautology) $\Leftrightarrow \psi$.

Contradiction Law

1. $\psi \lor$ (a contradiction) $\Leftrightarrow \psi$.

Conditional Laws (CL)

1. $(\psi \to \varphi) \Leftrightarrow (\neg\psi \lor \varphi)$.
2. $(\psi \to \varphi) \Leftrightarrow \neg(\psi \land \neg\varphi)$.

Contrapositive Law

1. $(\psi \to \varphi) \Leftrightarrow (\neg\varphi \to \neg\psi)$.

Biconditional Law

1. $(\psi \leftrightarrow \varphi) \Leftrightarrow (\psi \to \varphi) \land (\varphi \to \psi)$.

The Tautology Law and Contradiction Law can be easily illustrated. Observe that $\varphi \vee \neg\varphi$ is a tautology. From the Tautology Law we obtain the following logical equivalence: $\psi \wedge (\varphi \vee \neg\varphi) \Leftrightarrow \psi$. On the other hand, because $\varphi \wedge \neg\varphi$ is a contradiction, it follows that $\psi \vee (\varphi \wedge \neg\varphi) \Leftrightarrow \psi$ by the Contradiction Law.

Let α and β be two propositional sentences that are logically equivalent. Now, suppose that α appears in a given propositional sentence Θ. If we replace occurrences of α in Θ with β, then the resulting new sentence will be logically equivalent to Θ. To illustrate this substitution principle, suppose that we have the propositional sentence $\neg Q \vee \alpha$ and we also know that $\alpha \Leftrightarrow \beta$. Then we can conclude that $(\neg Q \vee \alpha) \Leftrightarrow (\neg Q \vee \beta)$. Now, using this substitution principle and the propositional logic laws, we will establish a new logic law without the use of truth tables.

Problem 4. Show that $(P \rightarrow R) \wedge (Q \rightarrow R) \Leftrightarrow (P \vee Q) \rightarrow R$, using logic laws.

Solution. We first start with the more complicated side $(P \rightarrow R) \wedge (Q \rightarrow R)$ and derive the simpler side as follows:

$$(P \rightarrow R) \wedge (Q \rightarrow R) \Leftrightarrow (\neg P \vee R) \wedge (\neg Q \vee R) \quad \text{by Conditional Law (1)}$$
$$\Leftrightarrow (\neg P \wedge \neg Q) \vee R \quad \text{by Distribution Law (4)}$$
$$\Leftrightarrow \neg(P \vee Q) \vee R \quad \text{by De Morgan's Law (1)}$$
$$\Leftrightarrow (P \vee Q) \rightarrow R \quad \text{by Conditional Law (1).}$$

Therefore, $(P \rightarrow R) \wedge (Q \rightarrow R) \Leftrightarrow (P \vee Q) \rightarrow R$. ⓢ

Using a list of propositional components, say $A, B, C, D, \ldots$, and the logical connectives $\wedge, \vee, \neg, \rightarrow, \leftrightarrow$, we can form a variety of propositional sentences. For example,

$$(P \rightarrow R) \wedge \neg(Q \leftrightarrow (S \vee T)).$$

The logical connectives are also used to tie together a variety of mathematical statements. A good understanding of these connectives and propositional logic will allow us to more easily understand and define set-theoretic concepts. The following problem and solution illustrate this observation.

Problem 5. Let A and B be any two sets. Show that $x \notin A \setminus B$ is equivalent to the statement $x \notin A$ or $x \in B$.

Solution. We shall show that $x \notin A \setminus B \Leftrightarrow (x \notin A \vee x \in B)$ as follows:

$x \notin A \setminus B \Leftrightarrow \neg x \in A \setminus B$ by definition of $\notin$

      $\Leftrightarrow \neg(x \in A \wedge x \notin B)$ by definition of $\setminus$

      $\Leftrightarrow x \notin A \vee x \in B$ by De Morgan's Law and Double Negation.

Therefore, $x \notin A \setminus B$ is equivalent to the assertion $(x \notin A \vee x \in B)$.   Ⓢ

Exercises 1.2

1. Using truth tables, show that $\neg(P \rightarrow Q) \Leftrightarrow (P \wedge \neg Q)$.

2. Construct truth tables to show that $(P \leftrightarrow Q) \Leftrightarrow (P \rightarrow Q) \wedge (Q \rightarrow P)$.

3. Using truth tables, show that $P \Leftrightarrow (\neg P \rightarrow (Q \wedge \neg Q))$.

4. Using truth tables, show that $(P \vee Q) \wedge R \Leftrightarrow (P \wedge R) \vee (P \wedge Q)$.

5. Show that $(P \rightarrow Q) \wedge (P \rightarrow R) \Leftrightarrow P \rightarrow (Q \wedge R)$, using logic laws.

6. Show that $(P \rightarrow R) \vee (Q \rightarrow R) \Leftrightarrow (P \wedge Q) \rightarrow R$, using logic laws.

7. Using propositional logic laws, show that $P \rightarrow (Q \rightarrow R) \Leftrightarrow (P \wedge Q) \rightarrow R$.

8. Show that $(P \rightarrow Q) \rightarrow R$ and $P \rightarrow (Q \rightarrow R)$ are not logically equivalent.

1.3   Predicates and Quantifiers

Variables, for instance, x and y, are used throughout mathematics to represent unspecified values. They are employed when we are interested in "properties" that may be true or false, depending on the values represented by the variables. A *predicate* is simply a statement that proclaims that certain variables satisfy a property. For example, "x is a number" is a predicate, and we can symbolize this predicate by $N(x)$. Of course, the truth or falsity of the expression $N(x)$ can be determined only when a value for x is given. For example, the expression $N(4)$, which means "4 is a number," is clearly true.

When our attention is to be focused on just the elements in a particular set, we shall then refer to that set as our **universe of discourse**. For example, if we were just talking about real numbers, then our universe of discourse would be the set of real numbers $\mathbb{R}$. Furthermore, every statement made in a specific universe of discourse applies to just the elements in that universe.

Given a statement $P(x)$, which says something about the variable x, we often want to assert that *every* element x in the universe of discourse satisfies $P(x)$. Moreover, there will be times when we want to express the fact that *at least one* element x in the universe makes $P(x)$ true. We will thus form sentences using the quantifiers $\forall$ and $\exists$. The quantifier $\forall$ means "for all" and is called the *universal quantifier*. The quantifier $\exists$ means "there exists," and it is identified as the *existential quantifier*. For example, we can form the sentences

1. $\forall x P(x)$ [means "for all x, $P(x)$"].
2. $\exists x P(x)$ [means "there exists an x such that $P(x)$"].

Any statement of the form $\forall x P(x)$ is called a *universal statement*. A statement having the form $\exists x P(x)$ is called an *existential statement*. Quantifiers offer us a valuable tool for clear thinking in mathematics, where many concepts begin with the expression "for every" or "there exists." Of course, the truth or falsity of a quantified statement depends on the universe of discourse.

Suppose that a variable x appears in an assertion $P(x)$. In the two statements $\forall x P(x)$ and $\exists x P(x)$, we say that x is a *bound variable* because x is *bound* by a quantifier. In other words, when a variable in a statement is immediately used by a quantifier, then that variable is referred to as being a bound variable. If a variable in a statement is not bound by a quantifier, then we shall say that the variable is a *free variable*. When a variable is free, then substitution may take place, that is, one can replace a free variable with any particular value from the universe of discourse–perhaps 1 or $\emptyset$. For example, the assertion $\forall x(P(x) \rightarrow x = y)$ has the one free variable y. Therefore, we can perform a substitution to obtain $\forall x(P(x) \rightarrow x = 2)$. In a given context, if all of the free variables in a statement are replaced with values, then one can determine the truth or falsity of the resulting statement.

There are times in mathematics when one is required to prove that there is exactly one value that satisfies a property. There is another quantifier that is sometimes used, though not very often. It is called the *uniqueness quantifier*. This quantifier is written as $\exists! x P(x)$, and it means that "there exists a unique x satisfying $P(x)$." This is in contrast with $\exists x P(x)$, which simply means that "at least one x satisfies $P(x)$."

As already noted, the quantifier $\exists!$ is rarely used. One reason for this is that the assertion $\exists! x P(x)$ can be expressed in terms of the other quantifiers $\exists$ and $\forall$. In particular, the statement $\exists! x P(x)$ is equivalent to

$$\exists x P(x) \wedge \forall x \forall y([P(x) \wedge P(y)] \rightarrow x = y).$$

The above statement is equivalent to $\exists!xP(x)$ because it means that "there is an x such that $P(x)$ holds, and any individuals x and y that satisfy $P(x)$ and $P(y)$ must be the same individual."

In addition to the quantifiers $\forall$ and $\exists$, *bounded set quantifiers* are often used when one wants to restrict a quantifier to a specific set of values. For example, to state that every real number x satisfies a property $P(x)$, we can simply write $(\forall x \in \mathbb{R})P(x)$. Similarly, to say that some real number x satisfies $P(x)$, we can write $(\exists x \in \mathbb{R})P(x)$.

Definition 1.3.1. (Bounded Set Quantifiers) For each set A, we shall write $(\forall x \in A)P(x)$ to mean that *for every x in A, $P(x)$ is true*. Similarly, we will write $(\exists x \in A)P(x)$ to signify that *for some x in A, $P(x)$ is true*.

The assertion $(\forall x \in A)P(x)$ means that for every x, if $x \in A$, then $P(x)$ is true. Similarly, the statement $(\exists x \in A)P(x)$ means that there is an x such that $x \in A$ and $P(x)$ is true. Thus, we have the logical equivalences:

1. $(\forall x \in A)P(x) \Leftrightarrow \forall x(x \in A \rightarrow P(x))$.
2. $(\exists x \in A)P(x) \Leftrightarrow \exists x(x \in A \land P(x))$.

Quantifier Negation Laws (QNL)

We now introduce logic laws that involve the negation of a quantified assertion. Let $P(x)$ be any predicate. The statement $\forall xP(x)$ means that "for every x, $P(x)$ is true." Thus, the assertion $\neg\forall xP(x)$ means that "it is not the case that every x makes $P(x)$ true." Therefore, $\neg\forall xP(x)$ means there is an x that does not make $P(x)$ true, which can be expressed as $\exists x\neg P(x)$. This reasoning is reversible as we will now show. The assertion $\exists x\neg P(x)$ means that "there is an x that makes $P(x)$ false." Hence, $P(x)$ is not true for every x; that is, $\neg\forall xP(x)$. Therefore, $\neg\forall xP(x)$ and $\exists x\neg P(x)$ are logically equivalent. Similar reasoning will show that $\neg\exists xP(x)$ and $\forall x\neg P(x)$ are also equivalent. We now formally state these important logic laws that connect quantifiers with negation.

Quantifier Negation Laws 1.3.2. For any predicate $P(x)$, we have the logical equivalences:

1. $\neg\forall xP(x) \Leftrightarrow \exists x\neg P(x)$.
2. $\neg\exists xP(x) \Leftrightarrow \forall x\neg P(x)$.

The above reasoning used to justify the quantifier negation laws can also be used to verify two negation laws for bounded set quantifiers. Thus, given a set A

and predicate $P(x)$, the following two logic laws show us how statements of the form $(\forall x \in A)P(x)$ and $(\exists x \in A)P(x)$ interact with negation. Notice that when you push the negation symbol through a bounded set quantifier, the quantifier changes and the negation symbol passes over "$x \in A$."

Bounded Quantifier Negation Laws 1.3.3. For every predicate $P(x)$, we have the logical equivalences:

1. $\neg(\forall x \in A)P(x) \Leftrightarrow (\exists x \in A)\neg P(x)$.
2. $\neg(\exists x \in A)P(x) \Leftrightarrow (\forall x \in A)\neg P(x)$.

Quantifier Interchange Laws (QIL)

Adjacent quantifiers have the form $\exists x \exists y$, $\forall x \forall y$, $\forall x \exists y$, and $\exists x \forall y$. In this section, we will see how to interpret statements that contain adjacent quantifiers. When a statement contains adjacent quantifiers, one should address the quantifiers, one at a time, in the order in which they are presented.

Problem 1. Let the universe of discourse be a group of people and let $L(x, y)$ mean "x likes y." What do the following formulas mean?

1. $\exists x \exists y L(x, y)$.
2. $\exists y \exists x L(x, y)$.

Solution. Note that "x likes y" also means that "y is liked by x." We will now translate each of these formulas from "left to right" as follows:

1. $\exists x \exists y L(x, y)$ means "there is a person x such that $\exists y L(x, y)$," that is, "there is a person x who likes some person y." Therefore, $\exists x \exists y L(x, y)$ means that "someone likes someone."
2. $\exists y \exists x L(x, y)$ states that "there is a person y such that $\exists x L(x, y)$," that is, "there is a person y who is liked by some person x." Thus, $\exists y \exists x L(x, y)$ means that "someone is liked by someone."

Hence, the statements $\exists x \exists y L(x, y)$ and $\exists y \exists x L(x, y)$ mean the same thing. Ⓢ

Problem 2. Let the universe be a group of people and $L(x, y)$ mean "x likes y." What do the following formulas mean in English?

1. $\forall x \forall y L(x, y)$.
2. $\forall y \forall x L(x, y)$.

Solution. We will work again from "left to right" as follows:

1. $\forall x \forall y L(x, y)$ means "for every person x, we have that $\forall y L(x, y)$," that is, "for every person x, we have that x likes every person y." Hence, $\forall x \forall y L(x, y)$ means that "everyone likes everyone."

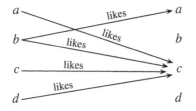

Figure 1.3. A world where $\exists y \forall x L(x, y)$ is true, since someone is liked by everyone.

2. $\forall y \forall x L(x, y)$ proclaims that "for each person y, we have that $\forall x L(x, y)$," that is, "for each person y, we have that y is liked by every person x." Thus, $\forall y \forall x L(x, y)$ means "everyone is liked by everyone."

So the statements $\forall x \forall y L(x, y)$ and $\forall y \forall x L(x, y)$ mean the same thing.   Ⓢ

Adjacent quantifiers of a different type are referred to as **mixed quantifiers**.

Problem 3. Let the universe be a group of people and $L(x, y)$ mean "x likes y." What do the following mixed quantifier formulas mean in English?

1. $\forall x \exists y L(x, y)$.
2. $\exists y \forall x L(x, y)$.

Solution. We will translate the formulas as follows:

1. $\forall x \exists y L(x, y)$ asserts that "for every person x we have that $\exists y L(x, y)$," that is, "for every person x there is a person y such that x likes y." Thus, $\forall x \exists y L(x, y)$ means that "everyone likes someone."
2. $\exists y \forall x L(x, y)$ states that "there is a person y such that $\forall x L(x, y)$," that is, "there is a person y who is liked by every person x." In other words, $\exists y \forall x L(x, y)$ means "someone is liked by everyone."

We conclude that the mixed quantifier statements $\forall x \exists y L(x, y)$ and $\exists y \forall x L(x, y)$ are not logically equivalent, that is, they do **not** mean the same thing.   Ⓢ

To clarify the conclusion obtained in our solution of Problem 3, consider the universe $U = \{a, b, c, d\}$ consisting of just four individuals with names as given. For this universe, Figure 1.3 identifies a world where $\exists y \forall x L(x, y)$ is true, where we portray the property $L(x, y)$ using the "arrow notation" $x \xrightarrow{\text{likes}} y$. Figure 1.3 illustrates a world where there is an individual who is very popular because everyone likes this person; that is, "someone is liked by everyone."

Figure 1.4 presents a slightly different world in which $\forall x \exists y L(x, y)$ is true. So, in this new world, "everyone likes someone."

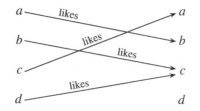

Figure 1.4. A world where $\forall x\exists y L(x, y)$ is true, because everyone likes someone.

Let us focus our attention on Figure 1.4. Clearly, the statement $\forall x\exists y L(x, y)$ is true in the world depicted in this figure. Moreover, notice that $\exists y\forall x L(x, y)$ is actually false in this world. Thus, $\forall x\exists y L(x, y)$ is true and $\exists y\forall x L(x, y)$ is false in the world presented in Figure 1.4. We can now conclude that $\forall x\exists y L(x, y)$ and $\exists y\forall x L(x, y)$ do not mean the same thing.

Our solution to Problem 1 shows that $\exists x\exists y L(x, y)$ and $\exists y\exists x L(x, y)$ both mean "someone likes someone." This supports the true logical equivalence:

$$\exists x\exists y L(x, y) \Leftrightarrow \exists y\exists x L(x, y).$$

Similarly, Problem 2 confirms the true logical equivalence:

$$\forall x\forall y L(x, y) \Leftrightarrow \forall y\forall x L(x, y).$$

Therefore, interchanging adjacent quantifiers of the same kind does not change the meaning. Problem 3, however, verifies that the two statements $\forall y\exists x L(x, y)$ and $\exists x\forall y L(x, y)$ are not logically equivalent. We conclude this discussion with a summary of the above observations:

- *Adjacent quantifiers of the same type are interchangeable.*
- *Adjacent quantifiers of a different type may not be interchangeable.*

We offer another example, involving the real numbers, which shows that the interchange of mixed quantifiers can change the meaning of a statement.

Example 4. Let the universe of discourse be $\mathbb{R}$, the set of real numbers.

1. $\forall x\exists y(x + y = 0)$ means that for every real number x there is a real number y such that $x + y = 0$. We see that the sentence $\forall x\exists y(x + y = 0)$ is true.
2. $\exists y\forall x(x + y = 0)$ states there is a $y \in \mathbb{R}$ such that $\forall x(x + y = 0)$. This is false.

Quantifier Interchange Laws 1.3.4. For every predicate $P(x, y)$, the following three statements are valid:

1. $\exists x\exists y P(x, y) \Leftrightarrow \exists y\exists x P(x, y).$
2. $\forall x\forall y P(x, y) \Leftrightarrow \forall y\forall x P(x, y).$
3. $\exists x\forall y P(x, y) \Rightarrow \forall y\exists x P(x, y).$

We will be using the arrow $\Rightarrow$ as an abbreviation for the word "implies." The conditional connective $\rightarrow$ shall be reserved for formal logical formulas. It should be noted that the implication in item 3 cannot, in general, be reversed.

The quantifier interchange laws also hold for bounded set quantifiers; for example, we have that

$$\exists x(\exists y \in A)P(x, y) \Leftrightarrow (\exists y \in A)\exists x P(x, y),$$

$$\exists x(\forall y \in A)P(x, y) \Rightarrow (\forall y \in A)\exists x P(x, y).$$

Quantifier Distribution Laws (QDL)

A quantifier can sometimes "distribute" over a particular logical connective. The quantifier distribution laws, given below, capture relationships that hold between a quantifier and the two logical connectives $\vee$ and $\wedge$. In particular, the existential quantifier distributes over disjunction (see 1.3.5(1)), and the universal quantifier distributes over conjunction (see 1.3.6(1)). The following quantifier distribution laws can be useful when proving certain set identities.

Existential Quantifier Distribution Laws 1.3.5. For any predicates $P(x)$ and $Q(x)$ we have the following distribution laws:

1. $\exists x P(x) \vee \exists x Q(x) \Leftrightarrow \exists x(P(x) \vee Q(x))$.
2. $(\exists x \in A)P(x) \vee (\exists x \in A)Q(x) \Leftrightarrow (\exists x \in A)(P(x) \vee Q(x))$.
3. $(\exists x \in A)P(x) \vee (\exists x \in B)P(x) \Leftrightarrow (\exists x \in A \cup B)P(x)$.
4. $(\exists x \in A \cap B)P(x) \Rightarrow (\exists x \in A)P(x) \wedge (\exists x \in B)P(x)$.

If R is a statement that does not involve the variable x, then we have:

5. $R \wedge \exists x Q(x) \Leftrightarrow \exists x(R \wedge Q(x))$.
6. $R \wedge (\exists x \in A)Q(x) \Leftrightarrow (\exists x \in A)(R \wedge Q(x))$.

Universal Quantifier Distribution Laws 1.3.6. For any predicates $P(x)$ and $Q(x)$ we have the following equivalences:

1. $\forall x P(x) \wedge \forall x Q(x) \Leftrightarrow \forall x(P(x) \wedge Q(x))$.
2. $(\forall x \in A)P(x) \wedge (\forall x \in A)Q(x) \Leftrightarrow (\forall x \in A)(P(x) \wedge Q(x))$.
3. $(\forall x \in A)P(x) \wedge (\forall x \in B)P(x) \Leftrightarrow (\forall x \in A \cup B)P(x)$.

If R is a statement that does not involve the variable x, then we have:

4. $R \vee \forall x Q(x) \Leftrightarrow \forall x(R \vee Q(x))$.
5. $R \vee (\forall x \in A)Q(x) \Leftrightarrow (\forall x \in A)(R \vee Q(x))$.

1.4 A Formal Language for Set Theory

Cantor employed an informal approach in his development of set theory. For example, Cantor regularly used the **Comprehension Principle**: *The collection of all objects that share a property forms a set.* Thus, given a property $P(x)$, the comprehension principle asserts that the collection $\{x : P(x)\}$ is a set. Using this principle, one can construct the *intersection* of two sets A and B via the property "$x \in A$ and $x \in B$"; namely, the intersection of A and B is the set $\{x : x \in A \text{ and } x \in B\}$. Similarly, we can form the *union* of A and B to be the set $\{x : x \in A \text{ or } x \in B\}$. In addition, we obtain the *power set* of A, denoted by $\mathcal{P}(A)$, which is the set whose elements are all of the subsets of A; that is, $\mathcal{P}(A) = \{X : X \subseteq A\}$. The comprehension principle allowed Cantor to establish the existence of many important sets. Today Cantor's approach to set theory is referred to as *naive set theory*.

Cantor's set theory soon became an indispensable tool for the development of new mathematics. For example, using fundamental set theoretic concepts, the mathematicians Émile Borel, René-Louis Baire, and Henri Lebesgue in the early 1900s created modern measure theory and function theory. The work of these mathematicians (and others) demonstrated the great mathematical utility of set theory.

Relying on Cantor's naive set theory, mathematicians discovered and proved many significant theorems. Then a devastating contradiction was announced by Bertrand Russell. This contradiction is now called *Russell's paradox*. Consider the property $x \notin x$, where x is understood to represent a set. The comprehension principle would allow us to conclude that $A = \{x : x \notin x\}$ is a set. Therefore,

the set A consists of all the sets x that satisfy $x \notin x$.   (▲)

Clearly, either $A \in A$ or $A \notin A$. Suppose $A \in A$. Then, as noted in (▲), A must satisfy the property $A \notin A$, which is a contradiction. Suppose $A \notin A$. Since A satisfies $A \notin A$, we infer from (▲) that $A \in A$, which is also a contradiction.

Russell's paradox thus threatened the very foundations of mathematics and set theory. If one can deduce a contradiction from the comprehension principle, then one can derive anything; in particular, one can prove that $1 = 2$. Cantor's set theory is therefore inconsistent, and the validity of the very important work of Borel and Lebesgue then became questionable. It soon became clear that the comprehension principle needed to be restricted in some way and the following question needed to be addressed: *How can one correctly construct a set?*

Ernst Zermelo resolved the problems discovered with the comprehension principle by producing a collection of axioms for set theory. Shortly afterward, Abraham Fraenkel amended Zermelo's axioms to obtain the Zermelo–Fraenkel axioms that have now become the accepted formulation of Cantor's ideas about

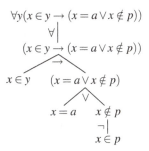

Figure 1.5. Construction of the formula $\forall y(x \in y \rightarrow (x = a \vee x \notin p))$

the nature of sets. In particular, these axioms will allow us to construct a power set and to form the intersection and union of two sets. These axioms also offer a highly versatile tool for exploring deeper topics in mathematics, such as infinity and the nature of infinite sets.

Before presenting the axioms of set theory, we must first describe a formal language for set theory. This formal language involves the logical connectives $\wedge, \vee, \neg, \rightarrow, \leftrightarrow$ together with the quantifier symbols $\forall$ and $\exists$. In addition, this formal language uses the relation symbols $=$ and $\in$ (also $\neq$ and $\notin$).

What is a formula in the *language of set theory*? An **atomic formula** is one that has the form $x \in y$ or $x = z$, where x, y, z can be replaced with any other variables, say, $a, b, x', y', z', x_1, x_2, y_5, X, Y, \ldots$. We say that χ is a **formula** (in the language of set theory) if χ is an atomic formula, or it can be constructed from atomic formulas by repeatedly applying the following **recursive rule**: If φ and ψ are formulas, then the next seven items are also formulas:

$$\forall x \varphi, \ \exists y \psi, \ (\neg \varphi), \ (\varphi \wedge \psi), \ (\varphi \vee \psi), \ (\varphi \rightarrow \psi), \ (\varphi \leftrightarrow \psi).$$

Hence, $\forall y(x \in y \rightarrow (x = a \vee x \notin p))$ is a formula in the language of set theory because it can be constructed from the atomic formulas $x \in y$, $x = a$, $x \in p$ and repeated applications of the above recursive rule. Figure 1.5 illustrates this construction, where the statement $x \notin p$ is used to abbreviate $(\neg x \in p)$.

Formulas are viewed as "grammatically correct" statements in the language of set theory. Moreover, the expression $\forall x((\rightarrow xy$ is not a formula because it cannot be constructed from the atomic formulas and the above recursive rule. In practice, we shall use parentheses so that our formulas are clear and readable. We will also be using, for any formulas φ and ψ, the following three conventions:

1. The *outermost* parentheses need not be explicitly written; that is, one can write $\varphi \wedge \psi$ to denote $(\varphi \wedge \psi)$.
2. The negation symbol will apply to as little as possible. We can therefore use $\neg \varphi \wedge \psi$ to denote $(\neg \varphi) \wedge \psi$.

3. Bounded set quantifiers shall be used. Thus, we can abbreviate the formula $\forall x(x \in y \rightarrow x \notin a)$ by the more readable $(\forall x \in y)(x \notin a)$.

We will also use symbols that are designed to make things easier to understand. For example, we may write $A \subseteq B$ rather than $\forall x(x \in A \rightarrow x \in B)$.

Throughout the book, we will be using the notation $\varphi(x, \ldots, z)$ to identify $x, \ldots, z$ as being free variables (see page 14) that appear in the formula φ. If the variables $x, \ldots, z$ are free, then substitution may take place. Thus, we can replace all occurrences of x, appearing in φ, with a particular set x_0 and obtain $\varphi(x_0, \ldots, z)$. Moreover, a formula $\varphi(x, \ldots, z)$ may contain *parameters*, that is, free variables other than $x, \ldots, z$ that represent unspecified (arbitrary) sets. Parameters denote "unassigned fixed sets." For an example, let $\varphi(x)$ be the formula

$$\forall y(x \in y \rightarrow (x = \varnothing \vee x \notin p)).$$

So, $\varphi(x)$ has x as an identified free variable, $\varnothing$ as a constant, and a parameter p (an unassigned set). To *replace* a parameter p in a formula $\varphi(x, \ldots, z)$ with an specific set p_0 means that *every* occurrence of p, in φ, is replaced with p_0.

We will now explore the expressive power of this set theoretic language. For example, the formula $\exists x(x \in y)$ asserts that the set y is nonempty. Moreover, $\neg \exists y \forall x(x \in y)$ states that "it is not the case that there is a set that contains all sets as elements." In addition, one can translate statements in English, which concern sets, into the language of set theory. Consider the English sentence "the set x contains at least two elements." This sentence can be translated into the language of set theory by $\exists y \exists z((y \in x \wedge z \in x) \wedge y \neq z)$.

Let $\varphi(z)$ be a formula with free variable z and let a be a set. The sentence "there is a set x whose members are just those y's that satisfy $y \in a$ and $\varphi(y)$," is represented by the formula $\exists x \forall y(y \in x \leftrightarrow (y \in a \wedge \varphi(y)))$.

Let $\varphi(x)$ and $\psi(x)$ be formulas. Now consider the relationship

$$y = \begin{cases} a, & \text{if } \varphi(x); \\ b, & \text{if } \psi(x) \text{ and not } \varphi(x); \\ \varnothing, & \text{if none of the above hold.} \end{cases} \tag{1.1}$$

This relationship can be translated into the language of set theory by

$$(\varphi(x) \wedge y = a) \vee ((\psi(x) \wedge \neg\varphi(x)) \wedge y = b) \vee ((\neg\psi(x) \wedge \neg\varphi(x)) \wedge y = \varnothing). \tag{1.2}$$

Let $\gamma(x, y)$ be the formula in (1.2). One can verify that $\gamma(x, y)$ holds if and only if (1.1) holds. Note that for all x there is a unique y such that $\gamma(x, y)$.

1. What does the formula $\exists x \forall y (y \notin x)$ say in English?

2. What does the formula $\forall y \exists x (y \notin x)$ say in English?

3. What does the formula $\forall y \exists x (x \notin y)$ say in English?

4. What does the formula $\forall y \neg \exists x (x \notin y)$ say in English?

5. What does the formula $\forall z \exists x \exists y (x \in y \wedge y \in z)$ say in English?

6. Let $\varphi(x)$ be a formula. What does $\forall z \forall y ((\varphi(x) \wedge \varphi(y)) \rightarrow z = y)$ assert?

7. Translate each of the following into the language of set theory.
   (a) x is the union of a and b.
   (b) x is not a subset of y.
   (c) x is the intersection of a and b.
   (d) a and b have no elements in common.

8. Let $a, b, C,$ and D be sets. Show that the relationship

$$y = \begin{cases} a, & \text{if } x \in C \setminus D; \\ b, & \text{if } x \notin C \setminus D, \end{cases}$$

can be translated into the language of set theory.

1.5 The Zermelo–Fraenkel Axioms

The axiomatic approach to mathematics was pioneered by the Greeks well over 2000 years ago. The Greek mathematician Euclid formally introduced, in the *Elements*, an axiomatic system for proving theorems in plane geometry. Ever since Euclid's success, mathematicians have developed a variety of axiomatic systems. The axiomatic method has now been applied in virtually every branch of mathematics. In this book, we will show how the axiomatic method can be applied to prove theorems in set theory.

We shall now present the Zermelo–Fraenkel axioms. Each of these axioms is first stated in English and then written in logical form. After the presentation, we will then discuss these axioms and some of their consequences; however, throughout the book we shall more carefully examine each of these axioms, beginning in Chapter 2. While reading these axioms, keep in mind that in set theory everything is a set, including the elements of a set. Also, recall that the notation $\theta(x, \ldots, z)$ means that $x, \ldots, z$ are free variables in the formula θ and that θ is allowed to contain parameters (free variables other than $x, \ldots, z$).

1. **Extensionality Axiom.** *Two sets are equal if and only if they have the same elements.*

$$\forall A \forall B(A = B \leftrightarrow \forall x(x \in A \leftrightarrow x \in B)).$$

2. **Empty Set Axiom.** *There is a set with no elements.*

$$\exists A \forall x(x \notin A).$$

3. **Subset Axiom.** *Let $\varphi(x)$ be a formula. For every set A there is a set S that consists of all the elements $x \in A$ such that $\varphi(x)$ holds.* [2]

$$\forall A \, \exists S \, \forall x(x \in S \leftrightarrow (x \in A \wedge \varphi(x))).$$

4. **Pairing Axiom.** *For every u and v there is a set that consists of just u and v.*

$$\forall u \forall v \exists A \forall x(x \in A \leftrightarrow (x = u \vee x = v)).$$

5. **Union Axiom.** *For every set $\mathcal{F}$ there exists a set U that consists of all the elements that belong to at least one set in $\mathcal{F}$.*

$$\forall \mathcal{F} \, \exists U \, \forall x(x \in U \leftrightarrow \exists C(x \in C \wedge C \in \mathcal{F})).$$

6. **Power Set Axiom.** *For every set A there is a set P that consists of all the sets that are subsets of A.*

$$\forall A \, \exists P \, \forall x(x \in P \leftrightarrow \forall y(y \in x \rightarrow y \in A)).$$

7. **Infinity Axiom.** *There is a set I that contains the empty set as an element and whenever $x \in I$, then $x \cup \{x\} \in I$.*

$$\exists I \, (\varnothing \in I \wedge \forall x(x \in I \rightarrow x \cup \{x\} \in I)).$$

8. **Replacement Axiom.** *Let $\psi(x, y)$ be a formula. For every set A, if for each $x \in A$ there is a unique y such that $\psi(x, y)$, then there is a set S that consists of all the elements y such that $\psi(x, y)$ for some $x \in A$.* (See endnote 2.)

$$\forall A \, ((\forall x \in A) \exists! y \psi(x, y) \rightarrow \exists S \forall y(y \in S \leftrightarrow (\exists x \in A) \psi(x, y))).$$

9. **Regularity Axiom.** *Every nonempty set A has an element that is disjoint from A.*

$$\forall A \, (A \neq \varnothing \rightarrow \exists x(x \in A \wedge x \cap A = \varnothing)).$$

The extensionality axiom simply states that two sets are equal if and only if they have exactly the same elements (see Definition 1.1.1(1)). The empty set axiom asserts that there exists a set with no elements. Since the extensionality axiom implies that this set is unique, we let $\varnothing$ denote the empty set.

The subset axiom proclaims that any definable subcollection of a set is itself a set. In other words, whenever we have a formula $\varphi(x)$ and a set A, we can then conclude that $\{x \in A : \varphi(x)\}$ is a set. Clearly, the subset axiom is a restricted form of the comprehension principle, but it does not lead to the contradiction that we encountered in Russell's paradox. The subset axiom, also called the *axiom of separation* (see page 3), is described as an *axiom schema*, because it yields infinitely many axioms–one for each formula φ. Similarly, the replacement axiom is also referred to as an axiom schema.

The pairing axiom states that for any two given sets, there is a set consisting of just those two sets. Therefore, for all sets u and v, the set $\{u, v\}$ exists. Since $\{u, u\} = \{u\}$, it follows that the set $\{u\}$ also exists for each u.

The union axiom asserts that for any set $\mathcal{F}$, there is a set U whose elements are precisely those elements that belong to at least one member of $\mathcal{F}$. More specifically, the union axiom proclaims that the union of any set $\mathcal{F}$ exists; that is, there is a set U so that $x \in U$ if and only if $x \in A$ for some $A \in \mathcal{F}$. As we will see, the set U is denoted by $\bigcup \mathcal{F}$.

The infinity axiom declares that there is a set I such that $\varnothing \in I$ and whenever $x \in I$, then $x \cup \{x\} \in I$. Since $\varnothing \in I$, we thus conclude that $\varnothing \cup \{\varnothing\} = \{\varnothing\} \in I$. Now, as $\{\varnothing\} \in I$, we also have that $\{\varnothing\} \cup \{\{\varnothing\}\} = \{\varnothing, \{\varnothing\}\} \in I$. Continuing in this manner, we see that the set I must contain all of the following sets:

$$\varnothing, \ \{\varnothing\}, \ \{\varnothing, \{\varnothing\}\}, \ \{\varnothing, \{\varnothing\}, \{\varnothing, \{\varnothing\}\}\}, \ \ldots.$$

Observe, by the extensionality axiom, that $\varnothing \neq \{\varnothing\}$. One can also show that any two of the sets in the above list are distinct. Therefore, the set I contains an infinite number of elements; that is, I is an infinite set.

The replacement axiom plays a crucial role in modern set theory (see [8]). Let A be a set and let $\psi(x, y)$ be a formula. Suppose that for each $x \in A$, there is a unique y such that $\psi(x, y)$. Thus, we shall say that y is "uniquely connected" to x. The replacement axiom can now be interpreted as asserting the following: If for each $x \in A$ there is an element y that is uniquely connected to x, then we can *replace* each $x \in A$ with its unique connection y and the result forms a new set. In the words of Paul Halmos [7], "anything intelligent that one can do to the elements of a set yields a set."

Given any nonempty set A, the regularity axiom asserts the $A \cap a = \varnothing$ for some $a \in A$. Can a set belong to itself? The regularity axiom rules out this possibility (see Exercise 3).

The formulas in the subset and replacement axioms may contain parameters. We will soon be proving theorems about formulas that may possess parameters. Because parameters represent arbitrary sets, any axiom/theorem that concerns a generic formula with parameters is applicable whenever the parameters are replaced with identified sets. As a result, such an axiom/theorem can be applied when a formula contains fixed sets, as these sets can be viewed as ones that have replaced parameters. For example, the subset axiom concerns a generic formula $\varphi(x)$. So this axiom can be applied when specific sets appear in $\varphi(x)$.

This completes our preliminary examination of the set-theoretic axioms that were first introduced by Ernst Zermelo and Abraham Fraenkel; however, we will more fully examine each of these axioms in the remainder of the book. Furthermore, before we make our first appeal to a particular axiom, it shall be reintroduced prior to its initial application. In addition, *we will not invoke an axiom before its time*; that is, *if we are able to prove a theorem without appealing to a specific axiom, then we shall do so.* Accordingly, we will not be using the regularity axiom to prove a theorem until the last section of Chapter 8.

It is a most remarkable fact that essentially all mathematical objects can be defined as sets. For example, the natural numbers and the real numbers can be constructed within set theory. Consequently, the theorems of mathematics can be viewed as statements about sets. These theorems can also be proven using the axioms of set theory. Thus, "mathematics can be embedded into set theory."

Exercises 1.5

1. Let u, v, and w be sets. By the pairing axiom, the sets $\{u\}$ and $\{v, w\}$ exist. Using the pairing and union axioms, show that the set $\{u, v, w\}$ exists.

2. Let A be a set. Show that the pairing axiom implies that the set $\{A\}$ exists.

*3. Let A be a set. The pairing axiom implies that the set $\{A\}$ exists. Using the regularity axiom, show that $A \cap \{A\} = \varnothing$. Conclude that $A \notin A$.

4. For sets A and B, the set $\{A, B\}$ exists by the pairing axiom. Let $A \in B$. Using the regularity axiom, show that $A \cap \{A, B\} = \varnothing$, and thus $B \notin A$.

5. Let A, B, and C be sets. Suppose that $A \in B$ and $B \in C$. Using the regularity axiom, show that $C \notin A$. [Hint: Consider the set $\{A, B, C\}$.]

6. Let A and B be sets. Using the subset and power set axioms, show that the set $\mathcal{P}(A) \cap B$ exists.

7. Let A and B be sets. Using the subset axiom, show that the set $A \setminus B$ exists.

8. Show that no two of the sets $\varnothing$, $\{\varnothing\}$, $\{\varnothing, \{\varnothing\}\}$ are equal to each other.

*9. Let A be a set with no elements. Show that for all x, we have that $x \in A$ if and only if $x \in \emptyset$. Using the extensionality axiom, conclude that $A = \emptyset$.

10. Let $\varphi(x, y)$ be the formula $\forall z(z \in y \leftrightarrow z = x)$ which asserts that $y = \{x\}$. As noted on page 25, for all x the set $\{x\}$ exists. So $\forall x \exists ! y \varphi(x, y)$. Let A be a set. Show that the collection $\{\{x\} : x \in A\}$ is a set.

2 Basic Set-Building Axioms and Operations

Nearly all mathematical fields apply concepts from set theory, and practically every mathematical statement can be phrased into one about sets. For these reasons, set theory is often viewed as a foundation for mathematics. In this chapter, we will begin to investigate some of the axioms of set theory that were introduced earlier. Moreover, we shall begin proving theorems about sets, and each of our proofs will be justified by the axioms. We will also show that the sets and operations discussed in Chapter 1 can be derived from these axioms.

2.1 The First Six Axioms

The first six axioms, taken together, will allow us to develop the operations on sets that were discussed in the previous chapter. Furthermore, they will allow us to define additional operations on sets that were not previously covered.

2.1.1 The Extensionality Axiom

Our first axiom is a very simple, but necessary, statement that proclaims that a set is uniquely determined by its elements. In other words, two sets are different if and only if one set contains an element that is not in the other set.

Extensionality Axiom. Two sets are equal if and only if they have exactly the same elements.

For sets A and B, the extensionality axiom yields two alternative strategies for proving that $A = B$:

(a) Prove that $A \subseteq B$ and $B \subseteq A$.
(b) Prove for all x, that $x \in A$ if and only if $x \in B$.

We will refer to strategy (a) as the "double-subset" strategy, and strategy (b) will be described as the "iff" strategy, where iff abbreviates the phrase "if and only if." One efficient way of executing strategy (b) is to derive a succession of *equivalences* starting with $x \in A$ and then ending with $x \in B$. This is usually done by citing appropriate definitions and logic laws (see Sections 1.2 and 1.3). We will illustrate the difference between these two strategies by presenting two proofs of Theorem 2.1.1 in Section 2.1.3.

The extensionality axiom does not imply that there exists a set. Our next five axioms do assert that certain sets exist. In addition, these axioms can be used to establish the existence of many of the sets that we typically take for granted in mathematics.

2.1.2 The Empty Set Axiom

Our next axiom explicitly implies that "a set exists."

Empty Set Axiom. There is a set with no elements.

Is there more than one set without any elements? The Extensionality Axiom implies that there is only one such set (see Exercise 9 on page 27), and we shall denote it by the symbol $\varnothing$.

The next four axioms will allow us to build new sets from given sets.

2.1.3 The Subset Axiom

Subset Axiom. Let $\varphi(x)$ be a formula. For every set A there exists a set S that consists of all the elements $x \in A$ such that $\varphi(x)$ holds.

The subset axiom states that for any set A and for any formula $\varphi(x)$, one can construct a set that consists of just the elements in A that satisfy the property $\varphi(x)$. In other words, the set $S = \{x \in A : \varphi(x)\}$ exists. Observe that $S \subseteq A$.

Let A and B be two sets. The subset axiom implies that the intersection and difference of these two sets exist, because

$$A \cap B = \{x \in A : x \in B\},$$

$$A \setminus B = \{x \in A : x \notin B\},$$

and thus, by the subset axiom, $A \cap B$ and $A \setminus B$ are sets. On the other hand, the subset axiom does not imply, in general, that the union $A \cup B$ is a set.

Our next theorem follows from the extensionality and subset axioms. We will give two proofs of this theorem. The first proof employs the double-subset strategy, and the second proof applies the iff strategy (see page 28).

Theorem 2.1.1. *Suppose A, B, and C are sets. Then $A \cap (B \setminus C) = (A \cap B) \setminus C$.*

First Proof. Let A, B, and C be sets. We prove that $A \cap (B \setminus C) = (A \cap B) \setminus C$.

($\subseteq$). Let $x \in A \cap (B \setminus C)$. Hence, $x \in A$ and $x \in B \setminus C$. Thus, $x \in A$, $x \in B$, and $x \notin C$. Because $x \in A$ and $x \in B$, we have $x \in A \cap B$. Since $x \notin C$, we conclude that $x \in (A \cap B) \setminus C$.

($\supseteq$). Let $x \in (A \cap B) \setminus C$. Thus, $x \in A \cap B$ and $x \notin C$. Because $x \in A \cap B$, we have that $x \in A$ and $x \in B$. We also have that $x \notin C$, and so we now conclude that $x \in B \setminus C$. Furthermore, we know that $x \in A$. Hence, $x \in A \cap (B \setminus C)$.

Therefore, $A \cap (B \setminus C) = (A \cap B) \setminus C$.   □

In our first proof of Theorem 2.1.1, the annotations ($\subseteq$) and ($\supseteq$) are added as a courtesy to the reader. The notation ($\subseteq$) is used to make it clear to the reader that we are proving that the first set[1] is a subset of the second set. The notation ($\supseteq$) indicates that we are proving that the second set is a subset of the first set. We shall now reprove Theorem 2.1.1 using the iff strategy.

Second Proof. Suppose that A, B, and C are sets. Let x be arbitrary. Then

$$
\begin{array}{ll}
x \in A \cap (B \setminus C) \text{ iff } x \in A \wedge x \in (B \setminus C) & \text{by definition of } \cap \\
\qquad\qquad \text{iff } x \in A \wedge (x \in B \wedge x \notin C) & \text{by definition of } \setminus \\
\qquad\qquad \text{iff } (x \in A \wedge x \in B) \wedge x \notin C & \text{by logical associativity} \\
\qquad\qquad \text{iff } x \in A \cap B \wedge x \notin C & \text{by definition of } \cap \\
\qquad\qquad \text{iff } x \in (A \cap B) \setminus C & \text{by definition of } \setminus.
\end{array}
$$

Therefore, $A \cap (B \setminus C) = (A \cap B) \setminus C$.   □

The argument used in Russell's paradox will be applied in the proof of our next theorem, which states that *there is no set of all sets.*

Theorem 2.1.2. *There is no set in which every set is a member.*

Proof. Suppose, for a contradiction, that there exists a set in which every set is a member. Let A denote this set. Thus $\forall x (x \in A)$; that is, the set A contains *all* sets as members. By the subset axiom there exists a set B such that

$$B = \{x \in A : x \notin x\}. \tag{2.1}$$

Observe from the definition of B we have that

$$
\begin{array}{ll}
B \in B \text{ iff } B \in \{x \in A : x \notin x\} & \text{by (2.1)} \\
\qquad \text{iff } B \in A \text{ and } B \notin B & \text{by definition of } \{x \in A : x \notin x\}.
\end{array}
\tag{$\star$}
$$

Because B is a set and A contains all sets as members, we conclude that $B \in A$. Now that $B \in A$ holds, the above $(\star)$ implies that $B \in B$ if and only if $B \notin B$, which is clearly a contradiction.   □

Since every set is equal to itself, we see that each set belongs in the collection $\{x : x = x\}$. Theorem 2.1.2 implies that $\{x : x = x\}$ is not a set. Thus, given a formula $\varphi(x)$, we cannot immediately conclude that the collection $\{x : \varphi(x)\}$ is a set. Hence, we shall refer to any collection of the form $\{x : \varphi(x)\}$ as a **class**. Thus, $\{x : x = x\}$ is a class that is not a set. If a class is not a set, then it is a **proper class**, and hence, it is an "unbounded collection" (see Exercise 34).

In the future, we will sometimes be required to prove that some classes are actually sets (see the proof of Theorem 2.1.7). When can we prove that a given class $\{x : \varphi(x)\}$ is also a set? The following theorem addresses this question.

Theorem 2.1.3. *Let $\varphi(x)$ be a formula. Suppose that there is a set A such that for all x, if $\varphi(x)$, then $x \in A$. Then there is a unique set $\mathcal{D}$ such that for all x,*

$x \in \mathcal{D}$ *if and only if $\varphi(x)$.*

In other words, the class $\{x : \varphi(x)\}$ is, in fact, equal to the set $\mathcal{D}$.

Proof. Let A be a set such that ($\blacktriangle$) for all x, if $\varphi(x)$, then $x \in A$. By the subset axiom, let $\mathcal{D}$ be the set $\{x \in A : \varphi(x)\}$. Clearly, ($\blacktriangle$) implies that $x \in \mathcal{D}$ if and only if $\varphi(x)$, for all x. The set $\mathcal{D}$ is unique by the extensionality axiom.   □

Remark 2.1.4. The regularity axiom implies that no set can be a member of itself (see Exercise 3 on page 26). Thus, in the proof of Theorem 2.1.2, the axioms of set theory imply that the collection B, defined in (2.1), is equal to A.

Remark 2.1.5. When applying the subset axiom, we can use a formula that is expressed in English, using appropriate symbols (e.g., $\varnothing$, $\cap$). Such English formulas can actually be written as formulas in the language of set theory.

2.1.4 The Pairing Axiom

The pairing axiom states that whenever we have two sets u and v, there is a set P whose only members are u and v. We can use the extensionality axiom to show that this set P is unique. We shall denote the set P by $\{u, v\}$.

Pairing Axiom. For every u and v there is a set that consists of just u and v.

Given any sets u and v, the pairing and extensionality axioms justify the following definition.

Definition 2.1.6. For sets u and v, the **pair set** $\{u, v\}$ is the unique set whose only members are u and v.

Thus, if $a \in \{u, v\}$, then $a = u$ or $a = v$. The pairing axiom implies that for any set u, we also have the set $\{u, u\}$, which, of course, is equal to the set $\{u\}$ by the extensionality axiom. So, given any set u, we can construct the set $\{u\}$. Since the set $\{u\}$ has only one element, such a set is referred to as a *singleton*.

Suppose that we have three sets u, v, and w. The pairing axiom yields the sets $\{u, v\}$, $\{\{u, v\}, w\}$, and $\{\{u, v\}, \{w\}\}$. Can we conclude that there is a set containing only u, v, and w? We will be able to do this once we understand the union axiom.

2.1.5 The Union Axiom

Let $\mathcal{F}$ be a set. The union axiom allows us to construct a set that consists of all the elements that belong to at least one of the sets in $\mathcal{F}$.[2]

Union Axiom. For every set $\mathcal{F}$ there is a set U that consists of all the elements that belong to at least one set in $\mathcal{F}$.

The union axiom states that, for any given set $\mathcal{F}$, there exists a set U whose members are precisely the elements of members in $\mathcal{F}$. The set U given by the union axiom is denoted by $\bigcup \mathcal{F}$. Thus, $\bigcup \mathcal{F}$ is the set of elements x such that $x \in C$ for some $C \in \mathcal{F}$; that is,

$$\bigcup \mathcal{F} = \{x : x \in C \text{ for some } C \in \mathcal{F}\}.$$

We shall say that the set $\bigcup \mathcal{F}$ is the **union** of $\mathcal{F}$. So for any set x we have that

$$x \in \bigcup \mathcal{F} \text{ if and only if } (\exists C \in \mathcal{F})(x \in C).$$

Example 1. Let $\mathcal{F} = \{\{a, b, c\}, \{e, f\}, \{e, c, d\}\}$. Then $\bigcup \mathcal{F} = \{a, b, c, d, e, f\}$.

The union axiom, together with the pairing axiom, allows us to construct the union of any two sets A and B. Let $\mathcal{F} = \{A, B\}$. Then

$$\bigcup \{A, B\} = \{x : x \text{ belongs to some member of } \{A, B\}\}$$

$$= \{x : x \in A \text{ or } x \in B\}$$

$$= A \cup B.$$

Similarly, we have that $\bigcup \{A, B, C, D\} = A \cup B \cup C \cup D$ and $\bigcup \{A\} = A$. In addition, we have $\bigcup \emptyset = \emptyset$.

Example 2. Consider the set $\mathcal{F} = \{\{a, b, c\}, \{e, f\}, \{e, c, d\}\}$. Then

$$\bigcup\bigcup\mathcal{F} = \bigcup\{a, b, c, d, e, f\} = a \cup b \cup c \cup d \cup e \cup f.$$

Suppose that we have three sets u, v, and w. We can now use the pairing and union axioms to construct a set that consists of only u, v, and w. We have the sets $\{u, v\}$ and $\{w\}$ by the pairing axiom. By the union axiom we have the set

$$\{u, v\} \cup \{w\} = \{u, v, w\},$$

which consists of the three elements u, v, and w. Similarly, one can assemble the set $\{x_1, x_2, x_3, x_4\}$ consisting of the four elements x_1, x_2, x_3, x_4, and so on.

We saw that the subset axiom allows us to construct the intersection of two sets. Our next theorem will allow us to take the intersection of all the sets that belong to a given nonempty set.

Theorem 2.1.7. *Let $\mathcal{F}$ be a nonempty set. There is a unique set I such that for all x,*

$$x \in I \text{ if and only if } x \text{ belongs to every member of } \mathcal{F}. \tag{2.2}$$

Proof. Let $\mathcal{F}$ be a nonempty set. So let $A \in \mathcal{F}$. For each x, if x belongs to every member of $\mathcal{F}$, then clearly $x \in A$. Since A is a set, Theorem 2.1.3 implies that there is a unique set I satisfying (2.2). $\qquad\square$

Definition 2.1.8. Let $\mathcal{F}$ be a nonempty set. We denote the unique set I given in the above Theorem 2.1.7 by $\bigcap\mathcal{F}$, which is said to be the **intersection** of $\mathcal{F}$.

Thus, if $\mathcal{F}$ is a nonempty set, then $x \in \bigcap\mathcal{F}$ if and only if $(\forall C \in \mathcal{F})(x \in C)$.

Example 3. Suppose $\mathcal{F} = \{\{a, b, c, e\}, \{e, f\}, \{e, c, d\}\}$. Then $\bigcap\mathcal{F} = \{e\}$.

We see that $\bigcap\{A, B, C, D\} = A \cap B \cap C \cap D$ and $\bigcap\{A\} = A$. Furthermore, if $\varnothing \in \mathcal{F}$, then $\bigcap\mathcal{F} = \varnothing$.

Why does Definition 2.1.8 require that $\mathcal{F} \neq \varnothing$? Observe that the statement "x belongs to every member of $\varnothing$" is vacuously true. To see this, suppose that this statement is false. Then x does not belong to some member of $\varnothing$, but the set $\varnothing$ has no elements! Therefore, the statement is true for all sets x.

Now, suppose $\bigcap\varnothing$ is a set. Thus,

$$\bigcap\varnothing = \{x : x \text{ belongs to every member of } \varnothing\}.$$

Because the statement "x belongs to every member of $\varnothing$" is true for every x, we conclude that $\bigcap \varnothing$ must contain every set, and this contradicts Theorem 2.1.2. Therefore, the notation $\bigcap \varnothing$ is undefined.

Example 4. Let $\mathcal{F} = \{\{c, e, f\}, \{e, f\}, \{e, f, c, d\}\}$. Then

1. $\bigcap \bigcap \mathcal{F} = \bigcap \{e, f\} = e \cap f$.
2. $\bigcup \bigcap \mathcal{F} = \bigcup \{e, f\} = e \cup f$.
3. $\bigcap \bigcup \mathcal{F} = \bigcap \{c, e, d, f\} = c \cap d \cap e \cap f$.
4. $\bigcup \bigcup \mathcal{F} = \bigcup \{c, e, d, f\} = c \cup d \cup e \cup f$.

The next example clearly states what it means for an element to be in, and not in, a union and intersection.

Example 5. Suppose that $\mathcal{F}$ is set. Then the following statements are true:

(1) $x \in \bigcup \mathcal{F}$ means that $x \in C$ for some $C \in \mathcal{F}$.
(2) $x \notin \bigcup \mathcal{F}$ means that $x \notin C$ for every $C \in \mathcal{F}$.
(3) $x \in \bigcap \mathcal{F}$ means that $x \in C$ for every $C \in \mathcal{F}$.
(4) $x \notin \bigcap \mathcal{F}$ means that $x \notin C$ for some $C \in \mathcal{F}$.

We now demonstrate how one works with our generalized notions of union and intersection in the following theorem and proof.

Theorem 2.1.9. *Suppose that $\mathcal{F} \subseteq \mathcal{G}$. Then $\bigcup \mathcal{F} \subseteq \bigcup \mathcal{G}$, and if $\mathcal{F}$ is nonempty, then $\bigcap \mathcal{G} \subseteq \bigcap \mathcal{F}$.*

Proof. Let $\mathcal{F} \subseteq \mathcal{G}$. To prove that $\bigcup \mathcal{F} \subseteq \bigcup \mathcal{G}$, let $x \in \bigcup \mathcal{F}$. Thus, $x \in C$ for some $C \in \mathcal{F}$. Since $\mathcal{F} \subseteq \mathcal{G}$, we have that $C \in \mathcal{G}$. So $x \in C$ for some $C \in \mathcal{G}$. Hence, $x \in \bigcup \mathcal{G}$. Now, suppose that $\mathcal{F}$ is nonempty. Because $\mathcal{F} \subseteq \mathcal{G}$, we see that $\mathcal{G}$ is also nonempty. Let $x \in \bigcap \mathcal{G}$. Thus, $x \in C$ for every $C \in \mathcal{G}$. Since $\mathcal{F} \subseteq \mathcal{G}$, we conclude that $x \in C$ for every $C \in \mathcal{F}$ as well. Therefore, $x \in \bigcap \mathcal{F}$. □

2.1.6   The Power Set Axiom

Recall that $X \subseteq A$ means that X is a subset of A; that is, every element in X is also an element in A. Given any set A, the power set axiom ensures that one can form a set that consists of every subset of A.

Power Set Axiom. For every set A there is a set P that consists of all the sets that are subsets of A.

When A is a set, the power set axiom states that there is a set P such that if $X \subseteq A$, then $X \in P$. In addition, this axiom implies that if $X \in P$, then $X \subseteq A$.

Definition 2.1.10. Let A be a set. The **power set** of A, denoted by $\mathcal{P}(A)$, is the set whose elements are all of the subsets of A; that is, $\mathcal{P}(A) = \{X : X \subseteq A\}$.

Thus, as noted above, $X \in \mathcal{P}(A)$ if and only if $X \subseteq A$. Recall that $\varnothing \subseteq A$ and $A \subseteq A$ (see page 2). Therefore, $\varnothing \in \mathcal{P}(A)$ and $A \in \mathcal{P}(A)$ for every set A. The power set of a set A has more elements than A. For example,

1. $\mathcal{P}(\varnothing) = \{\varnothing\}$.
2. $\mathcal{P}(\{a\}) = \{\varnothing, \{a\}\}$.
3. $\mathcal{P}(\{\varnothing, a\}) = \{\varnothing, \{\varnothing\}, \{a\}, \{\varnothing, a\}\}$.
4. $\mathcal{P}(\{1, 2, 3\}) = \{\varnothing, \{1\}, \{2\}, \{3\}, \{1, 2\}, \{1, 3\}, \{2, 3\}, \{1, 2, 3\}\}$.

The power set operation distributes over the intersection of two sets (for a generalization, see Exercise 14 on page 40).

Theorem 2.1.11. *Let A and B be sets. Then $\mathcal{P}(A \cap B) = \mathcal{P}(A) \cap \mathcal{P}(B)$.*

Proof. Let A and B be sets. We shall prove that $\mathcal{P}(A \cap B) = \mathcal{P}(A) \cap \mathcal{P}(B)$.

($\subseteq$). Let $X \in \mathcal{P}(A \cap B)$. Hence, $X \subseteq A \cap B$. Thus, $X \subseteq A$ and $X \subseteq B$ (see Exercise 4). So $X \in \mathcal{P}(A)$ and $X \in \mathcal{P}(B)$. Therefore, $X \in \mathcal{P}(A) \cap \mathcal{P}(B)$.

($\supseteq$). Let $X \in \mathcal{P}(A) \cap \mathcal{P}(B)$. Thus, $X \in \mathcal{P}(A)$ and $X \in \mathcal{P}(B)$. Hence, $X \subseteq A$ and $X \subseteq B$. Therefore, $X \subseteq A \cap B$ (see Exercise 4), and we now conclude that $X \in \mathcal{P}(A \cap B)$. □

Theorem 2.1.11 motivates the following simple question: Can one prove the equality $\mathcal{P}(A \cup B) = \mathcal{P}(A) \cup \mathcal{P}(B)$ for any two sets A and B? The answer is no. Let $A = \{1, 2\}$ and $B = \{2, 3\}$. Clearly, the set $X = \{1, 3\}$ is subset of $A \cup B$, and thus, $X \in \mathcal{P}(A \cup B)$. Because X is not a subset of A and is also not a subset of B, we see that $X \notin \mathcal{P}(A) \cup \mathcal{P}(B)$. So $X \in \mathcal{P}(A \cup B)$ and $X \notin \mathcal{P}(A) \cup \mathcal{P}(B)$. Therefore, $\mathcal{P}(A \cup B) \neq \mathcal{P}(A) \cup \mathcal{P}(B)$.

Exercises 2.1

Prove the following theorems, where A, B, C, and D are sets:

1. **Theorem.** If $A \subseteq B$, then $A \subseteq A \cup B$ and $A \cap B \subseteq A$.

*2. **Theorem.** If $A \subseteq B$ and $B \subseteq C$, then $A \subseteq C$.

*3. **Theorem.** If $B \subseteq C$, then $A \setminus C \subseteq A \setminus B$.

*4. **Theorem.** $C \subseteq A$ and $C \subseteq B$ if and only if $C \subseteq A \cap B$.

5. **Theorem.** There exists an x such that $x \notin A$.

6. **Theorem.** $A \cap B = B \cap A$.

7. **Theorem.** $A \cup B = B \cup A$.

8. **Theorem.** $A \cap (B \cup C) = (A \cap B) \cup (A \cap C)$.

9. **Theorem.** $A \cup (B \cap C) = (A \cup B) \cap (A \cup C)$.

10. **Theorem.** $A \cup (B \cup C) = (A \cup B) \cup C$.

11. **Theorem.** $A \cap (B \cap C) = (A \cap B) \cap C$.

12. **Theorem.** $C \setminus (A \cap B) = (C \setminus A) \cup (C \setminus B)$.

13. **Theorem.** $C \setminus (A \cup B) = (C \setminus A) \cap (C \setminus B)$.

14. **Theorem.** $(A \setminus B) \cap (C \setminus B) = (A \cap C) \setminus B$.

15. **Theorem.** $A \cap (B \cap C) = (A \cap B) \cap C$ and $A \cup (B \cup C) = (A \cup B) \cup C$.

16. **Theorem.** $(A \cup B) \setminus (A \cap B) = (A \setminus B) \cup (B \setminus A)$.

17. **Theorem.** If $A \setminus B \subseteq C$, then $A \setminus C \subseteq B$.

18. **Theorem.** If $A \subseteq B$ and $B \cap C = \varnothing$, then $A \subseteq B \setminus C$.

19. **Theorem.** If $A \setminus B \subseteq C$ and $A \nsubseteq C$, then $A \cap B \neq \varnothing$.

20. **Theorem.** $A \subseteq B$ if and only if $\mathcal{P}(A) \subseteq \mathcal{P}(B)$.

21. **Theorem.** $\mathcal{P}(A) \cup \mathcal{P}(B) \subseteq \mathcal{P}(A \cup B)$.

22. **Theorem.** If $\mathcal{P}(A \cup B) \subseteq \mathcal{P}(A) \cup \mathcal{P}(B)$, then $A \subseteq B$ or $B \subseteq A$.

23. **Theorem.** Let A be a set. Then $\{A\} \subseteq \mathcal{P}(A)$.

24. **Theorem.** Let A be a set. Then $\mathcal{P}(A \setminus B) \neq \mathcal{P}(A) \setminus \mathcal{P}(B)$.

25. **Theorem.** Let $\mathcal{F}$ be a set and let $C \in \mathcal{F}$. Then $C \subseteq \bigcup \mathcal{F}$.

26. **Theorem.** Let $\mathcal{F}$ be a set and let $C \in \mathcal{F}$. Then $\bigcap \mathcal{F} \subseteq C$.

27. **Theorem.** Let $\mathcal{F}$ and A be sets. If $A \subseteq C$ for some $C \in \mathcal{F}$, then we have that $A \subseteq \bigcup \mathcal{F}$.

28. **Theorem.** Let $\mathcal{F} \neq \varnothing$ and A be sets. If $A \subseteq C$ for all $C \in \mathcal{F}$, then we have that $A \subseteq \bigcap \mathcal{F}$.

29. Theorem. Let $\mathcal{F}$ and A be sets. Suppose $C \subseteq A$ for every $C \in \mathcal{F}$. Then $\bigcup \mathcal{F} \subseteq A$.

30. Theorem. Let A be a set. Then $\bigcup \mathcal{P}(A) = A$.

31. Theorem. Let A be a set. Then $A \subseteq \mathcal{P}(\bigcup A)$.

*32. Theorem. Let $C \in \mathcal{F}$. Then $\mathcal{P}(C) \in \mathcal{P}(\mathcal{P}(\bigcup \mathcal{F}))$.

33. Theorem. The collection $\{x : \exists y(x \in y)\}$ is not a set.

*34. Theorem. Let $\varphi(x)$ be a formula. Then $\{x : \varphi(x)\}$ is a proper class if and only if for every set A there is an x such that $\varphi(x)$ and $x \notin A$.

35. Theorem. Define the *symmetric difference* of any sets X and Y to be the set $X \triangle Y = (X \setminus Y) \cup (Y \setminus X)$. Then for sets A, B, C we have
(a) $A \cap (B \triangle C) = (A \cap B) \triangle (A \cap C)$.
(b) $A \triangle (B \triangle C) = (A \triangle B) \triangle C$.

Exercise Notes: Exercises **8–9** are referred to as the distributive laws for sets. Exercises **10–11** are called the associative laws for sets.

2.2 Operations on Sets

Sets can be combined in a number of ways to construct another set. We will now see how the first six axioms of Zermelo–Fraenkel set theory can be used to prove a number of theorems concerning several important methods, not identified in the previous section, for constructing a new set from given sets.

2.2.1 De Morgan's Laws for Sets

Lemma 2.2.1. *Let A and $\mathcal{F}$ be sets. Then there is a unique set $\mathcal{D}$ such that for all Y,*

$$Y \in \mathcal{D} \text{ iff } Y = A \setminus C \text{ for some } C \in \mathcal{F}. \tag{2.3}$$

We shall let $\{A \setminus C : C \in \mathcal{F}\}$ denote the set $\mathcal{D}$.

Proof. Let A and $\mathcal{F}$ be sets. If $Y = A \setminus C$ for some $C \in \mathcal{F}$, then $Y \subseteq A$ and so $Y \in \mathcal{P}(A)$. Since $\mathcal{P}(A)$ is a set, Theorem 2.1.3 implies that there exists a unique set $\mathcal{D}$ satisfying (2.3). $\square$

The "double-subset" proof strategy will be used to prove our next theorem. Before reading this proof, one should review items (1)–(4) of Example 5 on page 34 and note that $x \in \bigcap\{A \setminus C : C \in \mathcal{F}\}$ iff $x \in A \setminus C$ for every $C \in \mathcal{F}$.

Theorem 2.2.2 (De Morgan's Laws). *If A is a set and $\mathcal{F}$ is nonempty, then*

(1) $A \setminus \bigcup \mathcal{F} = \bigcap\{A \setminus C : C \in \mathcal{F}\}$,
(2) $A \setminus \bigcap \mathcal{F} = \bigcup\{A \setminus C : C \in \mathcal{F}\}$.

Proof. We prove that $A \setminus \bigcup \mathcal{F} = \bigcap\{A \setminus C : C \in \mathcal{F}\}$ and leave (2) as an exercise.

($\subseteq$). Let $x \in A \setminus \bigcup \mathcal{F}$. We prove that $x \in \bigcap\{A \setminus C : C \in \mathcal{F}\}$ as follows:[3]

$x \in A \setminus \bigcup \mathcal{F} \Rightarrow x \in A$ and $x \notin \bigcup \mathcal{F}$ by definition of $\setminus$

$\Rightarrow x \in A$ and $x \notin C$ for every $C \in \mathcal{F}$ by definition of $\bigcup$

$\Rightarrow x \in A \setminus C$ for every $C \in \mathcal{F}$ by definition of $\setminus$

$\Rightarrow x \in \bigcap\{A \setminus C : C \in \mathcal{F}\}$ by definition of $\bigcap$.

Therefore, $A \setminus \bigcup \mathcal{F} \subseteq \bigcap\{A \setminus C : C \in \mathcal{F}\}$.

($\supseteq$). Let $x \in \bigcap\{A \setminus C : C \in \mathcal{F}\}$. We prove that $x \in A \setminus \bigcup \mathcal{F}$ as follows:

$x \in \bigcap\{A \setminus C : C \in \mathcal{F}\}$

$\Rightarrow x \in A \setminus C$ for every $C \in \mathcal{F}$ by definition of $\bigcap$

$\Rightarrow x \in A$ and $x \notin C$ for every $C \in \mathcal{F}$ by definition of $\setminus$

$\Rightarrow x \in A$ and $x \notin \bigcup \mathcal{F}$ by definition of $\bigcup$

$\Rightarrow x \in A \setminus \bigcup \mathcal{F}$ by definition of $\setminus$.

Therefore, $\bigcap\{A \setminus C : C \in \mathcal{F}\} \subseteq A \setminus \bigcup \mathcal{F}$. The proof of (1) is complete. $\square$

2.2.2 Distributive Laws for Sets

Lemma 2.2.3. *For all sets A and $\mathcal{F}$, there exist unique sets $\mathcal{D}$ and $\mathcal{E}$ such that for all Y:*

$Y \in \mathcal{D}$ iff $Y = A \cup C$ for some $C \in \mathcal{F}$,   (2.4)

$Y \in \mathcal{E}$ iff $Y = A \cap C$ for some $C \in \mathcal{F}$.   (2.5)

We let $\{A \cup C : C \in \mathcal{F}\}$ denote the set $\mathcal{D}$ and $\{A \cap C : C \in \mathcal{F}\}$ denote the set $\mathcal{E}$.

Proof. Let A and $\mathcal{F}$ be sets. If $Y = A \cup C$ for a $C \in \mathcal{F}$, then $Y \subseteq A \cup (\bigcup \mathcal{F})$. Thus, $Y \in \mathcal{P}(A \cup (\bigcup \mathcal{F}))$. As $\mathcal{P}(A \cup (\bigcup \mathcal{F}))$ is a set, Theorem 2.1.3 asserts that there exists a unique set $\mathcal{D}$ satisfying (2.4). In a similar manner, one can show that there is a unique set $\mathcal{E}$ satisfying (2.5) (see Exercise 5). $\square$

Theorem 2.2.4 (Distributive Laws). *If A and $\mathcal{F}$ are sets, then*

(1) $A \cup (\bigcap \mathcal{F}) = \bigcap \{A \cup C : C \in \mathcal{F}\}$ *for $\mathcal{F} \neq \varnothing$,*
(2) $A \cap (\bigcup \mathcal{F}) = \bigcup \{A \cap C : C \in \mathcal{F}\}$.

Proof. We prove (1) using the "iff" strategy and leave (2) as an exercise. Let x be given. We prove that $x \in A \cup (\bigcap \mathcal{F})$ iff $x \in \bigcap \{A \cup C : C \in \mathcal{F}\}$ as follows:

$$x \in A \cup \left(\bigcap \mathcal{F} \right) \text{ iff } x \in A \text{ or } x \in \bigcap \mathcal{F} \qquad \text{by def. of } \cup$$
$$\text{iff } x \in A \text{ or } x \in C \text{ for every } C \in \mathcal{F} \quad \text{by def. of } \bigcap$$
$$\text{iff } x \in A \cup C \text{ for every } C \in \mathcal{F} \qquad \text{by def. of } \cup$$
$$\text{iff } x \in \bigcap \{A \cup C : C \in \mathcal{F}\} \qquad \text{by Lemma 2.2.3.} \quad \square$$

The above proof implicitly applied the quantifier distribution law (QDL) 1.3.6(5). We demonstrate this in the following more "symbolic" proof:

$$x \in A \cup \left(\bigcap \mathcal{F} \right) \text{ iff } x \in A \vee x \in \bigcap \mathcal{F} \qquad \text{by def. of } \cup$$
$$\text{iff } x \in A \vee (\forall C \in \mathcal{F})x \in C \quad \text{by def. of } \bigcap$$
$$\text{iff } (\forall C \in \mathcal{F})(x \in A \vee x \in C) \quad \text{by QDL 1.3.6(5)}$$
$$\text{iff } (\forall C \in \mathcal{F})(x \in A \cup C) \qquad \text{by def. of } \cup$$
$$\text{iff } x \in \bigcap \{A \cup C : C \in \mathcal{F}\} \qquad \text{by Lemma 2.2.3.} \quad \square$$

We now prove a lemma that will justify some of the exercises of this section.

Lemma 2.2.5. *If $\mathcal{F}$ is a set, then there exists a unique set $\mathcal{D}$ such that for all Y*

$$Y \in \mathcal{D} \text{ iff } Y = \mathcal{P}(C) \text{ for some } C \in \mathcal{F}. \tag{2.6}$$

The set $\mathcal{D}$ is denoted by $\{\mathcal{P}(C) : C \in \mathcal{F}\}$.

Proof. Let $\mathcal{F}$ be any set. If $Y = \mathcal{P}(C)$ for some $C \in \mathcal{F}$, then $Y \in \mathcal{P}(\mathcal{P}(\bigcup \mathcal{F}))$ by Exercise 32 on page 37. As $\mathcal{P}(\mathcal{P}(\bigcup \mathcal{F}))$ is a set, there exists a unique set $\mathcal{D}$ satisfying (2.6) by Theorem 2.1.3. $\square$

Exercises 2.2

Prove the following theorems:

1. **Theorem.** Let A be a set and $\mathcal{F} \neq \varnothing$. Then $A \setminus \bigcap \mathcal{F} = \bigcup\{A \setminus C : C \in \mathcal{F}\}$.

2. **Theorem.** Let A and $\mathcal{F}$ be sets. Then $A \cup (\bigcup \mathcal{F}) = \bigcup\{A \cup C : C \in \mathcal{F}\}$.

3. **Theorem.** Let A and $\mathcal{F}$ be sets. Then $A \cap (\bigcup \mathcal{F}) = \bigcup\{A \cap C : C \in \mathcal{F}\}$.

4. **Theorem.** Let $\mathcal{F} \neq \varnothing$. Then $A \cap (\bigcap \mathcal{F}) = \bigcap\{A \cap C : C \in \mathcal{F}\}$ for any set A.

*5. **Theorem.** Let A and $\mathcal{F}$ be sets. Then there exists a unique set $\mathcal{E}$ such that for all Y we have that $Y \in \mathcal{E}$ if and only if $Y = A \cap C$ for some $C \in \mathcal{F}$.

6. **Theorem.** Let $\mathcal{F}$ and $\mathcal{G}$ be sets. Then there is a unique set $\mathcal{D}$ such that for all Y, we have that $Y \in \mathcal{D}$ if and only if $Y = A \cap B$ for some $A \in \mathcal{F}$ and some $B \in \mathcal{G}$.

7. **Theorem.** $(\bigcup \mathcal{F}) \cap (\bigcup \mathcal{G}) = \bigcup\{A \cap B : A \in \mathcal{F} \text{ and } B \in \mathcal{G}\}$ for any sets $\mathcal{F}$ and $\mathcal{G}$.

8. **Theorem.** Let $\mathcal{F}$ and $\mathcal{G}$ be sets. Then there is a unique set $\mathcal{D}$ such that for all Y, we have that $Y \in \mathcal{D}$ if and only if $Y = A \cup B$ for some $A \in \mathcal{F}$ and some $B \in \mathcal{G}$.

9. **Theorem.** $(\bigcap \mathcal{F}) \cup (\bigcap \mathcal{G}) = \bigcap\{A \cup B : A \in \mathcal{F} \text{ and } B \in \mathcal{G}\}$ when $\mathcal{F}$ and $\mathcal{G}$ are nonempty.

10. **Theorem.** $(\bigcup \mathcal{F}) \cap (\bigcup \mathcal{G}) = \bigcup\{A \cap B : A \in \mathcal{F} \text{ and } B \in \mathcal{G}\}$ for any sets $\mathcal{F}$ and $\mathcal{G}$.

11. **Theorem.** Let $\mathcal{F}$ and $\mathcal{G}$ be sets. Then $\bigcup(\mathcal{F} \cup \mathcal{G}) = (\bigcup \mathcal{F}) \cup (\bigcup \mathcal{G})$.

*12. **Theorem.** If $\mathcal{F}$ and $\mathcal{G}$ are nonempty, then $\bigcap(\mathcal{F} \cup \mathcal{G}) = (\bigcap \mathcal{F}) \cap (\bigcap \mathcal{G})$.

13. **Theorem.** Let $\mathcal{F}$ and $\mathcal{G}$ be sets. Then $\bigcup(\mathcal{F} \cap \mathcal{G}) \subseteq (\bigcup \mathcal{F}) \cap (\bigcup \mathcal{G})$.

*14. **Theorem.** Let $\mathcal{F}$ be a nonempty set. Then $\mathcal{P}(\bigcap \mathcal{F}) = \bigcap\{\mathcal{P}(C) : C \in \mathcal{F}\}$.

15. **Theorem.** Let $\mathcal{F}$ be a set. Then $\bigcup\{\mathcal{P}(C) : C \in \mathcal{F}\} \subseteq \mathcal{P}(\bigcup \mathcal{F})$.

16. **Theorem.** Let $\mathcal{F}$ be a set. If there is an $A \in \mathcal{F}$ such that $C \subseteq A$ for all $C \in \mathcal{F}$, then $\mathcal{P}(\bigcup \mathcal{F}) \subseteq \bigcup\{\mathcal{P}(C) : C \in \mathcal{F}\}$.

3 Relations and Functions

In mathematics, relations and functions are usually defined in terms of ordered pairs. The concepts of an ordered pair, a relation, and a function are essential for mathematics. Is there a way to define these three notions in terms of sets? We will show that this, in fact, can be done.

3.1 Ordered Pairs in Set Theory

Before discussing relations or functions, we must first define an ordered pair. The set $\{2, 3\}$ is an unordered pair because $\{2, 3\} = \{3, 2\}$. We need to define an object of the form $\langle 2, 3 \rangle$ that captures the notion of the first component 2 and the second component 3. Furthermore, $\langle 2, 3 \rangle$ must be a set, and we must have that $\langle 2, 3 \rangle \neq \langle 3, 2 \rangle$.

Can an ordered pair be defined to be a set? If it can, then such a definition must satisfy the key property

$$\langle a, b \rangle = \langle c, d \rangle \text{ if and only if } a = c \text{ and } b = d. \tag{3.1}$$

The equivalence (3.1) ensures that an ordered pair has a unique first component and a unique second component. In 1921, Kazimierz Kuratowski answered the above question with the following definition.

Definition 3.1.1. The **ordered pair** $\langle x, y \rangle$, with *first component* x and *second component* y, is the set defined by

$$\langle x, y \rangle = \{\{x\}, \{x, y\}\}.$$

Clearly, for sets x and y, the above $\langle x, y \rangle$ is a set. To prove that Kuratowski's definition satisfies (3.1), we first establish a lemma.

Lemma 3.1.2. *Let u, v, x, y be given and suppose that $\langle x, y \rangle = \langle u, v \rangle$. If $x = y$ or $u = v$, then $x = u$ and $y = v$.*

Proof. Assume $\langle x, y \rangle = \langle u, v \rangle$. So (▲) $\{\{x\}, \{x, y\}\} = \{\{u\}, \{u, v\}\}$. If $x = y$, then $\{\{x\}, \{x, y\}\} = \{\{x\}, \{x, x\}\} = \{\{x\}, \{x\}\} = \{\{x\}\}$. Thus, by (▲),

$$\{\{x\}\} = \{\{u\}, \{u, v\}\}.$$

Hence, $\{x\} = \{u\} = \{u, v\}$. So $x = u = v$. Therefore, as $x = y$, we conclude that $x = u$ and $y = v$. If $u = v$, a similar argument shows that $x = u$ and $y = v$.   □

Theorem 3.1.3. *Let u, v, x, y be given. Then*

$$\langle x, y \rangle = \langle u, v \rangle \text{ if and only if } x = u \text{ and } y = v.$$

Proof. Assume that $\langle x, y \rangle = \langle u, v \rangle$. If either $x = y$ or $u = v$, then Lemma 3.1.2 implies that $x = u$ and $y = v$. Suppose that $x \neq y$ and $u \neq v$. Since

$$\{\{x\}, \{x, y\}\} = \{\{u\}, \{u, v\}\}, \tag{3.2}$$

we conclude that either $\{x\} = \{u\}$ or $\{x\} = \{u, v\}$. Since $u \neq v$, we must have $\{x\} = \{u\}$, and so $x = u$. Thus, $\{\{x\}, \{x, y\}\} = \{\{x\}, \{x, v\}\}$ by Equation (3.2). Since $x \neq y$, we infer that $\{x, y\} = \{x, v\}$, and since $x \neq y$, we deduce that $y = v$.

Conversely, if $x = u$ and $y = v$, then clearly $\langle x, y \rangle = \langle u, v \rangle$.   □

Let A and B be sets. Let $x \in A$ and $y \in B$. Note that the sets $\{x\}$ and $\{x, y\}$ are both elements in $\mathcal{P}(A \cup B)$. Therefore, $\{\{x\}, \{x, y\}\} \subseteq \mathcal{P}(A \cup B)$, and hence $\{\{x\}, \{x, y\}\} \in \mathcal{P}(\mathcal{P}(A \cup B))$. So $\langle x, y \rangle \in \mathcal{P}(\mathcal{P}(A \cup B))$ for all $x \in A$ and $y \in B$.

Lemma 3.1.4. *Let A and B be sets. There exists a unique set $\mathcal{D}$ such that for all Y,*

$$Y \in \mathcal{D} \text{ iff } Y = \langle x, y \rangle \text{ for some } x \in A \text{ and } y \in B. \tag{3.3}$$

We shall let $\{\langle x, y \rangle : x \in A \text{ and } y \in B\}$ denote the set $\mathcal{D}$.

Proof. Let A and B be sets. If $Y = \langle x, y \rangle$ for some $x \in A$ and some $y \in B$, then $Y \in \mathcal{P}(\mathcal{P}(A \cup B))$ (see above). Since $\mathcal{P}(\mathcal{P}(A \cup B))$ is a set, Theorem 2.1.3 now implies that there exists a unique set $\mathcal{D}$ satisfying (3.3).   □

Definition 3.1.5. Given sets A and B, the **Cartesian product** $A \times B$ is defined to be the set

$A \times B = \{\langle x, y \rangle : x \in A \text{ and } y \in B\}.$

In other words, $A \times B$ is the unique set, given by Lemma 3.1.4, that consists of the ordered pairs $\langle x, y \rangle$ such that $x \in A$ and $y \in B$.

Example 1. Let $A = \{2, 3\}$ and let $B = \{a, b, c, 3\}$. Then

$A \times B = \{\langle 2, a \rangle, \langle 2, b \rangle, \langle 2, c \rangle, \langle 2, 3 \rangle, \langle 3, a \rangle, \langle 3, b \rangle, \langle 3, c \rangle, \langle 3, 3 \rangle\}.$

Exercises 3.1

1. Define $\langle a, b, c \rangle = \langle \langle a, b \rangle, c \rangle$ for any sets a, b, c. Prove that this yields an *ordered triple*; that is, prove that if $\langle x, y, z \rangle = \langle a, b, c \rangle$, then $x = a$, $y = b$, and $z = c$.

2. Prove that $(A \cup B) \times C = (A \times C) \cup (B \times C)$.

3. Prove that $(A \setminus B) \times C = (A \times C) \setminus (B \times C)$.

4. Prove that $(\bigcup \mathcal{F}) \times C = \bigcup \{A \times C : A \in \mathcal{F}\}$.

3.2 Relations

We now discuss the mathematical theory of relations. We are already familiar with the three relations $a = b$ (equality), $a < b$ (less than), and $X \subseteq Y$ (subset). Many of the fundamental concepts of mathematics can be described in terms of relations. In this section, we shall identify a relation as a set of ordered pairs and explore various properties that a relation may possess.

Definition 3.2.1. A **relation** R is a set of ordered pairs.

Thus, a set is a relation if each of its elements is an ordered pair. In other words, a set R is a relation iff for each $x \in R$, there is an a and b such that

$x = \langle a, b \rangle$. Definition 3.2.1 does not put any restrictions on the ordered pairs that can appear in a relation.

Definition 3.2.2. Let A and B be sets. A *relation* R from A to B is a subset of $A \times B$; that is, $R \subseteq A \times B$.

Let $A = \{2, 3\}$ and let $B = \{a, b, c, 3\}$. Then

$$R = \{\langle 2, a \rangle, \langle 2, b \rangle, \langle 3, b \rangle, \langle 3, c \rangle, \langle 3, 3 \rangle\}$$

is an example of a relation from A to B. One can also show that the empty set is also a relation (see Exercise 1).

Given a formula $\varphi(x, y)$ and sets A and B, the subset axiom implies that one can construct a relation R by forming the set

$$R = \{\langle x, y \rangle \in A \times B : \varphi(x, y)\}.$$

Example 1. Let A and B be sets. Consider the following relations:

(1) $R = \{\langle x, y \rangle \in A \times B : x = y\}$.
(2) $R = \{\langle x, y \rangle \in A \times B : x \in y\}$.
(3) $R = \{\langle x, y \rangle \in A \times B : x \subseteq y\}$.

Definition 3.2.3. A *relation* R on A is a subset of $A \times A$; that is, $R \subseteq A \times A$.

Example 2. Let $A = \{2, 3\}$. Then the set $R = \{\langle 2, 2 \rangle, \langle 2, 3 \rangle, \langle 3, 3 \rangle\}$ is a relation on A.

Example 3. Let $\mathbb{N} = \{0, 1, 2, 3, \ldots\}$ and define the relation R on the set $\mathbb{N}$ by $R = \{\langle x, y \rangle \in \mathbb{N} \times \mathbb{N} : (\exists k \in \mathbb{N})(x = yk)\}$. So $\langle 6, 3 \rangle \in R$ and $\langle 9, 5 \rangle \notin R$.

For any set A, the relation

$$I_A = \{\langle x, y \rangle \in A \times A : x = y\}$$

is called the *identity relation* on A.

We know that every relation is just a set of ordered pairs. Can every relation also be viewed as a relation on a set? That is, given any relation R, can we find a set A and prove that $R \subseteq A \times A$? The following lemma answers this question.

Lemma 3.2.4. *Let R be a relation and let $A = \bigcup\bigcup R$. Then $R \subseteq A \times A$.*

Proof. Let R be a relation and let $A = \bigcup\bigcup R$. Clearly, A is a set by the union axiom. We prove that $R \subseteq A \times A$. Let $\langle a, b \rangle \in R$. So $\langle a, b \rangle = \{\{a\}, \{a, b\}\}$. Since $\{\{a\}, \{a, b\}\} \in R$, we see that $\{a\} \in \bigcup R$ and $\{a, b\} \in \bigcup R$. Now, since $\{a\} \in \bigcup R$, we conclude that $a \in \bigcup\bigcup R$. Similarly, we infer that $b \in \bigcup\bigcup R$. Hence, as $A = \bigcup\bigcup R$, we have that $a \in A$ and $b \in A$. So $\langle a, b \rangle \in A \times A$. $\square$

Let R be a relation and consider the class $\{x : \exists y(\langle x, y \rangle \in R)\}$. Is this class a set? If $\exists y(\langle x, y \rangle \in R)$, Lemma 3.2.4 implies that $x \in \bigcup\bigcup R$. Since $\bigcup\bigcup R$ is a set, Theorem 2.1.3 implies that $\{x : \exists y(\langle x, y \rangle \in R)\}$ is a set. A similar argument shows that $\{y : \exists x \langle x, y \rangle \in R\}$ is also a set. We can now define the sets $\mathrm{dom}(R)$ and $\mathrm{ran}(R)$ for any relation R.

Definition 3.2.5. Suppose R is a relation.

1. The **domain** of R, denoted by $\mathrm{dom}(R)$, is $\mathrm{dom}(R) = \{x : \exists y(\langle x, y \rangle \in R)\}$.
2. The **range** of R, denoted by $\mathrm{ran}(R)$, is the set $\mathrm{ran}(R) = \{y : \exists x(\langle x, y \rangle \in R)\}$.
3. The **field** of R is the set $\mathrm{fld}(R) = \mathrm{dom}(R) \cup \mathrm{ran}(R)$.

Consider the relation $R = \{\langle a, b \rangle \in A \times B : a \text{ evenly divides } b \text{ and } a < b\}$ where $A = \{0, 1, 2, 3, 4, 5\}$ and $B = \{4, 6, 7\}$. Let us evaluate the sets $\mathrm{dom}(R)$ and $\mathrm{ran}(R)$. We first observe that

$$R = \{\langle 1, 4 \rangle, \langle 1, 6 \rangle, \langle 1, 7 \rangle, \langle 2, 4 \rangle, \langle 2, 6 \rangle, \langle 3, 6 \rangle\}.$$

Hence, $\mathrm{dom}(R) = \{1, 2, 3\}$ and $\mathrm{ran}(R) = \{4, 6, 7\}$.

3.2.1 Operations on Relations

Let R be a relation. In our next definition, we define R^{-1} to be the class

$$R^{-1} = \{\langle v, u \rangle : \langle u, v \rangle \in R\}.$$

We now show that R^{-1} is a set. Since $\mathrm{ran}(R)$ and $\mathrm{dom}(R)$ are sets, we conclude from Lemma 3.1.4 that $\mathrm{ran}(R) \times \mathrm{dom}(R)$ is a set. Because $R^{-1} = \{x : \varphi(x)\}$ where $\varphi(x)$ is the formula $\exists u \exists v(x = \langle v, u \rangle \wedge \langle u, v \rangle \in R)$, it follows that for all x, if $\varphi(x)$, then $x \in \mathrm{ran}(R) \times \mathrm{dom}(R)$. So, by Theorem 2.1.3, R^{-1} is a set.

Our next definition is usually applied to functions, but it is also applicable to relations. We will see in Section 3.3 that functions are just relations that satisfy a special property.

Definition 3.2.6. Let R and S be relations.

(1) The **inverse** of R is the relation

$$R^{-1} = \{\langle v, u \rangle : \langle u, v \rangle \in R\}.$$

Thus, $\langle y, x \rangle \in R^{-1}$ iff $\langle x, y \rangle \in R$.

(2) For each set A, the **restriction** of R to A is the relation

$$R \restriction A = \{\langle u, v \rangle : \langle u, v \rangle \in R \wedge u \in A\}.$$

Consequently, $\langle x, y \rangle \in R \restriction A$ iff $\langle x, y \rangle \in R \wedge x \in A$.

(3) For any set A, the **image** of A under R is the set given by

$$R[A] = \{v : (\exists u \in A)(\langle u, v \rangle \in R)\}.$$

Therefore, $y \in R[A]$ iff $(\exists x \in A)(\langle x, y \rangle \in R)$.

(4) For a set B, the **inverse image** of B under R is the set

$$R^{-1}[B] = \{u : (\exists v \in B)(\langle u, v \rangle \in R)\}.$$

Hence, $x \in R^{-1}[B]$ iff $(\exists y \in B)(\langle x, y \rangle \in R)$.

(5) The **composition** of R and S is the relation

$$R \circ S = \{\langle u, v \rangle : \exists t(\langle u, t \rangle \in S \wedge \langle t, v \rangle \in R)\}.$$

Consequently, $\langle x, y \rangle \in R \circ S$ iff $\exists t(\langle x, t \rangle \in S \wedge \langle t, y \rangle \in R)$.

As noted earlier, R^{-1} is a set for every relation R. Similar reasoning shows that the above $R \restriction A$, $R[A]$, $R^{-1}[B]$, and $R \circ S$ are also sets (see Exercise 7).

Example 4. Consider the relation C on the set P of all people defined by

$$C = \{\langle x, y \rangle \in P \times P : x \text{ is a child of } y\}.$$

Then $C^{-1} = \{\langle y, x \rangle \in P \times P : y \text{ is a parent of } x\}$ and

$$C \circ C = \{\langle x, z \rangle \in P \times P : x \text{ is a grandchild of } z\}.$$

If A is the set of all people born in Australia, then

$$C^{-1}[A] = \{x \in P : x \text{ has a parent who was born in Australia}\}.$$

Now consider the relation L on P given by $L = \{\langle u, v \rangle \in P \times P : u \text{ loves } v\}$. To evaluate $L \circ C^{-1}$, we obtain

$$\langle a, b \rangle \in L \circ C^{-1} \quad \text{iff} \quad \langle a, t \rangle \in C^{-1} \wedge \langle t, b \rangle \in L \qquad \text{for some } t$$

$$\text{iff} \quad \langle t, a \rangle \in C \wedge \langle t, b \rangle \in L \qquad \text{for some } t$$

$$\text{iff} \quad a \text{ has a child who loves } b.$$

Theorem 3.2.7. *Let R, S, and T be relations. Then*

(1) $\text{dom}(R^{-1}) = \text{ran}(R)$.
(2) $\text{ran}(R^{-1}) = \text{dom}(R)$.
(3) $(R^{-1})^{-1} = R$.
(4) $(R \circ S)^{-1} = S^{-1} \circ R^{-1}$.
(5) $R \circ (S \circ T) = (R \circ S) \circ T$.

Proof. The proofs of items (1)–(3) are reserved for the exercises. We will prove items (4) and (5) by appealing to propositional logic laws.

(4) We shall prove that $(R \circ S)^{-1} = S^{-1} \circ R^{-1}$. Let x and y be given. Then

$\langle x, y \rangle \in (R \circ S)^{-1}$	iff $\langle y, x \rangle \in R \circ S$	Def. 3.2.6(1)
	iff $\langle y, t \rangle \in S \wedge \langle t, x \rangle \in R$	for some t, by Def. 3.2.6(5)
	iff $\langle t, x \rangle \in R \wedge \langle y, t \rangle \in S$	commutative logic law
	iff $\langle x, t \rangle \in R^{-1} \wedge \langle t, y \rangle \in S^{-1}$	Def. 3.2.6(1)
	iff $\langle x, y \rangle \in S^{-1} \circ R^{-1}$	Def. 3.2.6(5).

Therefore, $(R \circ S)^{-1} = S^{-1} \circ R^{-1}$.

(5) To prove that $R \circ (S \circ T) = (R \circ S) \circ T$, let x and y be given. Then

$\langle x, y \rangle \in R \circ (S \circ T)$	iff $\langle x, u \rangle \in S \circ T \wedge \langle u, y \rangle \in R$	for some u
	iff $(\langle x, w \rangle \in T \wedge \langle w, u \rangle \in S) \wedge \langle u, y \rangle \in R$	for some w
	iff $\langle x, w \rangle \in T \wedge (\langle w, u \rangle \in S) \wedge \langle u, y \rangle \in R)$	associative law
	iff $\langle x, w \rangle \in T \wedge \langle w, y \rangle \in R \circ S$	Def. 3.2.6(5)
	iff $\langle x, y \rangle \in (R \circ S) \circ T$	Def. 3.2.6(5).

Therefore, $R \circ (S \circ T) = (R \circ S) \circ T$. $\qquad\square$

Let R be a relation and let $\mathcal{G}$ be a set (of sets). So $R[\bigcup \mathcal{G}]$ can be viewed as the image of a union. In Exercise 8, one is asked to prove that the collection of images $\{R[C] : C \in \mathcal{G}\}$ is a set. Thus, $\bigcup \{R[C] : C \in \mathcal{G}\}$ is the union of images, where $y \in \bigcup \{R[C] : C \in \mathcal{G}\}$ iff $y \in R[C]$ for some $C \in \mathcal{G}$. In our next theorem, items (1) and (2) show that "the image of a union is the union of the images," whereas items (3) and (4) proclaim that "the image of an intersection is a subset of the intersection of the images." In the proof of this theorem, we will be using the quantifier distribution laws (QDL), namely, 1.3.5(3) and 1.3.5(4) on page 19.

Theorem 3.2.8. *Let R be a relation. Suppose that A and B are sets. In addition, suppose that $\mathcal{G}$ is a set (of sets). Then*

(1) $R[A \cup B] = R[A] \cup R[B]$.
(2) $R[\bigcup \mathcal{G}] = \bigcup \{R[C] : C \in \mathcal{G}\}$.
(3) $R[A \cap B] \subseteq R[A] \cap R[B]$.
(4) $R[\bigcap \mathcal{G}] \subseteq \bigcap \{R[C] : C \in \mathcal{G}\}$, when $\mathcal{G} \neq \varnothing$.
(5) $R[A] \setminus R[B] \subseteq R[A \setminus B]$.

Proof. We prove only items (1) and (3). The proofs of (2), (4), and (5) will be left to the reader (see Exercises 8, 19, 20, and 21). Let R be a relation and let A and B be sets.

(1) We shall prove that $R[A \cup B] = R[A] \cup R[B]$. Let y be arbitrary. Then

$$y \in R[A \cup B] \text{ iff } (\exists x \in A \cup B)(\langle x, y \rangle \in R) \qquad \text{Def. 3.2.6}$$

$$\text{iff } (\exists x \in A)(\langle x, y \rangle \in R) \vee (\exists x \in B)(\langle x, y \rangle \in R) \quad \text{QDL 1.3.5(3)}$$

$$\text{iff } y \in R[A] \vee y \in R[B] \qquad \text{Def. 3.2.6}$$

$$\text{iff } y \in R[A] \cup R[B] \qquad \text{definition of } \cup.$$

Therefore, $R[A \cup B] = R[A] \cup R[B]$.

(3) To prove that $R[A \cap B] \subseteq R[A] \cap R[B]$, let $y \in R[A \cap B]$. Observe that

$$y \in R[A \cap B] \text{ iff } (\exists x \in A \cap B)(\langle x, y \rangle \in R) \qquad \text{Def. 3.2.6}$$

$$\Rightarrow (\exists x \in A)(\langle x, y \rangle \in R) \wedge (\exists x \in B)(\langle x, y \rangle \in R) \quad \text{QDL 1.3.5(4)}$$

$$\text{iff } y \in R[A] \wedge y \in R[B] \qquad \text{Def. 3.2.6}$$

$$\text{iff } y \in R[A] \cap R[B] \qquad \text{definition of } \cap.$$

Hence, $y \in R[A] \cap R[B]$. Therefore, $R[A \cap B] \subseteq R[A] \cap R[B]$.   □

Is there a particular property so that if a relation R satisfies this property, then equality will hold for items (4) and (5) of Theorem 3.2.8? Our next corollary shows that these two items can be expressed as equalities whenever R happens to be a single-rooted relation.

Definition 3.2.9. A relation R is **single-rooted** if for each $y \in \text{ran}(R)$ there is exactly one x such that $\langle x, y \rangle \in R$.

Thus, if R is a single-rooted relation, then whenever $\langle x, y \rangle \in R$ and $\langle z, y \rangle \in R$, we can conclude that $x = z$.

Corollary 3.2.10. *Let R be a single-rooted relation. Suppose that A and B are sets and that $\mathcal{G}$ is a nonempty set. Then*

(1) $R[A] \setminus R[B] = R[A \setminus B]$.
(2) $R[\bigcap \mathcal{G}] = \bigcap \{R[C] : C \in \mathcal{G}\}$.

Proof. Let R be a single-rooted relation.

To prove (1), let A and B be sets. By Theorem 3.2.8(5), we need to prove only that $R[A \setminus B] \subseteq R[A] \setminus R[B]$. Let $y \in R[A \setminus B]$. Thus, $\langle x, y \rangle \in R$ for some $x \in A \setminus B$. So $x \in A$ and $x \notin B$. Since $x \in A$ and $\langle x, y \rangle \in R$, we have $y \in R[A]$. We now show that $y \notin R[B]$. Assume, to the contrary, that $y \in R[B]$. Therefore, $\langle b, y \rangle \in R$ for some $b \in B$. Hence, $\langle b, y \rangle \in R$ and $\langle x, y \rangle \in R$. Because R is single-rooted, we must have that $x = b$ and so $x \in B$, which is a contradiction. Thus, $y \notin R[B]$. Therefore, $y \in R[A]$ and $y \notin R[B]$. We conclude that $y \in R[A] \setminus R[B]$.

To prove (2), let $\mathcal{G}$ be a nonempty set. By Theorem 3.2.8(4), we just need to prove that $\bigcap \{R[C] : C \in \mathcal{G}\} \subseteq R[\bigcap \mathcal{G}]$. Let $y \in \bigcap \{R[C] : C \in \mathcal{G}\}$. Thus, we have that $y \in R[C]$ for every $C \in \mathcal{G}$. In other words,

$$\text{for every } C \in \mathcal{G}, \text{ there is an } x \in C \text{ such that } \langle x, y \rangle \in R. \tag{3.4}$$

Claim. There is a d such that $d \in \bigcap \mathcal{G}$ and $\langle d, y \rangle \in R$.

Proof of Claim. As $\mathcal{G}$ is nonempty, let $D \in \mathcal{G}$. By (3.4) there is a $d \in D$ so that $\langle d, y \rangle \in R$. We now show that this particular d is an element of every $C \in \mathcal{G}$. Let $C \in \mathcal{G}$. By (3.4) there exists an $x \in C$ so that $\langle x, y \rangle \in R$. Thus, $\langle x, y \rangle \in R$ and $\langle d, y \rangle \in R$. Because R is single-rooted, we conclude that $x = d$. Hence, $d \in C$ and therefore, $d \in \bigcap \mathcal{G}$ and $\langle d, y \rangle \in R$. (Claim) $\square$

The above claim implies that $y \in R[\bigcap \mathcal{G}]$ and this concludes the proof of (2). The proof of the corollary is now complete. $\square$

3.2.2   Reflexive, Symmetric, and Transitive Relations

The symbols R and $\sim$ shall be used to denote relations. Let R and $\sim$ be relations on a set A. If $a \in A$ and $b \in A$, we will write $a R b$ to mean that $\langle a, b \rangle \in R$. When we use the notation $a R b$, we shall say that "a is related to b." Similarly, we write $a \sim b$ to mean that $\langle a, b \rangle \in \sim$ and say that a is related to b. We also write $a \nsim b$ to assert that "a is not related to b."

Relations that have properties similar to those of equality appear frequently in mathematics. We now recognize three fundamental properties of equality. For quantities x, y, and z, we have the following:

1. $x = x$ (reflexive).
2. If $x = y$, then $y = x$ (symmetric).
3. If $x = y$ and $y = z$, then $x = z$ (transitive).

In this section, we will investigate relations that share some, or all, of the above three properties. In our next definition, we shall identify what it means for a relation to be reflexive, symmetric, and transitive. Relations that possess one or more of these properties occur naturally in many areas of mathematics, including set theory.

Definition 3.2.11. Let $\sim$ be a relation on A. Then

(1) $\sim$ is **reflexive** if $(\forall x \in A)(x \sim x)$; that is, $x \sim x$ for every $x \in A$.
(2) $\sim$ is **symmetric** when

$$(\forall x \in A)(\forall y \in A)(x \sim y \to y \sim x).$$

In other words, if $x \sim y$, then $y \sim x$ whenever $x \in A$ and $y \in A$.
(3) $\sim$ is **transitive** if

$$(\forall x \in A)(\forall y \in A)(\forall z \in A)[(x \sim y \land y \sim z) \to x \sim z].$$

That is, if $x \sim y$ and $y \sim z$, then $x \sim z$ whenever x, y, z are elements in A.

For a relation on A, the above properties can be expressed as follows:

1. The relation is *reflexive* if every element in the set A is related to itself.
2. The relation is *symmetric* if whenever x is related to y, then y is related to x.
3. The relation is *transitive* if whenever x is related to y and y is related to z, then x is also related to z.

3.2.3   Equivalence Relations and Partitions

Because the equality relation is so useful and essential, mathematicians have generalized this concept. A relation shall be called an *equivalence relation* if it satisfies the three key properties that are normally associated with equality.

Definition 3.2.12. A relation on a set A is an **equivalence relation on** A (or just an **equivalence relation** when A is clear from the context) if it is reflexive, symmetric, and transitive.

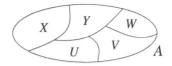

Figure 3.1. A partition of the set A.

An equivalence relation allows one to connect those elements of a set that have a particular property in common. One can show that each of the relations below is an equivalence relation:

1. The relation $\sim$ on the set $\mathbb{Z}$ defined by $m \sim n$ if and only if $m - n$ is even.
2. The relation $\sim$ on the set $\mathbb{R}$ defined by $x \sim y$ if and only if $|x| = |y|$.
3. The relation $\sim$ on the set $\mathbb{Z}$ defined by $m \sim n$ iff 3 evenly divides $(m - n)$.

A *partition* of a set breaks up the set into disjoint and nonempty subsets. In other words, a partition of a set A is a set of nonempty disjoint subsets of A whose union is equal to A. Figure 3.1 portrays a set A that is broken up into five subsets X, Y, U, V, W. Thus, $P = \{X, Y, U, V, W\}$ is a partition of A because every element of A is in one of the nonempty sets X, Y, U, V, W and any two of these sets are disjoint. The following definition formalizes this notion of breaking up a set into disjoint parts.

Definition 3.2.13. Let A be a set. Let P be a set of nonempty subsets of A. We say that P is a **partition** of A if the following two conditions hold:

1. For every element $a \in A$, there is a set $S \in P$ such that $a \in S$.
2. For all S and T in P, if $S \neq T$, then $S \cap T = \varnothing$.

Item 1 of Definition 3.2.13 asserts that every element in A belongs to a set in the partition P. Item 2 states that any two different sets in P are disjoint. When this occurs we say that the sets in P are *pairwise disjoint.* Thus, a partition of a set A divides the set into nonoverlapping parts that, together, cover all of A.

The main result that we will soon establish is that an equivalence relation on a set A induces a partition of A. Therefore, an equivalence relation can be used to create a new set from an old one. For each $a \in A$, we must first form the set of all those elements in A that are related to a.

Definition 3.2.14. Let $\sim$ be an equivalence relation on A. Let a be an element in A. The **equivalence class** of a, denoted by $[a]_\sim$, is the set of all elements in A that are related to a, namely, $[a]_\sim = \{x \in A : x \sim a\}$.

By the subset axiom, an equivalence class is a set. In Definition 3.2.14, we shall write $[a] = [a]_\sim$ whenever the relation $\sim$ is understood.

Example 5. Let $\sim$ be the equivalence relation on $\mathbb{R}$ defined by $x \sim y$ if and only if $|x| = |y|$. Then the equivalence classes $[1]$, $[-1]$, and $[2]$ are

$$[1] = \{x \in \mathbb{R} : x \sim 1\} = \{x \in \mathbb{R} : |x| = |1|\} = \{1, -1\},$$

$$[-1] = \{x \in \mathbb{R} : x \sim -1\} = \{x \in \mathbb{R} : |x| = |-1|\} = \{-1, 1\},$$

$$[2] = \{x \in \mathbb{R} : x \sim 2\} = \{x \in \mathbb{R} : |x| = |2|\} = \{2, -2\}.$$

Hence, $[1] = [-1]$ and $[1] \neq [2]$. In addition, observe that $[1] \cap [2] = \varnothing$.

Remark 3.2.15. Let $\sim$ be an equivalence relation on A and let $a \in A$. Then $[a] \subseteq A$, and moreover $x \in [a]$ if and only if $x \sim a$.

Example 6. Let $\sim$ be the equivalence relation on $\mathbb{Z}$ defined by $m \sim n$ if and only if $m - n$ is even. Thus, $[k] = \{m \in \mathbb{Z} : m \sim k\} = \{m \in \mathbb{Z} : m - k \text{ is even}\}$, for each integer k. We evaluate the equivalence classes $[1]$, $[2]$, and $[3]$ as follows:

$$[1] = \{m \in \mathbb{Z} : m - 1 \text{ is even}\} = \{\ldots, -5, -3, -1, 1, 3, 5, 7, \ldots\}.$$

$$[2] = \{m \in \mathbb{Z} : m - 2 \text{ is even}\} = \{\ldots, -6, -4, -2, 0, 2, 4, 6, 8, \ldots\}.$$

$$[3] = \{m \in \mathbb{Z} : m - 3 \text{ is even}\} = \{\ldots, -5, -3, -1, 1, 3, 5, 7, \ldots\}.$$

Therefore, $[1] = [3]$ and $[3] \neq [2]$. Furthermore, $[3] \cap [2] = \varnothing$.

For any equivalence relation, our next theorem shows that two elements are related if and only if their equivalence classes are equal.

Theorem 3.2.16. *Let $\sim$ be an equivalence relation on A. Then for all $a \in A$ and $b \in A$,*

$$a \sim b \text{ if and only if } [a] = [b].$$

Proof. Suppose that $\sim$ is an equivalence relation on A. Let $a \in A$ and $b \in A$. We shall prove that $a \sim b$ if and only if $[a] = [b]$. Assume $a \sim b$. We prove that $[a] = [b]$; that is, we prove that these two sets are equal. Let $x \in [a]$. We shall show that $x \in [b]$. Since $x \in [a]$, it follows that $x \sim a$. By assumption, we also have that $a \sim b$. Thus, $x \sim a$ and $a \sim b$. Because $\sim$ is transitive, we conclude that $x \sim b$ and hence, $x \in [b]$. So, $[a] \subseteq [b]$. A very similar argument shows that $[b] \subseteq [a]$. Therefore, $[a] = [b]$. To prove the converse, assume $[a] = [b]$. Since $a \in [a]$, we see that $a \in [b]$. Hence, $a \sim b$. $\qquad\square$

We now present a useful corollary.

Corollary 3.2.17. *Let $\sim$ be an equivalence relation on a set A. For any a and b in A, we have that $a \in [b]$ if and only if $[a] = [b]$.*

Let $\sim$ be an equivalence relation on a set A. Theorem 2.1.3 implies that the collection of all equivalence classes $\{x : (\exists a \in A)(x = [a])\}$ is a set. This set is denoted by $\{[a] : a \in A\}$. In our next theorem, we show that the set of all equivalence classes forms a partition of A. Thus, an equivalence relation on a set can be used to break up the set into nonempty subsets that do not overlap.

Theorem 3.2.18 (Fundamental Theorem of Equivalence Relations). *Let $\sim$ be an equivalence relation on A. Then the set $P = \{[a] : a \in A\}$ is a partition of A.*

Proof. Let $\sim$ be an equivalence relation on A. We shall prove that the collection $P = \{[a] : a \in A\}$ is a partition of A. To do this, we show that (1) for every element $x \in A$, we have that $x \in [x]$, and (2) for all $x \in A$ and $y \in A$, if $[x] \neq [y]$, then $[x] \cap [y] = \varnothing$. For (1), let $x \in A$. Clearly, $[x] \in P$ and $x \in [x]$, because $\sim$ is reflexive. Now, to prove (2), let $x \in A$, $y \in A$, and assume $[x] \cap [y] \neq \varnothing$ (we are using proof by contraposition). Thus, there is a $z \in A$ such that $z \in [x]$ and $z \in [y]$. Therefore, $[z] = [x]$ and $[z] = [y]$ by Corollary 3.2.17. So $[x] = [y]$. □

Definition 3.2.19. Let $\sim$ be an equivalence relation on A. Then $A/\!\!\sim$ denotes the partition $\{[a] : a \in A\}$ of A and is called the **quotient set** induced by $\sim$.

An equivalence relation $\sim$ on A breaks up the set A into disjoint subsets, as illustrated in the following figure:

$$A = \begin{array}{|c|c|c|c|c} \vdots & \vdots & \vdots & \vdots & \\ a & x & c & d & \cdots \\ \vdots & \vdots & \vdots & \vdots & \end{array}$$
$$\qquad \uparrow \ \ \uparrow \ \ \uparrow \ \ \uparrow$$
$$\quad [a]\ [x]\ [c]\ [d]\ \cdots$$

Mathematicians frequently use the quotient set $A/\!\!\sim$ to learn something new about the set A itself.

Example 7. Consider the equivalence relation $\sim$ on $\mathbb{Z}$ defined by $m \sim n$ if and only if $(m - n)$ is evenly divisible by 3. For all integers m and n, it follows that $m \sim n$ if and only if $m = 3k + n$ for some $k \in \mathbb{Z}$. Therefore,

$$[n] = \{m \in \mathbb{Z} : m \sim n\} = \{3k + n : k \in \mathbb{Z}\}$$

for each integer n. We can now evaluate the equivalence classes $[0]$, $[1]$, and $[2]$:

$$[0] = \{3k : k \in \mathbb{Z}\} = \{\ldots, -9, -6, -3, 0, 3, 6, 9, \ldots\}.$$

$$[1] = \{3k + 1 : k \in \mathbb{Z}\} = \{\ldots, -8, -5, -2, 1, 4, 7, 10, \ldots\}.$$

$$[2] = \{3k + 2 : k \in \mathbb{Z}\} = \{\ldots, -7, -4, -1, 2, 5, 8, 11, \ldots\}.$$

The quotient set $\mathbb{Z}/\sim$ is $\{[n] : n \in \mathbb{Z}\} = \{[0], [1], [2]\}$ and is illustrated in the figure:

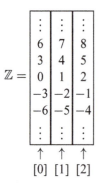

Exercises 3.2

*1. Explain why the empty set is a relation. [Hint: Suppose $\varnothing$ is not a relation and derive a contradiction.]

2. Prove items 1–3 of Theorem 3.2.7.

3. Let R be a relation. Prove that if $A \subseteq B$, then $R[A] \subseteq R[B]$.

4. Let R be the relation defined by

$$R = \{\langle 0, 1 \rangle, \langle 0, 2 \rangle, \langle 2, 0 \rangle, \langle 2, 3 \rangle, \langle 4, 4 \rangle, \langle 2, 4 \rangle, \langle 3, 3 \rangle, \langle 1, 2 \rangle, \langle 1, 3 \rangle, \langle 3, 2 \rangle\}.$$

Evaluate $\mathrm{dom}(R)$, $\mathrm{ran}(R)$, $R \circ R$, $R \upharpoonright \{1\}$, $R^{-1} \upharpoonright \{1\}$, $R[\{1\}]$, and $R^{-1}[\{1\}]$.

5. Suppose that R is a relation. Prove that $R \upharpoonright (A \cup B) = (R \upharpoonright A) \cup (R \upharpoonright B)$ for any sets A and B.

6. Let R be a relation. Prove that $\text{fld}(R) = \bigcup \bigcup R$.

*7. Let R and S be two relations and let A, B, and C be sets. Prove that $R \upharpoonright A$, $R^{-1}[B]$, $R[C]$, and $R \circ S$ are sets.

*8. Let R be a relation and $\mathcal{G}$ be a set. Prove that $\{R[C] : C \in \mathcal{G}\}$ is a set. Prove that if $\mathcal{G}$ is nonempty, then $\{R[C] : C \in \mathcal{G}\}$ is also nonempty.

9. Let R be a relation on A. Suppose that $\text{dom}(R) = A$ and $R^{-1} \circ R \subseteq R$. Prove that R is reflexive on A.

10. Let R be a relation on A. Prove that R is symmetric if and only if $R^{-1} \subseteq R$.

11. Let R be a relation on A. Prove that R is transitive if and only if $R \circ R \subseteq R$.

12. Let R be a relation on A. Prove that R is symmetric and transitive if and only if $R^{-1} \circ R = R$.

13. A relation R on A is said to be *antisymmetric* if, for all $x \in A$ and $y \in A$, whenever $\langle x, y \rangle \in R$ and $\langle y, x \rangle \in R$, then $x = y$. Prove that a relation R on A is antisymmetric if and only if $R \circ R^{-1} \subseteq I_A$ (see page 44).

14. Let $\mathcal{G}$ be a nonempty set of transitive relations on A. Prove that the relation $\bigcap \mathcal{G}$ is transitive.

15. Let R be a relation on A. Prove that if R is an equivalence relation on A, then R^{-1} is also an equivalence relation on A.

16. Let R be an equivalence relation on A. Prove that $R \circ R = R$.

17. Let R and S be relations on A. Suppose that R is reflexive on A. Prove that $S \subseteq R \circ S$ and $S \subseteq S \circ R$.

18. Let R and S be a reflexive relations on A. Suppose that R is also transitive. Prove $S \subseteq R$ if and only if $S \circ R = R$.

*19. Prove item (2) of Theorem 3.2.8 (see Exercise Notes).

*20. Prove item (4) of Theorem 3.2.8. Observe that $y \in \bigcap\{R[C] : C \in \mathcal{G}\}$ iff $y \in R[C]$ for every $C \in \mathcal{G}$.

*21. Prove item (5) of Theorem 3.2.8.

22. Let R and S be single-rooted relations. Prove that $R \circ S$ is single-rooted.

23. Let R and S be relations. If $\text{dom}(R) = \text{ran}(S)$ and $R \circ S$ is single-rooted, then prove that S is single-rooted.

24. Let P be a partition of a set A. Now define the relation $\sim$ on A as follows: $x \sim y$ if and only if $x \in C$ and $y \in C$ for some $C \in P$. Prove that $\sim$ is an equivalence relation on A.

Exercise Notes: For Exercise **19**, note that $y \in \bigcup\{R[C] : C \in \mathcal{G}\}$ iff $y \in R[C]$ for some $C \in \mathcal{G}$, and

$$y \in R[\textstyle\bigcup\mathcal{G}] \text{ iff } (\exists x \in \textstyle\bigcup\mathcal{G}) \langle x, y \rangle \in R$$

$$\text{iff } \exists x(x \in \textstyle\bigcup\mathcal{G} \wedge \langle x, y \rangle \in R)$$

$$\text{iff } \exists x(\exists C \in \mathcal{G})(x \in C \wedge \langle x, y \rangle \in R)$$

$$\text{iff } (\exists C \in \mathcal{G})\exists x(x \in C \wedge \langle x, y \rangle \in R),$$

recalling the quantifier interchange law 1.3.4(1). For Exercise **20**, note that

$$y \in R[\textstyle\bigcap\mathcal{G}] \text{ iff } (\exists x \in \textstyle\bigcap\mathcal{G}) \langle x, y \rangle \in R$$

$$\text{iff } \exists x(x \in \textstyle\bigcap\mathcal{G} \wedge \langle x, y \rangle \in R)$$

$$\text{iff } \exists x(\forall C \in \mathcal{G})(x \in C \wedge \langle x, y \rangle \in R)$$

$$\Rightarrow (\forall C \in \mathcal{G})\exists x(x \in C \wedge \langle x, y \rangle \in R),$$

recalling the quantifier interchange law 1.3.4(3). For Exercise **21**, observe that

$$y \in R[A] \setminus R[B] \text{ iff } y \in R[A] \wedge y \notin R[B]$$

$$\text{iff } (\exists x \in A) \langle x, y \rangle \in R \wedge \neg(\exists x \in B) \langle x, y \rangle \in R$$

$$\Rightarrow (\exists x \in A \setminus B) \langle x, y \rangle \in R.$$

Exercise **24** and Theorem 3.2.18 connect the seemingly unrelated concepts of an equivalence relation on a set A and a partition on A.

3.3 Functions

One of the most important ideas in modern mathematics is the concept of a *function*. Moreover, the function concept is one that is used extensively outside of mathematics. A function is a way of associating each element of a set A with exactly one element in another set B. For example, $f(x) = x^2$ is a function that associates each real number x with the real number x^2. The functions studied in a first calculus course have the set $\mathbb{R}$ of real numbers, or subsets of $\mathbb{R}$, as domain and codomain. In this chapter, we will look at functions in a more general context and will examine some important properties that a function may possess. To do this, we must first give a precise set-theoretic definition of a function.

Definition 3.3.1. A relation R is **single-valued** if for each $x \in \text{dom}(R)$ there is exactly one y such that $\langle x, y \rangle \in R$.

Thus, if R is a single-valued relation, then whenever $\langle x, y \rangle \in R$ and $\langle x, z \rangle \in R$, we can conclude that $y = z$.

Definition 3.3.2. A **function** F is any single-valued relation; that is, for each $x \in \text{dom}(F)$ there is only one y such that $\langle x, y \rangle \in F$.

Since a function is also a relation, all of the results and exercises presented in Section 3.2 (excluding equivalence relations) apply to functions as well.

Let A and B be sets. A *function F from A to B* is a subset of $A \times B$ such that for each $x \in A$ there is exactly one $y \in B$ so that $\langle x, y \rangle \in F$. We now express this notion in terms of a formal definition.

Definition 3.3.3. Let A and B be sets, and let F be a relation from A to B. Then F is said to be a **function from A to B** if the following two conditions hold:

(1) $\text{dom}(F) = A$; that is, for each $x \in A$, there is a $y \in B$ such that $\langle x, y \rangle \in F$.
(2) F is single-valued; that is, if $\langle x, y \rangle \in F$ and $\langle x, z \rangle \in F$, then $y = z$.

The set A is the **domain** of F and the set B is called the **codomain** of F.

Example 1. Let $A = \{a, b, c, d, e\}$ and $B = \{5, 6, 7, 8, 9\}$. Then

$$F = \{\langle a, 8 \rangle, \langle b, 7 \rangle, \langle c, 9 \rangle, \langle d, 6 \rangle, \langle e, 5 \rangle\}$$

is a function from A to B because for each $x \in A$, there is exactly one $y \in B$ such that $\langle x, y \rangle \in F$. On the other hand, the set of ordered pairs

$$G = \{\langle a, 8 \rangle, \langle b, 7 \rangle, \langle c, 9 \rangle, \langle d, 6 \rangle, \langle b, 8 \rangle, \langle e, 5 \rangle\}$$

is not a function from A to B because $\langle b, 7 \rangle \in g$ and $\langle b, 8 \rangle \in g$, but $7 \neq 8$. Hence, item (2) of Definition 3.3.3 fails to hold. In addition, the set

$$H = \{\langle a, 8 \rangle, \langle b, 7 \rangle, \langle c, 9 \rangle, \langle e, 5 \rangle\}$$

is not a function from A to B because $d \in A$ and there is no $y \in B$ such that $\langle d, y \rangle \in H$. So item (1) of Definition 3.3.3 does not hold.

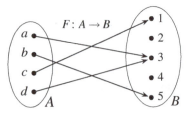

Figure 3.2. Arrow diagram of the function given in Example 2.

We write $F : A \to B$ to denote that F is a function from the set A to the set B. Thus, for each $x \in A$, there is exactly one $y \in B$ such that $\langle x, y \rangle \in F$. This unique y is called "the value of F at x" and is denoted by $F(x)$. Therefore, $\langle x, y \rangle \in F$ if and only if $F(x) = y$. The value $F(x)$ is called "F of x" or "the image of x under F." In addition, one can say that $x \in A$ is an *input* to the function F and that $F(x)$ is the resulting *output*. Moreover, one can also say that the function F maps x to $F(x)$, denoted by $x \mapsto F(x)$.

When we are given a function $F : A \to B$, we know that each $x \in A$ is mapped to exactly one element $F(x)$ in B, because F is single-valued.

Remark 3.3.4. Technically speaking, when F is a set of ordered pairs, one can use the notation $F(x)$ only when it is known, or it is clear, that F is a function.

Consider the set of ordered pairs $F = \{\langle x, y \rangle \in \mathbb{R} \times \mathbb{R} : y = 2x^3 + 1\}$. One can easily show that F satisfies the conditions of Definition 3.3.2. Thus, F is a function, and we can write $F(x) = 2x^3 + 1$, for all $x \in \mathbb{R}$.

When A and B are finite sets, then a function $F : A \to B$ can be represented by drawing an arrow from each element $x \in A$ to the corresponding element $F(x) \in B$ (see Figure 3.2). Such a drawing is called an *arrow diagram*. These diagrams can help one to gain a better understanding of the function concept and the concepts of domain and codomain.

Example 2. Let $A = \{a, b, c, d\}$ and $B = \{1, 2, 3, 4, 5\}$. Consider the function $F : A \to B$ defined by $F = \{\langle a, 3 \rangle, \langle b, 5 \rangle, \langle c, 1 \rangle, \langle d, 3 \rangle\}$. Thus,

$$F(a) = 3, \; F(b) = 5, \; F(c) = 1, \; \text{and} \; F(d) = 3.$$

Since A and B are finite, we can illustrate the function F by means of the arrow diagram in Figure 3.2. Clearly, each element $x \in A$ is mapped to exactly one element $F(x)$ in B. Observe that $a \neq d$ and $F(a) = F(d)$. So it is possible for distinct elements in the domain to produce the same value under a function. In fact, many functions have this "repeated value" property. Such a function is not *single-rooted*.

If F and G are relations, then $F = G$ if and only if F and G have exactly the same ordered pairs. The following lemma offers a useful tool for showing that two functions are equal.

Lemma 3.3.5. *Let F and G be functions such that* $\text{dom}(F) = \text{dom}(G)$. *Then* $F = G$ *if and only if* $F(x) = G(x)$ *for all x in their common domain.*

Proof. See Exercise 1. $\qquad\qquad\qquad\qquad\qquad\qquad\qquad\qquad\qquad\square$

Let A and B be sets. Consider the class $\{F : F$ is a function from A to $B\}$. Since a function $F : A \to B$ is a subset of $A \times B$, we conclude that every such function is an element in $\mathcal{P}(A \times B)$; it thereby follows from Theorem 2.1.3 that this class is a set.

Definition 3.3.6. Let A and B be sets. The set of all functions from A to B, denoted by ^{A}B, is defined by $^{A}B = \{F : F$ is a function from A to $B\}$.

Some authors use the notation B^A for the set of all functions from A to B. The notation ^{A}B is commonly used in set theory because the set A is read before B, emphasizing the fact that the set consists of all functions of the form $f : A \to B$.

Example 3. Let $\omega = \{0, 1, 2, 3, \ldots\}$ be the set of natural numbers.

1. $^{\omega}\{0, 1\}$ is the set of all functions $f : \omega \to \{0, 1\}$. One can think of $^{\omega}\{0, 1\}$ as the set of all infinite binary sequences.
2. $^{\omega}\omega$ is the set of all functions $f : \omega \to \omega$. One can think of $^{\omega}\omega$ as the set of all infinite sequences of natural numbers.

Remark 3.3.7. The **empty function** is the function whose domain is the empty set. For each set B, there is exactly one function $f : \varnothing \to B$. Since $f \subseteq \varnothing \times B$ and the product $\varnothing \times B = \varnothing$, this unique function f is just the empty set. In other words, the empty set $\varnothing$ vacuously satisfies the condition given in Definition 3.3.2. Thus, $\varnothing$ is a function from $\varnothing$ to B, for any set B. Hence, in Definition 3.3.6, $^{\varnothing}B = \{\varnothing\}$ for every set B, and $^{A}\varnothing = \varnothing$ when A is nonempty.

3.3.1 Operations on Functions

Let F be a function. Since F is a single-valued relation, we have that $\langle x, y \rangle \in F$ if and only if $F(x) = y$. Therefore, we can slightly modify Definition 3.2.6 to obtain the following definition.

Definition 3.3.8. Let F and G be functions.

(1) The **inverse** of F is the relation

$$F^{-1} = \{\langle v, u\rangle : \langle u, v\rangle \in F\}.$$

Thus, $\langle y, x\rangle \in F^{-1}$ iff $\langle x, y\rangle \in F$ iff $F(x) = y$.

(2) For each A, the **restriction** of F to A is the function

$$F \restriction A = \{\langle u, v\rangle : \langle u, v\rangle \in F \wedge u \in A\}$$

with domain $\mathrm{dom}(F) \cap A$. Therefore, $\langle x, y\rangle \in F \restriction A$ iff $F(x) = y \wedge x \in A$.

(3) For any $A \subseteq \mathrm{dom}(F)$, the **image** of A under F is the subset of $\mathrm{ran}(F)$ given by

$$F[A] = \{F(x) : x \in A\},$$

where $\{F(x) : x \in A\}$ is equal to the set $\{y : y = F(x) \text{ for some } x \in A\}$. Hence, $y \in F[A]$ iff $(\exists x \in A)(y = F(x))$. So, if $x \in A$, then $F(x) \in F[A]$.

(4) For a set B, the **inverse image** of B under F is the subset of $\mathrm{dom}(F)$ defined by

$$F^{-1}[B] = \{u : F(u) \in B\}.$$

Therefore, $x \in F^{-1}[B]$ iff $F(x) \in B$.

(5) The **composition** of F and G is the relation

$$F \circ G = \{\langle u, v\rangle : \exists t(\langle u, t\rangle \in G \wedge \langle t, v\rangle \in F)\}.$$

Consequently, $\langle x, y\rangle \in F \circ G$ iff $\exists t(\langle x, t\rangle \in G \wedge \langle t, y\rangle \in F)$.

Remark 3.3.9. We make a few notes concerning Definition 3.3.8.

(a) Let F be a function. In general, F^{-1} is not a function, but it is a relation.

(b) Suppose that F is a function and let $A \subseteq \mathrm{dom}(F)$. Then $F \restriction A$ is a function with domain A.

(c) Given functions F and G, we will prove below that the composition $F \circ G$ is a function.

(d) As a function F is a single-valued relation, the range of F is the set

$$\mathrm{ran}(F) = \{F(x) : x \in \mathrm{dom}(F)\}.$$

Theorem 3.3.10. *Let F and G be functions. Then*

(a) *$F \circ G$ is a function.*

(b) *$\mathrm{dom}(F \circ G) = \{x \in \mathrm{dom}(G) : G(x) \in \mathrm{dom}(F)\}$.*

(c) *For all $x \in \mathrm{dom}(F \circ G)$, we have $(F \circ G)(x) = F(G(x))$.*

Proof. Let F and G be functions. We shall prove statements (a), (b), and (c).

(a) To prove that the relation $F \circ G$ is a function, let $x \in \text{dom}(F \circ G)$. Assume $\langle x, y \rangle \in (F \circ G)$ and $\langle x, z \rangle \in (F \circ G)$. We shall prove that $y = z$. Since $\langle x, y \rangle \in (F \circ G)$ and $\langle x, z \rangle \in (F \circ G)$, there are s and t such that

$$\langle x, t \rangle \in G \text{ and } \langle t, y \rangle \in F,$$

$$\langle x, s \rangle \in G \text{ and } \langle s, z \rangle \in F.$$

Hence,

$$\langle x, t \rangle \in G \text{ and } \langle x, s \rangle \in G, \tag{3.5}$$

$$\langle t, y \rangle \in F \text{ and } \langle s, z \rangle \in F. \tag{3.6}$$

Since G is a function, we conclude from (3.5) that $t = s$. So from (3.6) we obtain $\langle t, y \rangle \in F$ and $\langle t, z \rangle \in F$. Because F is a function, we infer that $y = z$.

(b) Let u be given. Because G is a function, we see that $\exists t (\langle u, t \rangle \in G)$ if and only if $\langle u, G(u) \rangle \in G$. We have the following equivalences:

$$u \in \text{dom}(F \circ G) \text{ iff } \langle u, y \rangle \in F \circ G \qquad \text{for some } y$$

$$\text{iff } \langle u, t \rangle \in G \wedge \langle t, y \rangle \in F \qquad \text{for some } t$$

$$\text{iff } \langle u, G(u) \rangle \in G \wedge \langle G(u), y \rangle \in F \qquad t = G(u)$$

$$\text{iff } u \in \text{dom}(G) \wedge G(u) \in \text{dom}(F) \qquad \text{definition of domain}$$

$$\text{iff } u \in \{x \in \text{dom}(G) : G(x) \in \text{dom}(F)\} \qquad \text{definition of set.}$$

Therefore, $\text{dom}(F \circ G) = \{x \in \text{dom}(G) : G(x) \in \text{dom}(F)\}$.

(c) Let $x \in \text{dom}(F \circ G)$. Since $F \circ G$ is a function, we see that $\langle x, y \rangle \in F \circ G$ where $y = (F \circ G)(x)$. It follows, as in the proof of (b), that $\langle G(x), y \rangle \in F$. Thus, $y = F(G(x))$, as F is a function. So $(F \circ G)(x) = y = F(G(x))$. $\qquad \square$

Remark 3.3.11. If $G : A \to B$, $F : B \to C$, and $H : C \to D$, Theorem 3.3.10 implies that $(F \circ G) : A \to C$ and $(F \circ G)(x) = F(G(x))$ for all $x \in A$ and that $(H \circ F \circ G) : A \to D$ where

$$(H \circ F \circ G)(x) = ((H \circ F) \circ G)(x) = (H \circ (F \circ G))(x)$$

for all $x \in A$ (see Theorem 3.2.7(5)).

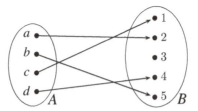

Figure 3.3. Arrow diagram of a one-to-one function $F : A \to B$.

3.3.2 One-to-One Functions

Some functions (see Figure 3.2) may have two inputs that are assigned to the same output and thus, such functions have a repeated value. If a function has no repeated values, then we will say that the function is *one-to-one*. For example, it is easy to see that the function in Figure 3.3 is one-to-one.

Definition 3.3.12. A function F is said to be **one-to-one** (or an **injection**), if and only if any two distinct elements in the domain of F are mapped to distinct elements, or equivalently

$$(\forall x \in \mathrm{dom}(F))(\forall y \in \mathrm{dom}(F))[F(x) = F(y) \to x = y].$$

The following lemma is straightforward to prove (see Exercise 1) and will be used to establish an important consequence of our next theorem.

Lemma 3.3.13. *A function F is one-to-one if and only if F is single-rooted.*

Theorem 3.3.14. *Let F be a relation. Then*

(a) *F is a function iff F^{-1} is single-rooted.*
(b) *F^{-1} is a function iff F is single-rooted.*

Proof. Since (a) and (b) are essentially the same statement, we just prove (a).

($\Rightarrow$). Assume F is a function. Thus, F is single-valued. So to prove that F^{-1} is single-rooted, assume that $\langle y, x \rangle \in F^{-1}$ and $\langle z, x \rangle \in F^{-1}$. We must prove $y = z$. Since $\langle y, x \rangle \in F^{-1}$ and $\langle z, x \rangle \in F^{-1}$, we have that $\langle x, y \rangle \in F$ and $\langle x, z \rangle \in F$, by the definition of F^{-1}. Because F is a function, we infer that $y = z$.

($\Leftarrow$). Suppose that F^{-1} is single-rooted. To prove that F is a function, assume that $\langle x, y \rangle \in F$ and $\langle x, z \rangle \in F$. We prove $y = z$. Since $\langle x, y \rangle \in F$ and $\langle x, z \rangle \in F$, we conclude that $\langle y, x \rangle \in F^{-1}$ and $\langle z, x \rangle \in F^{-1}$, by the definition of F^{-1}. Now, since F^{-1} is single-rooted, we conclude that $y = z$. □

Corollary 3.3.15. *If F is a one-to-one function, then F^{-1} is also a one-to-one function.*

Proof. Assume that the function F is one-to-one. Hence, F is single-rooted by Lemma 3.3.13. Theorem 3.3.14(b) now implies that F^{-1} is a function. As F is a function, Theorem 3.3.14(a) allows us to conclude that F^{-1} is single-rooted. Therefore, F^{-1} is one-to-one by Lemma 3.3.13. $\qquad\square$

Theorem 3.3.16. *Let F be a one-to-one function.*

(a) *If $x \in \mathrm{dom}(F)$, then $F^{-1}(F(x)) = x$.*
(b) *If $y \in \mathrm{ran}(F)$, then $F(F^{-1}(y)) = y$.*

Proof. Assume F is a one-to-one function. Corollary 3.3.15 implies that F^{-1} is a function. Since the proofs of (a) and (b) are similar, we just prove (a). Assume $x \in \mathrm{dom}(F)$. Thus, $\langle x, F(x)\rangle \in F$. Hence, $\langle F(x), x\rangle \in F^{-1}$. So $F^{-1}(F(x)) = x$, because F^{-1} is a function. $\qquad\square$

We note that the empty function vacuously satisfies Definition 3.3.12. Thus, $\varnothing$ is a one-to-one function from $\varnothing$ to B, for any set B.

Definition 3.3.17. A function $F: A \to B$ maps A **onto** B when $\mathrm{ran}(F) = B$, that is, if $(\forall y \in B)(\exists x \in A)[\, F(x) = y\,]$ (see Figure 3.4). Such a function is also referred to as a **surjection**.

Theorem 3.3.18. *If $F: A \to B$ is a one-to-one function that maps A onto B, then $F^{-1}: B \to A$ is a one-to-one function that maps B onto A, and*

(a) *$F^{-1}(F(x)) = x$ for all $x \in A$.*
(b) *$F(F^{-1}(y)) = y$ for all $y \in B$.*

Proof. Let $F: A \to B$ be a one-to-one function that maps A onto B. Hence, $\mathrm{dom}(F) = A$ and $\mathrm{ran}(F) = B$. Since F is one-to-one, Corollary 3.3.15 implies that F^{-1} is a one-to-one function, and, furthermore, Theorem 3.2.7 shows that $\mathrm{dom}(F^{-1}) = B$ and $\mathrm{ran}(F^{-1}) = A$. Therefore, $F^{-1}: B \to A$ is a one-to-one function that maps B onto A. Theorem 3.3.16 implies items (a) and (b). $\qquad\square$

A function $f: A \to B$ that is one-to-one and onto B is called a **bijection**.

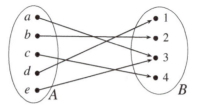

Figure 3.4. Arrow diagram of a function $F: A \to B$ that maps onto B.

Theorem 3.3.19. *If the functions $g: A \to B$ and $f: B \to C$ are one-to-one, then the composition $(f \circ g): A \to C$ is one-to-one.*

Proof. Suppose the functions $g: A \to B$ and $f: B \to C$ are one-to-one. To prove that the function $(f \circ g): A \to C$ is one-to-one, let $x \in A$ and $y \in A$. Assume that $(f \circ g)(x) = (f \circ g)(y)$. Thus, (i) $f(g(x)) = f(g(y))$ by Remark 3.3.11. Since f is one-to-one, we conclude from (i) that $g(x) = g(y)$. Because g is one-to-one, we see that $x = y$. This completes the proof. □

Theorem 3.3.20. *If $g: A \to B$ is onto B and $f: B \to C$ is onto C, then the composition $(f \circ g): A \to C$ is onto C.*

Proof. Assume $g: A \to B$ is onto B and $f: B \to C$ is onto C. We shall prove that the function $(f \circ g): A \to C$ is onto C. Let $z \in C$. Since $f: B \to C$ is onto C and $z \in C$, there is a $y \in B$ such that $f(y) = z$. Because $y \in B$ and $g: A \to B$ is onto B, there is an $x \in A$ such that $g(x) = y$. We will show that $(f \circ g)(x) = z$ as follows:

$$(f \circ g)(x) = f(g(x)) \quad \text{by definition of composition}$$
$$= f(y) \quad \text{because } g(x) = y$$
$$= z \quad \text{because } f(y) = z. \qquad \square$$

3.3.3   Indexed Sets

There are occasions when the range of a function is deemed to be more important than the function itself. When that is the case, both the terminology and the notation undergo radical alterations.

– Paul R. Halmos [7]

Let F be a function with domain I and range $\{F(i) : i \in I\}$. To simplify our notation, let us write $F_i = F(i)$, for each $i \in I$. It is sometimes useful to identify the function F by using the *indexed notation* $\langle F_i : i \in I \rangle$ where I is referred to as the **index set**, an element i in I is called an **index**, and each value F_i of the function at an index i is called a **term**. Thus, $\langle F_i : i \in I \rangle$ will be said to have *nonempty terms* when $F_i \neq \emptyset$ for all $i \in I$. We shall say that $\langle F_i : i \in I \rangle$ is an **indexed function**, and its range $\{F_i : i \in I\}$ shall be called an **indexed set**. An indexed set is also referred to as an **indexed family** of sets. Whenever $J \subseteq I$, then $\langle F_i : i \in J \rangle$ is the restriction of the function $\langle F_i : i \in I \rangle$ to J.

When we say that $\{F_i : i \in I\}$ is an indexed set, we just mean that it is the range of the indexed function $F = \langle F_i : i \in I \rangle$. Indexed sets appear frequently in mathematics. The union of an indexed set $\{F_i : i \in I\}$, written as $\bigcup_{i \in I} F_i$, appears routinely in mathematics. We now give a formal definition of this union.

Definition 3.3.21. Let $\{F_i : i \in I\}$ be an indexed set. Define

$$\bigcup_{i \in I} F_i = \bigcup \{F_i : i \in I\};$$

that is, $\bigcup_{i \in I} F_i = \{x : x \in F_i \text{ for some } i \in I\}$.

Thus, $\bigcup_{i \in I} F_i$ consists of those elements that are in at least one of the sets F_i. For example, suppose $I = \{1, 2, 3, 4\}$. Then $\{F_i : i \in I\} = \{F_1, F_2, F_3, F_4\}$, and

$$\bigcup_{i \in I} F_i = \bigcup \{F_1, F_2, F_3, F_4\} = F_1 \cup F_2 \cup F_3 \cup F_4.$$

Definition 3.3.22. Let $\{F_i : i \in I\}$ be an indexed set where $I \neq \emptyset$. Define

$$\bigcap_{i \in I} F_i = \bigcap \{F_i : i \in I\};$$

that is, $\bigcap_{i \in I} F_i = \{x : x \in F_i \text{ for every } i \in I\}$.

The intersection $\bigcap_{i \in I} F_i$ is composed of those elements that belong to each and every one of the sets F_i.

3.3.4 The Axiom of Choice

Suppose that we have a set S that contains only nonempty sets. Is it possible to uniformly select exactly one element from each set in S? In other words, is there a function F so that for each $A \in S$, we have that $F(A) \in A$? The following set theoretic principle will allow to answer this question.

Axiom of Choice. Let $\langle A_i : i \in I \rangle$ be an indexed function with nonempty terms. Then there is an indexed function $\langle x_i : i \in I \rangle$ such that $x_i \in A_i$, for all $i \in I$.

Definition 3.3.23. Let $\langle A_i : i \in I \rangle$ be an indexed function with nonempty terms. An indexed function $\langle x_i : i \in I \rangle$ shall be called a **choice function** for $\langle A_i : i \in I \rangle$, if $x_i \in A_i$ for all $i \in I$.

Let $\langle A_i : i \in I \rangle$ be a function where $A_i \neq \varnothing$ for all $i \in I$. Suppose that there exists an explicit method for identifying a single element in A_i and that this method works for each $i \in I$. Then one can construct a choice function without appealing to the axiom of choice. Moreover, if I is a finite set, then one can prove that there is a choice function (by induction) without using the axiom of choice (see Exercise 22 on page 116). On the other hand, there are times when the only way to obtain a choice function is by appealing to the axiom of choice. Mathematicians often use the axiom of choice when the index set is infinite and it is not clear how to construct a choice function.

To better understand the axiom of choice, let us look at three examples where we can define a choice function without using the axiom of choice:

1. $\langle A_i : i \in I \rangle$ is such that each A_i is a nonempty set of natural numbers. For each $i \in I$, define ℓ_i to be the least element in A_i. Then $\langle \ell_i : i \in I \rangle$ is a choice function for $\langle A_i : i \in I \rangle$.
2. $\langle A_i : i \in I \rangle$ is such that each A_i is a nonempty finite interval of real numbers. For each $i \in I$, let m_i be the midpoint of the interval A_i. Then $\langle m_i : i \in I \rangle$ is a choice function for $\langle A_i : i \in I \rangle$.
3. $\langle A_i : i \in I \rangle$ is such that each A_i contains one pair of shoes. For each $i \in I$, let r_i be the right-footed shoe in A_i. Then $\langle r_i : i \in I \rangle$ is a desired choice function.

In each of the above examples, we presented a method for identifying a single element in A_i, and this method worked for each $i \in I$. Thus, we were able to uniformly select exactly one element from each set A_i without using the axiom of choice.

Now let I be an infinite set and consider the following three examples where, in general, it is not possible to define a choice function:

1. $\langle A_i : i \in I \rangle$ is such that each A_i is an infinite set of real numbers.
2. $\langle A_i : i \in I \rangle$ is such that each A_i is a nonempty set of functions.
3. $\langle A_i : i \in I \rangle$ is such that each A_i contains one pair of identical socks.

It is not at all clear, in each of these three examples, how to uniformly select exactly one element from each set A_i. By the axiom of choice, however, there is such a choice function. (The socks/shoes examples are due to B. Russell.)

The Zermelo–Fraenkel system of axioms is usually denoted by ZF, and the axiom of choice is abbreviated by AC. The axiom of choice was formulated by Ernst Zermelo; however, Zermelo and Fraenkel did not include this axiom in ZF. For this reason, the result of adding the axiom of choice to the system ZF is denoted by ZFC. There were some early attempts to prove the axiom of choice assuming only the axioms in ZF. Since these attempts were not successful, mathematicians began to doubt the possibility of proving the axiom of choice, and eventually this was shown to be the case. The combined work of Kurt Gödel in 1940 and Paul Cohen in 1963 confirmed that the axiom of choice is independent of the Zermelo–Fraenkel axioms; that is, AC cannot be proven or refuted using just the axioms in ZF. Nevertheless, the axiom of choice is a powerful tool in mathematics, and some important theorems cannot be established without it. Consequently, mathematicians typically assume the axiom of choice and usually cite the axiom when they use it in a proof. Our next theorem uses the axiom of choice in its proof, and we alert the reader to this fact by using the notation (AC) just before the statement of the theorem.

Suppose that C is a set of nonempty sets. In the proof of our next theorem, we convert C into an indexed set via a simple change in notation. Using the set C itself as the index set, together with the identity function on C, we shall reexpress C as an indexed set.

Theorem 3.3.24 (AC). *Let C be a set of nonempty sets. Then there is a function* $H : C \to \bigcup C$ *such that* $H(A) \in A$ *for all* $A \in C$.

Proof. Let C be a set of nonempty sets. Let $F : C \to C$ be the identity function on C; that is, let $F(A) = A$ for all $A \in C$. Let $F_A = F(A) = A$, for each $A \in C$. We can now rewrite F as the following indexed function $\langle F_A : A \in C \rangle$. By the axiom of choice, there is an indexed function $\langle x_A : A \in C \rangle$ such that $x_A \in A$ for all $A \in C$. Define $H : C \to C$ by $H(A) = x_A$, for each $A \in C$. Therefore, $H(A) \in A$ for every $A \in C$. $\qquad\square$

The above function H shall be called a **choice function** for C. Informally, the axiom of choice declares that for any collection of nonempty sets, it is possible to uniformly choose exactly one element from each set in the collection.

The statement of Theorem 3.3.24 actually implies the axiom of choice, and thus, the theorem is equivalent to the axiom (see Exercise 21). We will revisit the axiom of choice in Chapter 7.

Exercises 3.3

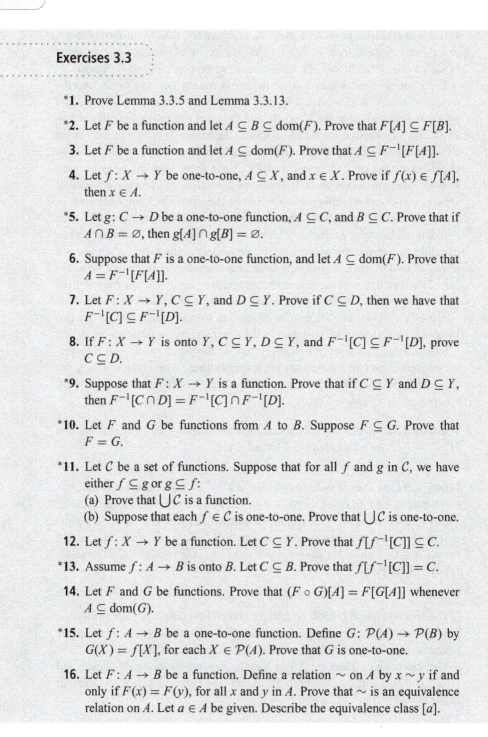

*1. Prove Lemma 3.3.5 and Lemma 3.3.13.

*2. Let F be a function and let $A \subseteq B \subseteq \mathrm{dom}(F)$. Prove that $F[A] \subseteq F[B]$.

3. Let F be a function and let $A \subseteq \mathrm{dom}(F)$. Prove that $A \subseteq F^{-1}[F[A]]$.

4. Let $f: X \to Y$ be one-to-one, $A \subseteq X$, and $x \in X$. Prove if $f(x) \in f[A]$, then $x \in A$.

*5. Let $g: C \to D$ be a one-to-one function, $A \subseteq C$, and $B \subseteq C$. Prove that if $A \cap B = \varnothing$, then $g[A] \cap g[B] = \varnothing$.

6. Suppose that F is a one-to-one function, and let $A \subseteq \mathrm{dom}(F)$. Prove that $A = F^{-1}[F[A]]$.

7. Let $F: X \to Y$, $C \subseteq Y$, and $D \subseteq Y$. Prove if $C \subseteq D$, then we have that $F^{-1}[C] \subseteq F^{-1}[D]$.

8. If $F: X \to Y$ is onto Y, $C \subseteq Y$, $D \subseteq Y$, and $F^{-1}[C] \subseteq F^{-1}[D]$, prove $C \subseteq D$.

*9. Suppose that $F: X \to Y$ is a function. Prove that if $C \subseteq Y$ and $D \subseteq Y$, then $F^{-1}[C \cap D] = F^{-1}[C] \cap F^{-1}[D]$.

*10. Let F and G be functions from A to B. Suppose $F \subseteq G$. Prove that $F = G$.

*11. Let $\mathcal{C}$ be a set of functions. Suppose that for all f and g in $\mathcal{C}$, we have either $f \subseteq g$ or $g \subseteq f$:
    (a) Prove that $\bigcup \mathcal{C}$ is a function.
    (b) Suppose that each $f \in \mathcal{C}$ is one-to-one. Prove that $\bigcup \mathcal{C}$ is one-to-one.

12. Let $f: X \to Y$ be a function. Let $C \subseteq Y$. Prove that $f[f^{-1}[C]] \subseteq C$.

*13. Assume $f: A \to B$ is onto B. Let $C \subseteq B$. Prove that $f[f^{-1}[C]] = C$.

14. Let F and G be functions. Prove that $(F \circ G)[A] = F[G[A]]$ whenever $A \subseteq \mathrm{dom}(G)$.

*15. Let $f: A \to B$ be a one-to-one function. Define $G: \mathcal{P}(A) \to \mathcal{P}(B)$ by $G(X) = f[X]$, for each $X \in \mathcal{P}(A)$. Prove that G is one-to-one.

16. Let $F: A \to B$ be a function. Define a relation $\sim$ on A by $x \sim y$ if and only if $F(x) = F(y)$, for all x and y in A. Prove that $\sim$ is an equivalence relation on A. Let $a \in A$ be given. Describe the equivalence class $[a]$.

17. Let $F: A \to A$ be a function. Suppose that F is a reflexive relation on A. Prove that $F(x) = x$ for all $x \in A$.

18. Let $F: A \to A$ be a function. Prove that F is a symmetric relation on A if and only if $F(F(x)) = x$ for all $x \in A$.

19. Let $F: A \to A$ be a function. Prove that F is a transitive relation on A if and only if $F(F(x)) = F(x)$ for all $x \in A$.

20. Prove the following theorems:
    (a) **Theorem.** Let $\{A_i : i \in I\}$ and $\{B_i : i \in I\}$ be indexed sets with the same index set I. Suppose $A_i \subseteq B_i$ for all $i \in I$. Then $\bigcup_{i \in I} A_i \subseteq \bigcup_{i \in I} B_i$.
    (b) **Theorem.** Let $\{A_i : i \in I\}$ and $\{B_i : i \in I\}$ be indexed sets with the same index set I. Suppose $A_i \subseteq B_i$ for all $i \in I$. Then $\bigcap_{i \in I} A_i \subseteq \bigcap_{i \in I} B_i$.
    (c) **Theorem.** Suppose that A is a set and that $\{B_i : i \in I\}$ is an indexed set. Then $A \cap \bigcup_{i \in I} B_i = \bigcup_{i \in I}(A \cap B_i)$.
    (d) **Theorem.** Suppose that A is a set and that $\{B_i : i \in I\}$ is an indexed set. Then $A \cup \bigcap_{i \in I} B_i = \bigcap_{i \in I}(A \cup B_i)$.
    (e) **Theorem.** Suppose that A is a set and that $\{B_i : i \in I\}$ is an indexed set. Then $A \setminus \bigcap_{i \in I} B_i = \bigcup_{i \in I}(A \setminus B_i)$.

*21. Let $\langle A_i : i \in I \rangle$ be an indexed function with nonempty terms. Prove that there is an indexed function $\langle x_i : i \in I \rangle$ so that $x_i \in A_i$ for all $i \in I$, using Theorem 3.3.24.

22. Let $A \subseteq \mathbb{R}$ and $\mathbb{N}^+ = \{1, 2, 3, \dots\}$. Suppose that for all $n \in \mathbb{N}^+$ there is an $x \in A$ so that $0 < x < \frac{1}{n}$. Using the axiom of choice, show that there is an indexed function $\langle x_n : n \in \mathbb{N}^+ \rangle$ such that $x_n \in A$ and $0 < x_n < \frac{1}{n}$, for all $n \in \mathbb{N}^+$.

3.4 Order Relations

The equality relation was generalized, in Section 3.2.3, by the introduction of an equivalence relation. An equivalence relation satisfies three key properties that are usually associated with the equality relation; namely, such a relation is reflexive, symmetric, and transitive. We shall now generalize the concept of "less than or equal to."

The familiar relation $\leq$ on the set $\mathbb{R}$ puts an order on the real numbers. This relation is reflexive and transitive; but, since $2 \leq 5$ and $5 \nleq 2$, the relation $\leq$ is not symmetric. Moreover, for x and y in $\mathbb{R}$, if $x \leq y$ and $y \leq x$, then $x = y$. We therefore say that $\leq$ is *antisymmetric*. We now present a general definition.

Definition 3.4.1. A relation R on a set A is **antisymmetric** when

$$(\forall x \in A)(\forall y \in A)((x\,R\,y \wedge y\,R\,x) \rightarrow x = y);$$

that is, if $x\,R\,y$ and $y\,R\,x$, then $x = y$ whenever x and y are in A.

Let us now review the three core properties of the relation $\leq$ on the set of real numbers. For all $a, b, c \in \mathbb{R}$ we have the following:

1. $a \leq a$ (reflexive).
2. If $a \leq b$ and $b \leq a$, then $a = b$ (antisymmetric).
3. If $a \leq b$ and $b \leq c$, then $a \leq c$ (transitive).

The above three properties are the ones that constitute the notion of *order*. We can now generalize the concept of "less than or equal to" and apply it to many different sets. To do this, we shall use the symbol $\preccurlyeq$ to denote a relation.

Definition 3.4.2. A relation $\preccurlyeq$ on a set A is a **partial order** if $\preccurlyeq$ is reflexive, antisymmetric, and transitive; that is, for x, y, and z in A, the following hold:

1. $x \preccurlyeq x$.
2. If $x \preccurlyeq y$ and $y \preccurlyeq x$, then $x = y$.
3. If $x \preccurlyeq y$ and $y \preccurlyeq z$, then $x \preccurlyeq z$.

If $\preccurlyeq$ is a partial order on the set A, then we shall say that the structure $(A, \preccurlyeq)$ is a **partially ordered set** or a **poset**.[1] For example, $(\mathbb{R}, \leq)$ is a poset.

Problem 1. Let $\mathcal{F}$ be a set (of sets) and let $\subseteq$ be the subset relation. Show that $(\mathcal{F}, \subseteq)$ is a partially ordered set.

Solution. Let A, B, and C be sets in $\mathcal{F}$. We verify the following three items:

(1) $A \subseteq A$.
(2) If $A \subseteq B$ and $B \subseteq A$, then $A = B$.
(3) If $A \subseteq B$ and $B \subseteq C$, then $A \subseteq C$.

Clearly, (1) is valid. If $A \subseteq B$ and $B \subseteq A$, then $A = B$ by the extensionality axiom. Thus, (2) holds. Finally, item (3) follows from Exercise 2 on page 35.

Ⓢ

Problem 2. For integers a and b, the *divisibility relation* $a \mid b$ means that $b = ai$ for some $i \in \mathbb{Z}$. Let A be a set of nonzero natural numbers. Show that $(A, \mid)$ is a partially ordered set.

Solution. We show that the divisibility relation is reflexive, antisymmetric, and transitive on the set A. Let a, b, c be in A. We need to verify the following three items:

(1) $a \mid a$.
(2) If $a \mid b$ and $b \mid a$, then $a = b$.
(3) If $a \mid b$ and $b \mid c$, then $a \mid c$.

Since $a = a \cdot 1$, we have that $a \mid a$. Items (2) and (3) can also be verified. ⓢ

Definition 3.4.3. Let $\preccurlyeq$ be a partial order on A. The relation $\preccurlyeq$ is a **total order** (or *linear order*) on A if $\preccurlyeq$ satisfies $(\forall x \in A)(\forall y \in A)(x \preccurlyeq y \lor y \preccurlyeq x)$; that is, either $x \preccurlyeq y$ or $y \preccurlyeq x$, for every x and y in A.

When $(A, \preccurlyeq)$ is a poset and $\preccurlyeq$ is a total order, then we shall say that $(A, \preccurlyeq)$ is a **totally ordered set**. The poset $(\mathbb{R}, \leq)$ is a totally ordered set. On the other hand, let $A = \{2, 5, 4, 8, 9\}$ and let $\mid$ be the divisibility relation. Then $(A, \mid)$ is a poset, but since $2 \nmid 5$ and $5 \nmid 2$, we see that $(A, \mid)$ is not a totally ordered set.

Definition 3.4.4. Let $(A, \preccurlyeq)$ be a poset. For x and y in A, we write $x \prec y$ if and only if $x \preccurlyeq y$ and $x \neq y$. The relation $\prec$ on A shall be referred to as the **strict order** corresponding to $\preccurlyeq$.

A strict order is not a partial order, because it is not reflexive. This is verified by our next lemma.

Lemma 3.4.5. *Let $(A, \preccurlyeq)$ be a poset, and let $\prec$ be the strict order corresponding to $\preccurlyeq$. Then for all x, y, and z in A, we have the following:*

1. $x \nprec x$.
2. *If $x \prec y$, then $y \nprec x$.*
3. *If $x \prec y$ and $y \prec z$, then $x \prec z$.*
4. *Let $\preccurlyeq$ be a total order on A. Then exactly one of the following holds: $x \prec y$, or $x = y$, or $y \prec x$.*

Proof. See Exercise 2. □

So, if $(A, \preccurlyeq)$ is a totally ordered set, then Lemma 3.4.5(4) implies that the strict order corresponding to $\preccurlyeq$ satisfies the **trichotomy law** on A; that is, for all x and y in A, exactly one of three alternatives holds: $x \prec y$, $x = y$, or $y \prec x$.

Let $\preccurlyeq$ be a partial order on a set A. If $x \prec y$, then we will say that x is **smaller** than y and that y is **larger** than x.

Definition 3.4.6. Let $\preccurlyeq$ be a partial order on a set A. An element $b \in A$ is called a **maximal element** if and only if $b \nprec x$ for all $x \in A$; that is, there is no element in A that is larger than b.

Definition 3.4.7. Let $\preccurlyeq$ be a partial order on a set A. An element $a \in A$ is said to be a **minimal element** if and only if $x \nprec a$ for all $x \in A$; that is, there is no element in A that is smaller than a.

Example 3. Let $A = \{2, 3, 4, 5, 6, 8, 9\}$. Thus, 2, 3, 5 are minimal elements in the poset $(A, \mid)$ (see Problem 2) and 5, 6, 8, 9 are maximal elements.

Definition 3.4.8. Let $\preccurlyeq$ be a partial order on A. Let $S \subseteq A$, $a \in A$, and $b \in A$.

• If b satisfies $(\forall x \in S)(x \preccurlyeq b)$, then b is called an **upper bound** for S.
• If a satisfies $(\forall x \in S)(a \preccurlyeq x)$, then a is called a **lower bound** for S.

Let $(A, \mid)$ be the poset given in Example 3 where $A = \{2, 3, 4, 5, 6, 9\}$. Hence, $S = \{2, 3\}$ is a subset of A. Observe that $6 \in A$ is an upper bound for S, but there is no lower bound for S in A.

Definition 3.4.9. Let $\preccurlyeq$ be a partial order on a set A, and let $S \subseteq A$.

• If ℓ is an upper bound for S and $\ell \preccurlyeq b$ whenever b is another upper bound for S, then ℓ is called the **least upper bound** for S.
• If g is a lower bound for S and $a \preccurlyeq g$ whenever a is another lower bound for S, then g is called the **greatest lower bound** for S.

Our next lemma shows that whenever a least upper bound for a set exists, it is unique. Similarly, if a greatest lower bound exists, then it is unique as well.

Lemma 3.4.10. *Let $\preccurlyeq$ be a poset on A. Let $S \subseteq A$. If ℓ and ℓ' are least upper bounds for S, then $\ell = \ell'$. If g and g' are greatest lower bounds for S, then $g = g'$.*

Proof. Let ℓ and ℓ' be least upper bounds for S. Since ℓ is a least upper bound and ℓ' is another upper bound for S, it follows from the definition of least upper bound that $\ell \preccurlyeq \ell'$. Similarly, it follows that $\ell' \preccurlyeq \ell$. By antisymmetry, we thus conclude that $\ell = \ell'$. An analogous argument shows that if g and g' are greatest lower bounds for S, then $g = g'$. □

Consider the poset $(A, \,|\,)$ where $A = \{1, 2, 3, 4, 5, \ldots\}$. Let $S = \{6, 9, 12\}$. So 36 is the least upper bound for S, and 3 is the greatest lower bound for S.

Definition 3.4.11. Let $\preccurlyeq$ be a partial order on A where $S \subseteq A$, $a \in A$, and $b \in A$.

- If $(\forall x \in S)(x \preccurlyeq b)$ and $b \in S$, then b is the **largest (greatest) element** in S.
- If $(\forall x \in S)(a \preccurlyeq x)$ and $a \in S$, then a is the **smallest (least) element** in S.

There is a simple distinction between an upper bound and a largest element. An upper bound for S need not be an element in S. A largest element of S is just an upper bound that is also an element in S (see Exercise 4). Similarly, a smallest element of S is just a lower bound that is also an element in the set S.

Lemma 3.4.12. *Let $\preccurlyeq$ be a partial order on A. Let $S \subseteq A$. If b and b' are largest elements of S, then $b = b'$. If a and a' are smallest elements of S, then $a = a'$.*

Proof. Let b and b' be largest elements of S. So, in particular, b and b' are both elements in S. Since b is a largest element of S and $b' \in S$, it follows that $b' \preccurlyeq b$. Similarly, it follows that $b \preccurlyeq b'$. By antisymmetry, we have that $b = b'$. If a and a' are smallest elements of S, an analogous argument shows that $a = a'$. □

Example 4. Consider the poset $(A, \,|\,)$, where $A = \{1, 2, 3, 4, 5, \ldots\}$ and $|$ is the divisibility relation. Let $S = \{4, 6, 12\}$. Then 12 is the largest element in S. The set S has no smallest element; however, 2 is the greatest lower bound for S.

Definition 3.4.13. Let $(A, \preccurlyeq)$ be a poset, and let x and y be elements in A. Then x and y are said to be **comparable** if either $x \preccurlyeq y$ or $y \preccurlyeq x$; otherwise they are said to be **incomparable**.

A partial order on a set A may not be a total order, but it is a total order on any subset of A in which any two elements are comparable.

Definition 3.4.14. Let $\preccurlyeq$ be a partial order on a set A, and let $C \subseteq A$. Then C is called a **chain** in A if for all x and y in C, either $x \preccurlyeq y$ or $y \preccurlyeq x$.

Clearly, if $D \subseteq C$ and C is a chain in A, then D is also a chain in A.

Example 5. Let $A = \{1, 2, 3, 4, 5, \dots\}$ and consider the poset $(A, \mid)$, where $\mid$ is the divisibility relation. Then the set $\{1, 3, 6, 12\}$ is a chain in A for which 24 is an upper bound. The set $\{3^n : n \in A\}$ is a chain in A that has no upper bound.

Now let $\preccurlyeq$ be a relation on A. Given any subset C of A, we can also view $\preccurlyeq$ as a relation on C (see Exercise 14).

Definition 3.4.15. Let $\preccurlyeq$ be a relation on A, and let $C \subseteq A$. Then the **induced relation** on C is defined to be $\preccurlyeq_C = \preccurlyeq \cap (C \times C)$. In other words,

$x \preccurlyeq_C y$ iff $x \in C$, $y \in C$, and $x \preccurlyeq y$.

Definition 3.4.16. Let $(A, \preccurlyeq)$ and $(B, \preccurlyeq^*)$ be partially ordered sets. A bijection $f : A \to B$ is called an **isomorphism** from $(A, \preccurlyeq)$ onto $(B, \preccurlyeq^*)$ if f satisfies

$x \preccurlyeq y$ if and only if $f(x) \preccurlyeq^* f(y)$,

for all x and y in A. If such an isomorphism exists, then $(A, \preccurlyeq)$ and $(B, \preccurlyeq^*)$ are said to be **isomorphic**.

When two posets are isomorphic, they are structurally the same poset.

Exercises 3.4

1. Define a relation $\preccurlyeq$ on the set of integers $\mathbb{Z}$ by $x \preccurlyeq y$ if and only if $x \leq y$ and $x + y$ is even, for all x, y in $\mathbb{Z}$. Prove that $\preccurlyeq$ is a partial order on $\mathbb{Z}$. Then answer the following questions about the poset $(\mathbb{Z}, \preccurlyeq)$:
   (a) Is $S = \{1, 2, 3, 4, 5, 6, \dots\}$ a chain in $\mathbb{Z}$?
   (b) Is $S = \{1, 3, 5, 7, \dots\}$ a chain in $\mathbb{Z}$?

(c) Does the set $S = \{1, 2, 3, 4, 5\}$ have a lower bound or an upper bound?

(d) Does $S = \{1, 2, 3, 4, 5\}$ have any maximal or minimal elements?

*2. Prove Lemma 3.4.5.

3. Find the greatest lower bound of the set $S = \{15, 20, 30\}$ in the poset $(A, \mid)$ where $A = \{n \in \mathbb{N} : n > 0\}$. Now find the least upper bound of S.

*4. Let $(A, \preccurlyeq)$ be a poset and let $S \subseteq A$. Suppose that b is the largest element of S. Prove that b is also the least upper bound of S.

5. Let $\preccurlyeq$ be a partial order on a set A, and let $S \subseteq A$. Suppose g and g' are both greatest lower bounds of S. Prove that $g = g'$.

6. Let $\mathcal{P}(A)$ be the set of all subsets of the set $A = \{a, b, c\}$. Thus, $(\mathcal{P}(A), \subseteq)$ is a partially ordered set, where

$$\mathcal{P}(A) = \{\varnothing, \{a\}, \{b\}, \{c\}, \{a, b\}, \{a, c\}, \{b, c\}, \{a, b, c\}\}.$$

Find an upper bound, the least upper bound, a lower bound, and the greatest lower bound of the following subsets of $\mathcal{P}(A)$:

(a) $S = \{\{a\}, \{a, b\}\}$.
(b) $S = \{\{a\}, \{b\}\}$.
(c) $S = \{\{a\}, \{a, b\}, \{a, b, c\}\}$.
(d) $S = \{\{a\}, \{c\}, \{a, c\}\}$.
(e) $S = \{\varnothing, \{a, b, c\}\}$.
(f) $S = \{\{a\}, \{b\}, \{c\}\}$.

7. Let $\preccurlyeq$ be a partial order on a set A and let $S \subseteq A$. Suppose a and a' are both smallest elements of S. Prove that $a = a'$.

8. Consider the poset $(\mathcal{P}(\mathbb{R}), \subseteq)$, where $\mathcal{P}(\mathbb{R}) = \{A : A \subseteq \mathbb{R}\}$; that is, $\mathcal{P}(\mathbb{R})$ is the set of all subsets of $\mathbb{R}$. Let C be the chain in $\mathcal{P}(\mathbb{R})$ defined by

$$C = \{\{1\}, \{1, 2, 3\}, \{1, 2, 3, 4, 5\}, \{1, 2, 3, 4, 5, 6, 7\}, \ldots\}.$$

Does C have an upper bound? A least upper bound?

9. Consider the poset $(P, \subseteq)$ where $P = \{A : A \text{ is a finite subset of } \mathbb{R}\}$; that is, P is the set of all finite subsets of $\mathbb{R}$. Let C be the chain in P defined by

$$C = \{\{1\}, \{1, 2, 3\}, \{1, 2, 3, 4, 5\}, \{1, 2, 3, 4, 5, 6, 7\}, \ldots\}.$$

Does C have an upper bound? A least upper bound?

10. Let $(A, \preccurlyeq)$ and $(B, \preccurlyeq')$ be posets. Suppose that the function $h\colon A \to B$ satisfies $x \preccurlyeq y$ if and only if $h(x) \preccurlyeq' h(y)$, for all x and y in A. Prove that h is one-to-one.

*11. Let $(B, \preccurlyeq')$ be a poset. Suppose $h\colon A \to B$ is a one-to-one function. Define the relation $\preccurlyeq$ on A by $x \preccurlyeq y$ if and only if $h(x) \preccurlyeq' h(y)$, for all x and y in A. Prove that $\preccurlyeq$ is a partial order on A.

*12. Let $(B, \preccurlyeq')$ be a totally ordered set. Suppose $h\colon A \to B$ is one-to-one. Define the relation $\preccurlyeq$ on A by $x \preccurlyeq y$ if and only if $h(x) \preccurlyeq' h(y)$, for all x and y in A. Prove that $\preccurlyeq$ is a total order on A.

13. Let $(A, \preccurlyeq)$ be a poset and let $S \subseteq A$. Suppose that a is the smallest element of S. Prove that a is also the greatest lower bound of S.

*14. Let $\preccurlyeq$ be a partial order on A. Let $C \subseteq A$. Show that $\preccurlyeq_C$ is a partial order on C. Show that if $\preccurlyeq$ is a total order on A, then $\preccurlyeq_C$ is a total order on C.

15. Let $(A, \preccurlyeq)$ be a poset. For each $x \in A$, let $P_x = \{a \in A : a \preccurlyeq x\}$. Let $\mathcal{F}$ be defined by $\mathcal{F} = \{P_x : x \in A\}$. Thus, $(\mathcal{F}, \subseteq)$ is a poset. Prove that

$$x \preccurlyeq y \text{ if and only if } P_x \subseteq P_y,$$

for all x and y in A. Hence, the posets $(A, \preccurlyeq)$ and $(\mathcal{F}, \subseteq)$ are isomorphic.

16. Let $(A, \preccurlyeq)$ and $(B, \preccurlyeq')$ be posets. Suppose that $h\colon A \to B$ satisfies

$$x \preccurlyeq y \text{ if and only if } h(x) \preccurlyeq' h(y),$$

for all x and y in A. Let $C \subseteq A$. Prove the following:
(a) If C is a chain in A, then the image $h[C]$ is a chain in B.
(b) If $h[C]$ is a chain in B, then C is a chain in A.
(c) If C has an upper bound, then $h[C]$ has an upper bound.
(d) If $h[C]$ has an upper bound and h is onto B, then C has an upper bound.

*17. Let $\preccurlyeq$ be a partial order on A. Show that $\mathrm{fld}(\preccurlyeq) = A$.

*18. Let $\mathcal{C}$ be a set where each $\preccurlyeq \,\in \mathcal{C}$ is a partial order on its field. Suppose that for any $\preccurlyeq$ and $\preccurlyeq'$ in $\mathcal{C}$, either $\preccurlyeq \,\subseteq\, \preccurlyeq'$ or $\preccurlyeq' \,\subseteq\, \preccurlyeq$.
(a) Show that for every $\preccurlyeq$ and $\preccurlyeq'$ in $\mathcal{C}$, if $\preccurlyeq \,\subseteq\, \preccurlyeq'$, then $\mathrm{fld}(\preccurlyeq) \subseteq \mathrm{fld}(\preccurlyeq')$.
(b) Let $\preccurlyeq^{\mathcal{C}}$ be the relation $\bigcup \mathcal{C}$ with field $\bigcup\{\mathrm{fld}(\preccurlyeq) : \preccurlyeq \,\in \mathcal{C}\}$. Prove that $\preccurlyeq^{\mathcal{C}}$ is a partial order on its field.

*19. Let $(A, \preccurlyeq)$ and $(B, \preccurlyeq')$ be posets. Define the relation $\preccurlyeq_\ell$ on $A \times B$ by

$$\langle a, b \rangle \preccurlyeq_\ell \langle x, y \rangle \ \text{ iff } \ a \prec x \vee (a = x \wedge b \preccurlyeq' y),$$

for all $\langle a, b \rangle$ and $\langle x, y \rangle$ in $A \times B$. The relation $\preccurlyeq_\ell$ is called the *lexicographic ordering* on $A \times B$.

(a) Prove that $\preccurlyeq_\ell$ is a partial ordering on $A \times B$.

(b) Suppose that $\preccurlyeq$ is a total order on A and that $\preccurlyeq'$ is a total order on B. Prove that $\preccurlyeq_\ell$ is a total order on $A \times B$.

3.5 Congruence and Preorder

Let $\sim$ be an equivalence relation on a set A. Recalling Definition 3.2.19, the quotient set $A/\!\sim$ denotes the partition $\{[a] : a \in A\}$ of A induced by $\sim$. By Theorem 3.2.16, we know that if $x \sim y$, then $[x] = [y]$. Let $f \colon A \to A$ be a function. Can we say that if $x \sim y$, then $[f(x)] = [f(y)]$? In this section, we will pursue such questions.

An equivalence relation that is preserved under certain identified operations is called a congruence. In this case, one can use these operations to construct similar operations on the quotient set and thereby address the above question.

A preorder is a relation that is reflexive and transitive. Thus, a preorder is "almost" a partial order. We will also show how one can take a preorder on a set and then use it to construct a partial order on a particular quotient set.

Congruence

Again, let $\sim$ be an equivalence relation on A. We now investigate the following question: Given a function $f \colon A \to A$, can we use f to define a function from $A/\!\sim$ to $A/\!\sim$? Specifically, (▲) is there a function $\hat{f} \colon A/\!\sim \to A/\!\sim$ such that

$$\hat{f}([x]) = [f(x)] \text{ for all } x \in A?$$

Is it obvious that such a function $\hat{f}$ exists? Given an equivalence class, is the resulting value of $\hat{f}$ independent of the element x chosen from the equivalence class? Suppose that $x \sim y$. Thus, $[x] = [y]$ by Theorem 3.2.16. If $f(x) \nsim f(y)$, then $[f(x)] \neq [f(y)]$. So $\hat{f}$ would not be single-valued, and hence, $\hat{f}$ will not be a function. To avoid this difficulty, we must have that $f(x) \sim f(y)$.

Definition 3.5.1. Let $\sim$ be an equivalence relation on A, and let $f \colon A \to A$ be a function. We say that f is **congruent with** $\sim$ if for all $x \in A$ and $y \in A$,

if $x \sim y$, then $f(x) \sim f(y)$.

We can now answer question (▲).

Theorem 3.5.2. *Let* $\sim$ *be an equivalence relation on A and let $f: A \to A$. There exists a function $\hat{f}: A/{\sim} \to A/{\sim}$ such that*

$$\hat{f}([x]) = [f(x)] \text{ for all } x \in A$$

if and only if f is congruent with $\sim$.

Proof. Let $\sim$ be an equivalence relation on A and let $f: A \to A$.

($\Rightarrow$). Let $\hat{f}: A/{\sim} \to A/{\sim}$ be such that $\hat{f}([x]) = [f(x)]$ for all $x \in A$. Let $x \in A$ and $y \in A$ be such that $x \sim y$. So $[x] = [y]$ by Theorem 3.2.16. Since $\hat{f}$ is a function, we conclude that $[f(x)] = [f(y)]$. Therefore, $f(x) \sim f(y)$, again by Theorem 3.2.16.

($\Leftarrow$). Suppose f is congruent with $\sim$. Let $\hat{f} = \{\langle [x], [f(x)]\rangle : x \in A\}$. To prove that the relation $\hat{f}$ is single-valued, let $\langle [x], [f(x)]\rangle \in \hat{f}$ and $\langle [y], [f(y)]\rangle \in \hat{f}$. Assume that $[x] = [y]$. We show that $[f(x)] = [f(y)]$ as follows:

$$[x] = [y] \Rightarrow x \sim y \qquad \text{by Theorem 3.2.16}$$
$$\Rightarrow f(x) \sim f(y) \qquad \text{as } f \text{ is a congruent with } \sim$$
$$\Rightarrow [f(x)] = [f(y)] \qquad \text{by Theorem 3.2.16.}$$

Hence, $\hat{f}$ is a function. Clearly, the domain of $\hat{f}$ equals $A/{\sim}$ and $\text{ran}(\hat{f}) \subseteq A/{\sim}$. Therefore, $\hat{f}: A/{\sim} \to A/{\sim}$ and $\hat{f}([x]) = [f(x)]$ for all $x \in A$. $\qquad \square$

We shall soon show how Theorem 3.5.2 can be extended to both relations and binary operations.

Definition 3.5.3. Any function of the form $f: A \times A \to A$ is called a **binary operation on A** (or just a **binary operation** whenever the context is clear). We shall write $f(x, y)$ for the value $f(\langle x, y\rangle)$ when $x \in A$ and $y \in A$.

Definition 3.5.1 above describes when a given function is "in harmony" with an equivalence relation. We now define what it means for a binary operation, and a relation, to be in harmony with an equivalence relation.

Definition 3.5.4. Let $\sim$ be an equivalence relation on A and $f: A \times A \to A$ be a binary operation. Then f is **congruent with** $\sim$ if for all x, y, w, z in A,

if $x \sim w$ and $y \sim z$, then $f(x, y) \sim f(w, z)$.

Definition 3.5.5. Let $\sim$ be an equivalence relation on A, and let R be a relation on A. Then R is **congruent with** $\sim$ if for all x, y, w, z in A,

if $x \sim w$ and $y \sim z$, then $x R y$ iff $w R z$.

One can easily adapt the proof of Theorem 3.5.2 to establish our next two theorems (see Exercises 2 and 3, respectively). The following theorem extends Theorem 3.5.2 to binary operations.

Theorem 3.5.6. *Let $\sim$ be an equivalence relation on A and let $f: A \times A \to A$. There exists a function $\hat{f}: A/\!\!\sim \times A/\!\!\sim \to A/\!\!\sim$ such that*

$$\hat{f}([x], [y]) = [f(x, y)] \text{ for all } x \text{ and } y \text{ in } A$$

if and only if f is congruent with $\sim$.

We now extend Theorem 3.5.2 to relations.

Theorem 3.5.7. *Let $\sim$ be an equivalence relation on A, and let R be a relation on A. There exists a relation $\hat{R}$ on $A/\!\!\sim$ such that*

$$\hat{R}([x], [y]) \text{ iff } R(x, y), \text{ for all } x \text{ and } y \text{ in } A$$

if and only if R is congruent with $\sim$.

Preorder

Definition 3.5.8. A relation $\preccurlyeq$ on a set A is a **preorder** if $\preccurlyeq$ is reflexive and transitive; that is, for any x, y, and z in A, the following hold:

1. $x \preccurlyeq x$.
2. If $x \preccurlyeq y$ and $y \preccurlyeq z$, then $x \preccurlyeq z$.

When $\preccurlyeq$ is a preorder on the set A, we shall say that $(A, \preccurlyeq)$ is a **preordered set** or a **proset**. Of course, if a preorder is also antisymmetric, then it is a partial order. On the other hand, there are preorders that are not partial orders.

Example 1. Let $X = \{1, 2, 3, 4\}$ and $A = \mathcal{P}(X)$. Define the relation $\preccurlyeq$ on A by

$a \preccurlyeq b$ if and only if $|a| \leq |b|$

for all $a \in A$ and $b \in A$, where $|c|$ denotes the number of elements in c. One can easily see that the relation $\preccurlyeq$ is reflexive and transitive; that is, $\preccurlyeq$ is a preorder on A. Clearly, $\{1, 2\} \preccurlyeq \{3, 4\}$ and $\{3, 4\} \preccurlyeq \{1, 2\}$. Since $\{1, 2\} \neq \{3, 4\}$, we see that $\preccurlyeq$ is not antisymmetric. So $\preccurlyeq$ is not a partial order on X.

Lemma 3.5.9. *Let $(A, \preccurlyeq)$ be a preordered set, and let $\sim$ be the relation on A defined by*

$a \sim b$ *if and only if $a \preccurlyeq b$ and $b \preccurlyeq a$.*

Then $\sim$ is an equivalence relation on A.

Proof. See Exercise 4. □

The relation $\sim$ in Lemma 3.5.9 is called the **derived equivalence relation**.

Lemma 3.5.10. *Suppose that $(A, \preccurlyeq)$ is a preordered set. Let $\sim$ be the derived equivalence relation on A. Then $\preccurlyeq$ is congruent with $\sim$; that is, for all for all a, b, c, d in A, if $a \sim c$ and $b \sim d$, then $a \preccurlyeq b$ if and only if $c \preccurlyeq d$.*

Proof. See Exercise 5. □

When $(A, \preccurlyeq)$ is a preordered set, Lemma 3.5.10 allows us to define a partial order on the quotient set $A/\sim$ when $\sim$ is the derived equivalence relation.

Theorem 3.5.11. *Suppose $(A, \preccurlyeq)$ is a preordered set, and let $\sim$ be the derived equivalence relation on A. Consider the relation $\hat{\preccurlyeq}$ on $A/\sim$ defined by*

$[a] \hat{\preccurlyeq} [b]$ *if and only if $a \preccurlyeq b$,* $\qquad\qquad$ (3.7)

for all $[a]$ and $[b]$ in $A/\sim$. Then $\hat{\preccurlyeq}$ is a partial order on $A/\sim$.

Proof. Let $(A, \preccurlyeq)$ be a proset, and let $\sim$ be the derived equivalence relation. By Lemma 3.5.10, $\preccurlyeq$ is congruent with $\sim$. Theorem 3.5.7 implies that there is a relation $\hat{\preccurlyeq}$ satisfying (3.7). Since $\preccurlyeq$ is a preorder on A, one can now show (using (3.7)) that $\hat{\preccurlyeq}$ is a partial order on $A/\sim$ (see Exercise 6). □

Definition 3.5.12. Let $\preccurlyeq$ be a preorder on A. Then $\preccurlyeq$ is a **total preorder** on A if $\preccurlyeq$ satisfies the property: $(\forall x \in A)(\forall y \in A)(x \preccurlyeq y \vee y \preccurlyeq x)$; that is, either $x \preccurlyeq y$ or $y \preccurlyeq x$, for any x and y in A.

Corollary 3.5.13. *Suppose that $(A, \preccurlyeq)$ is a preordered set. Let $\hat{\preccurlyeq}$ be the relation on $A/\sim$ defined by (3.7) where $\sim$ is the derived equivalence relation on A. If $\preccurlyeq$ is a total preorder on A, then $\hat{\preccurlyeq}$ is a total order on $A/\sim$.*

Proof. See Exercise 7. $\qquad\qquad\qquad\qquad\qquad\qquad\qquad\qquad\qquad$ $\square$

Exercises 3.5

1. Let $\sim$ be an equivalence relation on A. Let $f : A \to A/\sim$ be defined by

   $f(x) = [x]$, for all $x \in A$.

   Show that the function f is one-to-one if and only if each equivalence class in $A/\sim$ is a singleton.

*2. Prove Theorem 3.5.6.

*3. Prove Theorem 3.5.7.

*4. Prove Lemma 3.5.9.

*5. Prove Lemma 3.5.10.

*6. Prove Theorem 3.5.11.

*7. Prove Corollary 3.5.13.

8. Let $P = \{1, 2, 3, \ldots\}$. Define the equivalence relation $\sim$ on P by

   $m \sim n$ if and only if m and n have the same number of prime factors.

   For example, $25 \sim 15$ and $9 \not\sim 3$.
   (a) Is there an $h : P/\sim \to P/\sim$ so that $h([n]) = [3n]$ for all $n \in P$?
   (b) Is there an $h : P/\sim \times P/\sim \to P/\sim$ such that $h([m], [n]) = [m + n]$ for all $m, n \in P$?
   (c) Is there an $h : P/\sim \times P/\sim \to P/\sim$ such that $h([m], [n]) = [m \cdot n]$ for all $m, n \in P$?

9. Let $\mathbb{N} = \{0, 1, 2, 3, \ldots\}$ be the set of natural numbers and let $\leq$ be the standard order relation on $\mathbb{N}$. Let $\mathbb{N}^2 = \mathbb{N} \times \mathbb{N}$. Consider the relation $\preccurlyeq$ on $\mathbb{N}^2$ defined by

$$\langle m, n \rangle \preccurlyeq \langle i, j \rangle \text{ if and only if } m + j \leq n + i.$$

(a) Show that $\preccurlyeq$ not antisymmetric.

(b) Prove that $\preccurlyeq$ is a preorder on $\mathbb{N}^2$.

(c) Prove that $\preccurlyeq$ is a total preorder on $\mathbb{N}^2$.

(d) Let $\sim$ be the derived equivalence relation on $\mathbb{N}^2$. Is there a function $h: \mathbb{N}^2/\!\!\sim \; \to \mathbb{N}^2/\!\!\sim$ such that $h([\langle m, n \rangle]) = [\langle m + 2n, n + 2m \rangle]$?

(e) Define $f: \mathbb{N}^2 \times \mathbb{N}^2 \to \mathbb{N}^2$ by $f(\langle m, n \rangle, \langle i, j \rangle) = \langle m + i, n + j \rangle$. Now show that f is congruent with $\sim$.

Exercise Notes: Exercise **9** offers a first step in a set-theoretic construction of the integers using the natural numbers (see Remark 4.4.15).

4 The Natural Numbers

In this chapter, it shall be shown that number theory can be embedded within set theory. One consequence of this embedding is that all of the theorems in number theory can be proven from the axioms of set theory.

In order to show that number theory is, in fact, a special branch of set theory, we must first represent each natural number as a set. How can one define the natural numbers in set theory? To answer this question, we next consider a set theoretic construction that makes sense for any set.

Definition 4.0.1. For each set x, the **successor** x^+ is the set that is obtained by adjoining x to the elements of x, namely, $x^+ = x \cup \{x\}$.

We note the following three properties concerning the successor of a set x:

1. $a \in x^+$ iff $(a \in x \vee a = x)$.
2. $x \in x^+$.
3. $x \subseteq x^+$.

Using the successor operation, we can now construct, in set theory, the first few natural numbers as follows:

- $0 = \varnothing$.
- $1 = 0^+ = 0 \cup \{0\} = \{0\}$.
- $2 = 1^+ = 1 \cup \{1\} = \{0, 1\}$.
- $3 = 2^+ = 2 \cup \{2\} = \{0, 1, 2\}$.
- $4 = 3^+ = 3 \cup \{3\} = \{0, 1, 2, 3\}$.

We note some interesting properties that these "natural numbers" possess (see the above bulleted list):

1. $0 \in 1 \in 2 \in 3 \in 4 \in 5 \in \cdots$.
2. $0 \subseteq 1 \subseteq 2 \subseteq 3 \subseteq 4 \subseteq 5 \subseteq \cdots$.

4.1   Inductive Sets

The existence of an infinite set is crucial in modern set theory and mathematics. Zermelo realized that one cannot prove that an infinite set exists and thus found a fairly simple way to assert the existence of an infinite set.

Infinity Axiom. There is a set A that contains the empty set as an element and whenever $x \in A$, then $x^+ \in A$.

Using the infinity axiom, we shall prove that there is a set consisting of only the natural numbers; this set is denoted by ω. First, we define a property of a set that will ensure that each natural number belongs to such a set. A set that satisfies this property will contain the empty set and will be closed under the successor operation.

Definition 4.1.1. A set I is said to be **inductive** if and only if

1. $\varnothing \in I$,
2. $(\forall a \in I)(a^+ \in I)$, that is, I is "closed under successor."

An inductive set contains $\varnothing$ and also contains the successor of each one of its elements. Hence, an inductive set must contain all of the "natural numbers." Clearly, the infinity axiom declares that there is an inductive set.

Definition 4.1.2. A **natural number** is a set that belongs, as an element, to every inductive set.

Therefore, x is a natural number if and only if x is in every inductive set. We now want to prove that there is a set consisting of just the natural numbers. To do this, we need to prove that the class

$$\{x : x \text{ is in every inductive set}\} \tag{4.1}$$

is a set. By the infinity axiom, we know that there is an inductive set A. Hence, for each x, if x is in every inductive set, then $x \in A$. Theorem 2.1.3 therefore implies that (4.1) defines a set, and this set consists of the natural numbers.

Definition 4.1.3. The set consisting of the natural numbers is denoted by ω.

Of course, ω is the set $\{0, 1, 2, 3, \dots\}$, but we will identify ω as the set that satisfies

$$x \in \omega \text{ if and only if } x \text{ is in every inductive set.} \tag{4.2}$$

Theorem 4.1.4. *The set ω is inductive and ω is a subset of every inductive set. Thus, ω is the smallest inductive set.*

Proof. We first prove that ω is an inductive set; that is, we shall prove that

(1) $\varnothing \in \omega$,
(2) $(\forall a \in \omega)(a^+ \in \omega)$.

Since $\varnothing$ is in every inductive set, (4.2) implies that $\varnothing \in \omega$. To prove (2), let $a \in \omega$. We thus conclude that a is in every inductive set. So a^+ is also in every inductive set. Hence, $a^+ \in \omega$ and (2) holds. Now let I be an inductive set and let $x \in \omega$. The above (4.2) implies that x is in every inductive set. Therefore, $x \in I$ and thus, $\omega \subseteq I$. $\qquad\square$

We now prove the following simple corollary.

Corollary 4.1.5. *Let $I \subseteq \omega$ be an inductive set. Then $I = \omega$.*

Proof. Let I be an inductive set such that $I \subseteq \omega$. Since I is an inductive set, Theorem 4.1.4 implies that $\omega \subseteq I$. Hence, $I = \omega$. $\qquad\square$

Corollary 4.1.5 is, in fact, a restatement of the **principle of mathematical induction**. Suppose that $P(n)$ is some property. To prove *by induction* that

$$(\forall n \in \omega)P(n)$$

is true, let $I = \{n \in \omega : P(n)\}$ and so, $I \subseteq \omega$. If we can prove that

(1) $0 \in I$,
(2) $(\forall n \in \omega)(n \in I \rightarrow n^+ \in I)$,

then I is inductive and thus, $I = \omega$ by Corollary 4.1.5. Therefore, $(\forall n \in \omega)P(n)$ holds. Item (1) is often called the *base step*, and (2) is commonly refereed to as the *inductive step*. We apply this method of proof to establish our next theorem.

Theorem 4.1.6. *For every $n \in \omega$, either $n = 0$ or $n = k^+$ for some $k \in \omega$.*

Proof. Let $I = \{n \in \omega : n = 0 \vee (\exists k \in \omega)(n = k^+)\}$. Clearly, $0 \in I$. Let $n \in I$. So either $n = 0$ or $n = k^+$ for some $k \in \omega$. Therefore, $n^+ = 0^+$ or $n^+ = (k^+)^+$ where $k^+ \in \omega$. Thus, $n^+ \in I$. Hence, by Corollary 4.1.5, $I = \omega$. $\qquad\square$

Definition 4.1.7. A set A is said to be a **transitive set** if $(\forall a \in A)(a \subseteq A)$; that is, every member of A is also a subset of A.

Therefore, for any set A we have that

A is transitive iff $(\forall a)(\forall x)[(x \in a \wedge a \in A) \rightarrow x \in A]$ $\qquad$ (4.3)

$$\text{iff } \bigcup A \subseteq A. \qquad (4.4)$$

Example 1. The empty set $\varnothing$ is, vacuously, transitive. Each of the following sets is also transitive:

- $1 = 0^+ (= \{0\})$,
- $2 = 1^+ (= \{0, 1\})$,
- $3 = 2^+ (= \{0, 1, 2\})$,
- $4 = 3^+ (= \{0, 1, 2, 3\})$,
- $5 = 4^+ (= \{0, 1, 2, 3, 4\})$.

On the other hand, the set $A = \{\{0\}\}$ is not transitive because $\{0\} \in A$ and yet $\{0\} \not\subseteq A$. Similarly, the set $\{0, 1, 3\}$ is not transitive.

Remark 4.1.8. Regrettably, we are now using the word "transitive" to refer to two different concepts. A relation R on a set A is *transitive* provided that for all x, y, z in A, if $x R y$ and $y R z$, then $x R z$ (see Definition 3.2.11). A set A is *transitive*, if whenever $x \in A$, then $x \subseteq A$. If a set A is transitive, then one cannot infer that the membership relation $\in$ on A is a transitive relation; that is, one cannot always conclude that if $x \in y$ and $y \in z$, then $x \in z$, for all x, y, z in A.

Proposition 4.1.9. *Let A and B be sets. Then* $\bigcup(A \cup B) = (\bigcup A) \cup (\bigcup B)$.

Theorem 4.1.10. *Let a be a transitive set. Then* $\bigcup(a^+) = a$.

Proof. Suppose that a is a transitive set. Then

$$\bigcup(a^+) = \bigcup(a \cup \{a\}) \qquad \text{by def. successor}$$

$$= \bigcup a \cup \bigcup\{a\} \qquad \text{by Proposition 4.1.9}$$

$$= \bigcup a \cup a \qquad \text{because } \bigcup\{a\} = a$$

$$= a \qquad \text{as } \bigcup a \subseteq a \text{ by (4.4)}.$$

Therefore, $\bigcup(a^+) = a$. $\qquad\qquad\qquad\qquad\qquad\qquad\qquad \square$

Theorem 4.1.11. *Every natural number is a transitive set.*

Proof. We prove that $(\forall n \in \omega)(n$ is a transitive set$)$ by induction. Let

$I = \{n \in \omega : n$ is a transitive set$\}$.

Clearly, $0 \in I$, as $0 = \varnothing$ is a transitive set. Let $n \in I$. So n is transitive. To prove that n^+ is transitive, we will show that $\bigcup n^+ \subseteq n^+$ (see 4.4). By Theorem 4.1.10, $\bigcup n^+ = n$. Since $n \subseteq n^+$ (see item 3 on page 83), we conclude that $\bigcup n^+ \subseteq n^+$. Hence, n^+ is a transitive set and thus, $n^+ \in I$. By Corollary 4.1.5, we conclude that $I = \omega$, and this completes the proof. $\square$

Theorem 4.1.11 implies that if $a \in b$ and $b \in n$, then $a \in n$, whenever n is a natural number. We now prove that the successor function on ω is one-to-one.

Theorem 4.1.12. *For all m and n in ω, if $n^+ = m^+$, then $n = m$.*

Proof. Let $m \in \omega$ and $n \in \omega$. It follows from Theorem 4.1.11 that m and n are transitive sets. Assume that $n^+ = m^+$. Thus, $\bigcup n^+ = \bigcup m^+$. Therefore, $n = m$ by Theorem 4.1.10. $\square$

Theorem 4.1.13. *The set ω is a transitive set.*

Proof. We shall prove that $(\forall n \in \omega)(n \subseteq \omega)$ by mathematical induction. Let $I = \{n \in \omega : n \subseteq \omega\}$. Clearly, $0 \in I$, because $0 = \varnothing$ and $\varnothing \subseteq \omega$. Let $n \in I$. So (IH)[1] $n \subseteq \omega$. We shall prove that $n^+ \subseteq \omega$. To do this, let $k \in n^+$. We must prove that $k \in \omega$. Since $k \in n^+ = n \cup \{n\}$, we conclude that $k \in n$ or $k = n$. If $k \in n$, then $k \in \omega$ by (IH). If $k = n$, then $k \in \omega$ because $n \in \omega$. Hence, $n^+ \in I$. Therefore, $I = \omega$. $\square$

Exercises 4.1

1. Let I and J be inductive sets. Prove that $I \cap J$ is also inductive.

*2. Prove that if A is a transitive set, then A^+ is also a transitive set.

3. Prove that A is a transitive set if and only if $A \subseteq \mathcal{P}(A)$.

4. Prove that if A is a transitive set, then $\bigcup A$ is a transitive set.

5. Let $\mathcal{A}$ be a nonempty set such that B is a transitive set, for all $B \in \mathcal{A}$. Prove that $\bigcap \mathcal{A}$ and $\bigcup \mathcal{A}$ are transitive sets.

6. Prove Proposition 4.1.9.

7. Prove that $n \neq n^+$ for all $n \in \omega$.

8. Conclude from Exercise 7 that $n^+ \not\subseteq n$ for all $n \in \omega$.

9. Prove that for all $m \in \omega$ and all $n \in \omega$, if $m \in n$, then $n \not\subseteq m$.

*10. Conclude from Exercise 9 that $n \notin n$ for all $n \in \omega$.

11. Let A be a set and suppose that $\bigcup A = A$. Prove that A is transitive and for all $x \in A$, there is a $y \in A$ such that $x \in y$.

Exercise Notes: For Exercise 7, apply Theorem 4.1.12 for the inductive step. For Exercise 9, let $m \in \omega$ and let $I = \{n \in \omega : (m \in n \to n \not\subseteq m)\}$. Prove that I is inductive. Assume $n \in I$. To prove $n^+ \in I$, let $m \in n^+$. So $m \in n$ or $m = n$. If $m \in n$, then $n \in I$ implies that $n \not\subseteq m$ and thus, $n^+ \not\subseteq m$. If $m = n$, then $n^+ \not\subseteq m$ by Exercise 8.

4.2 The Recursion Theorem on ω

Induction is frequently applied in mathematical proofs. Moreover, one often defines a function by induction (recursion). A recursively defined function is one that is defined in terms of "previously evaluated values of the function." A function h on ω is defined recursively if its value at 0 is first specified and all of the remaining values are then defined using an earlier value, together with a given function. For example, let $f: A \to A$ be a function. Suppose that the initial value of a function h is given to be $h(0) = a$ for a fixed $a \in A$, and to get the next value $h(1)$ we use the initial value to obtain $h(1) = f(h(0)) = f(a)$. Similarly, to get the next value $h(2)$ we must evaluate $f(h(1))$. By continuing in this manner, we can evaluate more and more values of the function h. A more succinct way of describing this function h is to first define (i) $h(0) = a$ and then define the remaining values by (ii) $h(n^+) = f(h(n))$. Thus, one can evaluate $h(0), h(1), \ldots, h(100)$. Just because we can evaluate a finite number of values, do we really know that there is such a function h with domain ω? The existence of such a function seems quite reasonable. In mathematics, however, the acceptance of an existential statement requires a valid argument, that is, a proof. The proof of our next theorem shows that such a function does exist and that it is unique.

Recursion Theorem 4.2.1. *Let A be a set and $a \in A$. Suppose that $f: A \to A$ is a function. Then there exists a unique function $h: \omega \to A$ such that*

(1) $h(0) = a$,
(2) $h(n^+) = f(h(n))$, *for all $n \in \omega$.*

The function h in the recursion theorem, satisfying (1) and (2), is said to be *defined by recursion.* If a function has been defined by recursion, then proofs of statements about this function often use "proof by induction." Before we prove the recursion theorem, we make an observation. First, let us assume that the function h exists. When viewing h as a set of ordered pairs, conditions (1) and (2) imply that

- $\langle 0, a \rangle \in h$, and
- if $\langle n, u \rangle \in h$, then $\langle n^+, f(u) \rangle \in h$, for all $n \in \omega$.

In our proof, we will show that there exists a function h that satisfies the above bulleted conditions.

Proof. Let A be a set, $a \in A$, and let $f: A \to A$ be a function. Let us call a relation $R \subseteq \omega \times A$ *suitable* if the following two conditions hold:

(i) $\langle 0, a \rangle \in R$.
(ii) If $\langle n, u \rangle \in R$, then $\langle n^+, f(u) \rangle \in R$.

Our goal is to construct a suitable relation that is also a function.

Since every suitable relation is a subset of $\omega \times A$, Theorem 2.1.3 implies that the class $\mathcal{S} = \{R : R \text{ is suitable}\}$ is, in fact, a set. Because $\omega \times A$ is a suitable relation, $\mathcal{S}$ is nonempty. Let $h = \bigcap \mathcal{S}$. Clearly, $h \subseteq \omega \times A$, and so h is a relation. We now prove that h is suitable and that it is a function.

Claim 1. h is suitable.

Proof. We need to show that h satisfies items (i) and (ii). Every relation in $\mathcal{S}$ is suitable. So $\langle 0, a \rangle$ is an element in every relation in $\mathcal{S}$. Thus, $\langle 0, a \rangle \in \bigcap \mathcal{S}$; that is, $\langle 0, a \rangle \in h$. To prove (ii), assume that $\langle n, u \rangle \in h$. As $h = \bigcap \mathcal{S}$, it follows that $\langle n, u \rangle$ belongs to every relation in $\mathcal{S}$. Let $R \in \mathcal{S}$. Since $\langle n, u \rangle \in R$ and R is suitable, item (ii) implies that $\langle n^+, f(u) \rangle \in R$. So $\langle n^+, f(u) \rangle$ belongs to every relation in $\mathcal{S}$, and hence, $\langle n^+, f(u) \rangle \in h$. Thus, h is suitable. (Claim 1) $\square$

We now must prove that h is a function from ω to A.

Claim 2. h is a function from ω to A.

Proof. To prove that h is a function from ω to A, we must show that for each $n \in \omega$, there is exactly one $y \in A$ such that $\langle n, y \rangle \in h$. Let $I \subseteq \omega$ be defined by

$$I = \{n \in \omega : \text{there is exactly one } u \in A \text{ such that } \langle n, u \rangle \in h\}.$$

We now prove that I is an inductive set. To do this, we must first prove that $0 \in I$. We know that $\langle 0, a \rangle \in h$. To prove that $0 \in I$, we need to show there is no $\langle 0, b \rangle \in h$ where $a \neq b$. Suppose, for a contradiction, that $\langle 0, b \rangle \in h$ where $a \neq b$. Consider the relation

$$R = h \setminus \{\langle 0, b \rangle\}.$$

Clearly, (▲) $\langle 0, b \rangle \notin R$. As h is suitable, it easily follows that R is also suitable (because $n^+ \neq 0$ for all $n \in \omega$). Hence, $R \in \mathcal{S}$. Since $h = \bigcap \mathcal{S}$, we conclude that $h \subseteq R$, and thus, $\langle 0, b \rangle \in R$, which contradicts (▲). Consequently, a is the only element in A such that $\langle 0, a \rangle \in h$. Therefore, $0 \in I$.

Let $n \in I$; that is, assume that (**IH**) there is a unique $u \in A$ such that $\langle n, u \rangle \in h$. Because h is suitable, we conclude that $\langle n^+, f(u) \rangle \in h$. To show that $n^+ \in I$, we must show there is no $\langle n^+, b \rangle \in h$ where $f(u) \neq b$. Assume, for a contradiction, that $\langle n^+, b \rangle \in h$ where $f(u) \neq b$. Consider the relation

$$R = h \setminus \{\langle n^+, b \rangle\}.$$

Clearly, (♦) $\langle n^+, b \rangle \notin R$. We now show that R is suitable. Since h is suitable and $n^+ \neq 0$, we have only to show that if $\langle n, v \rangle \in R$, then $\langle n^+, f(v) \rangle \in R$. To do this, assume that $\langle n, v \rangle \in R$. Thus, $\langle n, v \rangle \in h$, and so $\langle n^+, f(v) \rangle \in h$. Since $\langle n, u \rangle \in h$, $\langle n, v \rangle \in h$, and $n \in I$, we conclude from (**IH**) that $u = v$, and thus, $f(u) = f(v)$. Because $f(v) = f(u) \neq b$ and $\langle n^+, f(v) \rangle \in h$, we infer that $\langle n^+, f(v) \rangle \in R$. So R is suitable. Hence, $R \in \mathcal{S}$. Since $h = \bigcap \mathcal{S}$, we conclude that $h \subseteq R$, and thus, $\langle n^+, b \rangle \in R$, which contradicts (♦). Therefore, $f(u)$ is the only element in A such that $\langle n^+, f(u) \rangle \in h$. So $n^+ \in I$, and thus, I is inductive. Corollary 4.1.5 implies that $I = \omega$, and as a result, h is a function from ω to A. (Claim 2) □

Because the function h is suitable, it satisfies conditions (1) and (2) given in the statement of the theorem. To prove that h is unique, suppose that $g \colon \omega \to A$ also satisfies properties (1) and (2). Thus, g is a suitable relation, and hence, $g \in \mathcal{S}$. Since $h = \bigcap \mathcal{S}$, we have that $h \subseteq g$. Exercise 10 on page 68 implies that $h = g$. Therefore, h is unique. (Theorem) □

Example 1. Let $f \colon \mathbb{R} \to \mathbb{R}$ and let $a = 2$. Theorem 4.2.1 implies that there is a function $h \colon \omega \to \mathbb{R}$ such that $h(0) = 2$ and $h(n^+) = f(h(n))$, for all $n \in \omega$.

When can we be assured that the function h in Theorem 4.2.1 is one-to-one? Our next theorem provides an answer.

Theorem 4.2.2. *Let A be a set, $a \in A$, $a \notin \text{ran}(f)$, and $f: A \to A$ be a function. Suppose that $h: \omega \to A$ satisfies*

(1) $h(0) = a$,
(2) $h(n^+) = f(h(n))$, *for all $n \in \omega$.*

If f is one-to-one, then h is also one-to-one.

Proof. Let $f: A \to A$ be one-to-one, $a \in A$, $a \notin \text{ran}(f)$, and let $h: \omega \to A$ satisfy (1) and (2). We prove that for all $n \in \omega$ and all $k \in \omega$, if $h(n) = h(k)$, then $n = k$. Let $I \subseteq \omega$ be defined by

$$I = \{n \in \omega : (\forall k \in \omega)(h(n) = h(k) \to n = k)\}.$$

We shall show that I is an inductive set. To prove that $0 \in I$, we must show that

$$(\forall k \in \omega)(h(0) = h(k) \to 0 = k). \tag{4.5}$$

So let $k \in \omega$ and assume that $h(0) = h(k)$. By (1), we have $h(0) = a$. Hence, $h(k) = a$. Suppose that $k \neq 0$. Then $k = j^+$ for some $j \in \omega$ by Theorem 4.1.6. Since $h(k) = a$, we conclude that $h(j^+) = a$. The above item (2) now implies that $f(h(j)) = a$. Thus, $a \in \text{ran}(f)$, which contradicts the fact that $a \notin \text{ran}(f)$. Hence, $k = 0$. Therefore, (4.5) holds, and so $0 \in I$.

Let $n \in I$. In other words, assume the induction hypothesis

$$(\forall k \in \omega)(h(n) = h(k) \to n = k). \tag{IH}$$

We shall prove that $n^+ \in I$; that is, we will prove that

$$(\forall k \in \omega)(h(n^+) = h(k) \to n^+ = k).$$

Let $k \in \omega$ and suppose that ($\blacktriangle$) $h(n^+) = h(k)$. Because $n^+ \neq 0$, the above (4.5) and ($\blacktriangle$) imply that $k \neq 0$. So $k = j^+$ for some $j \in \omega$. Hence, $h(n^+) = h(j^+)$. Item (2) implies that $f(h(n)) = f(h(j))$. Since f is one-to-one, we conclude that $h(n) = h(j)$. Our induction hypothesis (IH) implies that $n = j$, and thus, $n^+ = j^+$; that is, $n^+ = k$. Hence, $n^+ \in I$, and so I is an inductive set. Therefore, $I = \omega$ and h is a one-to-one function. $\qquad\square$

4.2.1 The Peano Postulates[2]

The Peano postulates offer an axiomatic foundation for the natural numbers. These postulates were first presented by the nineteenth-century Italian mathematician Giuseppe Peano. In 1888, Richard Dedekind proposed a set of axioms about the natural numbers. Peano then published a more precisely formulated

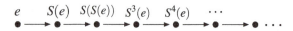

Figure 4.1. A Peano system (N, S, e)

version of these axioms in his 1889 book *The principles of arithmetic presented by a new method* [13]. The essence of the Peano postulates shall be summarized by means of the following two definitions.

Definition 4.2.3. Let $S: N \to N$ and let $A \subseteq N$. Then A is said to be **closed** under S if for all x, if $x \in A$, then $S(x) \in A$.

Let (x, y, z) denote an ordered **triple**, that is, $(x, y, z) = \langle\langle x, y\rangle, z\rangle$.

Definition 4.2.4. Let (N, S, e) be an ordered triple that consists of a set N, a function $S: N \to N$, and a particular element $e \in N$. Then (N, S, e) is a **Peano system** if the following three conditions hold:

(1) $e \notin \operatorname{ran}(S)$.
(2) S is one-to-one.
(3) For all $A \subseteq N$, if $e \in A$ and A is closed under S, then $A = N$.

The above condition (3) is referred to as the *induction postulate*. Figure 4.1 illustrates a Peano system.

Theorem 4.1.4 shows that ω is inductive. Thus, we can define the function $\sigma: \omega \to \omega$ by $\sigma(n) = n^+$ for all $n \in \omega$. We can now prove that a Peano system exists.

Theorem 4.2.5. *The triple $(\omega, \sigma, 0)$ is a Peano system.*

Proof. Since ω is inductive, we have that $0 \in \omega$ and that $\sigma(n) \in \omega$ for all $n \in \omega$. Clearly, $n^+ \neq 0$ for all $n \in \omega$, as $n \in n^+$ and $n \notin 0$. Thus, $0 \notin \operatorname{ran}(\sigma)$. Theorem 4.1.12 shows that σ is one-to-one. Finally, we need to show that $(\omega, \sigma, 0)$ satisfies the induction postulate. Let $A \subseteq \omega$ be such that $0 \in A$ and A is closed under σ. So $A \subseteq \omega$ is inductive. Corollary 4.1.5 implies that $A = \omega$. Therefore, $(\omega, \sigma, 0)$ is a Peano system because it satisfies conditions (1)–(3) of Definition 4.2.4. $\qquad\square$

Definition 4.2.6. Let (N, S, e) and (N', S', e') be Peano systems. Then (N, S, e) is **isomorphic** to (N', S', e') when there is a bijection $\pi : N \to N'$ such that

(1) $\pi(e) = e'$, and
(2) $\pi(S(x)) = S'(\pi(x))$ for all $x \in N$.

The function π is said to be an **isomorphism** from (N, S, e) onto (N', S', e').

Two Peano systems are isomorphic if they are essentially the same structure. We now prove that $(\omega, \sigma, 0)$ is isomorphic to any Peano system. Thus, the set of natural numbers ω is "the same" as the set $\mathbb{N}$ that is used in calculus.

Theorem 4.2.7. *Let (N, S, e) be a Peano system. Then $(\omega, \sigma, 0)$ is isomorphic to (N, S, e).*

Proof. Let (N, S, e) be a Peano system. So $S \colon N \to N$ is one-to-one, $e \in N$, and $e \notin \operatorname{ran}(S)$. Since $\sigma(n) = n^+$, Theorem 4.2.1 implies that there is a function $\pi : \omega \to N$ such that

(1) $\pi(0) = e$,
(2) $\pi(\sigma(n)) = S(\pi(n))$, for all $n \in \omega$.

As $e \notin \operatorname{ran}(S)$ and S is one-to-one, Theorem 4.2.2 implies that π is one-to-one. To show that π is onto N, we use the induction postulate (Definition 4.2.4(3)). Clearly, $\operatorname{ran}(\pi) \subseteq N$. Since $\pi(0) = e$, we see that $e \in \operatorname{ran}(\pi)$. Let $x \in \operatorname{ran}(\pi)$. Thus, $\pi(n) = x$ for some $n \in \omega$. Hence, $S(\pi(n)) = S(x)$. By (2), we conclude that $\pi(\sigma(n)) = \pi(n^+) = S(x)$. So $S(x) \in \operatorname{ran}(\pi)$, and $\operatorname{ran}(\pi)$ is thereby closed under S. The induction postulate implies that $\operatorname{ran}(\pi) = N$, that is, π is onto N. Therefore, π is an isomorphism from $(\omega, \sigma, 0)$ onto (N, S, e). $\qquad\square$

Theorem 4.2.7 implies that the Peano postulates characterize the system of natural numbers (up to isomorphism). Theorem 4.2.7 also implies that any two Peano systems are isomorphic to each other (see Exercises 7–8).

Exercises 4.2

1. How would one modify the proof of Theorem 4.2.1 in order to establish the following:

Theorem. *Let A be a set and let $a \in A$. Suppose that $f : \omega \times A \to A$ is a function. Then there exists a unique function $h : \omega \to A$ such that*
(1) $h(0) = a$,
(2) $h(n^+) = f(n, h(n))$, *for all $n \in \omega$.*

2. Let $a \in A$ and $f : A \to A$ be a function such that $a \notin \mathrm{ran}(f)$. Suppose that $h : \omega \to A$ satisfies
   (1) $h(0) = a$,
   (2) $h(n^+) = f(h(n))$, for all $n \in \omega$.

   Assume that h is one-to-one and onto A. Prove that f is one-to-one.

3. Let A be a set and let $a \in A$. Suppose that $f : A \to A$ satisfies
   (a) $a \in f(a)$,
   (b) if $x \in y$ then $f(x) \in f(y)$, for all x and y in A.

   Let $h : \omega \to A$ be as stated in Theorem 4.2.1. Prove that $h(n) \in h(n^+)$ for all $n \in \omega$.

4. Let $F : A \to A$ be a function and let $y \in A$.
   (a) Prove that the class $\mathcal{S} = \{B : B \subseteq A, \ y \in B, \text{ and } F[B] \subseteq B\}$ is a set.
   (b) Show that $\mathcal{S}$ is nonempty.
   (c) Let $C = \bigcap \mathcal{S}$. Prove that $y \in C$ and $F[C] \subseteq C$.
   (d) Prove that for all $B \subseteq A$, if $y \in B$ and $F[B] \subseteq B$, then $C \subseteq B$.
   (e) Prove that $y \in F[C]$ if and only if $F[C] = C$.

5. Let $F : A \to A$ be a function and let $Y \subseteq A$.
   (a) Prove that the class $\mathcal{S} = \{B : Y \subseteq B \subseteq A \text{ and } F[B] \subseteq B\}$ is a set.
   (b) Show that $\mathcal{S}$ is nonempty.
   (c) Let $C = \bigcap \mathcal{S}$. Prove that $Y \subseteq C$ and $F[C] \subseteq C$.
   (d) Prove that for all $B \subseteq A$, if $Y \subseteq B$ and $F[B] \subseteq B$, then $C \subseteq B$.
   (e) Prove that $Y \subseteq F[C]$ if and only if $F[C] = C$.

6. Let $F : A \to A$ be a function and let $Y \subseteq A$. The recursion theorem implies the existence of the function $h : \omega \to \mathcal{P}(A)$ that satisfies:
   (1) $h(0) = Y$,
   (2) $h(n^+) = h(n) \cup F[h(n)]$, for all $n \in \omega$.
   Now address the following:
   (a) Let $Y \subseteq B \subseteq A$ and $F[B] \subseteq B$. Prove by induction that $h(n) \subseteq B$ for all $n \in \omega$.
   (b) Let $C' = \bigcup_{n \in \omega} h(n)$. Prove that $Y \subseteq C' \subseteq A$ and $F\big[C'\big] \subseteq C'$.
   (c) Prove for all $B \subseteq A$, if $Y \subseteq B$ and $F[B] \subseteq B$, then $C' \subseteq B$.
   (d) Let C be as in the previous Exercise 5. Prove that $C' = C$.

*7. Let (N, S, e) and (N', S', e') be Peano systems. Let π be an isomorphism from (N, S, e) onto (N', S', e'). Prove that π^{-1} is thus an isomorphism from (N', S', e') onto (N, S, e).

*8. Let (N, S, e) and (N', S', e') be Peano systems. Let π be an isomorphism from (N, S, e) onto $(\omega, \sigma, 0)$, and let ρ be an isomorphism from $(\omega, \sigma, 0)$ onto (N', S', e'). Show that the composition $\rho \circ \pi$ is an isomorphism from (N, S, e) onto (N', S', e').

Exercise Notes: Exercises 4–5 do not involve Theorem 4.2.1, but they do apply some of the ideas in its proof. For Exercise 4(e) and for Exercise 5(e), in the direction ($\Rightarrow$), prove that $F[C] \in S$. For Exercise 6, item (a) implies (c); to prove (d), use (b) and (c) together with the results of Exercise 5(c)(d) to show that $C' \subseteq C$ and $C \subseteq C'$.

4.3 Arithmetic on ω

In this section, the recursion theorem will be used to define the operations of addition, multiplication, and exponentiation on the natural numbers. We will also prove the properties of arithmetic that we were taught as children. We first establish the following result.

Theorem 4.3.1. *Let* $g: \omega \to \omega$ *and* $f: \omega \times \omega \to \omega$ *be functions. Then there is a unique function* $h: \omega \times \omega \to \omega$ *such that for each* $m \in \omega$,

(1) $h(m, 0) = g(m)$,
(2) $h(m, n^+) = f(h(m, n), m)$, *for all* $n \in \omega$.

Proof. For each $m \in \omega$, by the recursion theorem (viewing m as a constant), there is a unique function $p_m: \omega \to \omega$ such that

(a) $p_m(0) = g(m)$,
(b) $p_m(n^+) = f(p_m(n), m)$, for all $n \in \omega$.

Since p_m is unique for each $m \in \omega$, the connection $m \mapsto p_m$ thus produces a function with domain ω. We can now define the function $h: \omega \times \omega \to \omega$ by $h(m, n) = p_m(n)$. The function h easily satisfies conditions (1) and (2) for all $m \in \omega$, and it is unique. $\square$

The function h in Theorem 4.3.1 is defined in terms of two known functions. Thus, before one can apply Theorem 4.3.1, one must first introduce the two relevant functions $g: \omega \to \omega$ and $f: \omega \times \omega \to \omega$.

Recalling Definition 3.5.3, a function of the form $h: B \times B \to B$ is called a binary operation on B. Our next definition is an application of Theorem 4.3.1; that is, using the functions $g(m) = m$ and $f(k, m) = k^+$, we define addition on the natural numbers.

Definition 4.3.2. Let $A: \omega \times \omega \to \omega$ be the unique function satisfying

(1) $A(m, 0) = m$,
(2) $A(m, n^+) = A(m, n)^+$

for all natural numbers m and n. The function A shall be referred to as **addition**, and we define the binary operation $+$ on ω to be

$$m + n = A(m, n)$$

for all m and n in ω.

Theorem 4.3.3. *For all natural numbers m and n,*

(A1) $m + 0 = m$,
(A2) $m + n^+ = (m + n)^+$.

Proof. Let m and n be elements in ω. Recalling that $A(m, n) = m + n$, we see that (1) and (2) in Definition 4.3.2 immediately produce (A1) and (A2). $\quad\square$

The next proposition shows that $m + 1$ is equal to m^+, the successor of m. After we prove this proposition, we will show that $2 + 2 = 4$.

Proposition 4.3.4. *For all $m \in \omega$, $m + 1 = m^+$.*

Proof. Let $m \in \omega$. Since $1 = 0^+$, we obtain

$$m + 1 = m + 0^+$$
$$= (m + 0)^+ \quad \text{by (A2)}$$
$$= m^+ \quad\quad\; \text{by (A1)}$$

and thus, $m + 1 = m^+$. $\quad\square$

We can now prove that $2 + 2 = 4$ as follows:

$$2 + 2 = 2 + 1^+ \qquad \text{because } 2 = 1^+$$
$$= (2 + 1)^+ \quad \text{by (A2)}$$
$$= (2^+)^+ \quad \text{by Prop. 4.3.4}$$
$$= 3^+ \qquad \text{as } 2^+ = 3$$
$$= 4 \qquad \text{as } 3^+ = 4.$$

Now that we have the operation $+$ of addition, we can apply Theorem 4.3.1 to define multiplication on the natural numbers using the functions $g(m) = 0$ and $f(k, m) = k + m$.

Definition 4.3.5. Let $M: \omega \times \omega \to \omega$ be the unique function satisfying

(1) $M(m, 0) = 0$,
(2) $M(m, n^+) = M(m, n) + m$

for all natural numbers m and n. The function M is called **multiplication**, and we define the binary operation $\cdot$ on ω to be

$$m \cdot n = M(m, n)$$

for all m and n in ω.

Theorem 4.3.6. *For all natural numbers m and n,*

(M1) $m \cdot 0 = 0$
(M2) $m \cdot n^+ = m \cdot n + m$.

Proof. Let m and n be elements in ω. Recalling that $M(m, n) = m \cdot n$, we see that (1) and (2) in Definition 4.3.5 yields (M1) and (M2). $\qquad \square$

Having presented the set-theoretic definitions of addition and multiplication, we will now show that some of the familiar laws of arithmetic are provable within set theory. Afterwards, we shall define exponentiation.

Theorem 4.3.7 (Associative Law for Addition). *For all m, n, and p in ω,*

$$m + (n + p) = (m + n) + p.$$

Proof. Let m and n be elements in ω, and let

$$I = \{p \in \omega : m + (n + p) = (m + n) + p\}.$$

We shall show that I is an inductive set. To show that $0 \in I$, observe that

$$m + (n + 0) = m + n \qquad \text{by (A1)}$$
$$= (m + n) + 0 \quad \text{by (A1)}$$

and thus, $m + (n + 0) = (m + n) + 0$, and so $0 \in I$. Assume that $p \in I$. Then

$$m + (n + p^+) = m + (n + p)^+ \qquad \text{by (A2)}$$
$$= (m + (n + p))^+ \quad \text{by (A2)}$$
$$= ((m + n) + p)^+ \quad \text{since } p \in I$$
$$= (m + n) + p^+ \qquad \text{by (A2).}$$

Therefore, $m + (n + p^+) = (m + n) + p^+$, and hence, $p^+ \in I$. $\qquad \square$

For $n \in \omega$, we know that $n + 0 = n$ by (A1). Since we have not proven that addition is commutative, we cannot immediately infer that $0 + n = n$, and thus, we must prove that this latter equation holds for all $n \in \omega$.

Lemma 4.3.8. *For all $n \in \omega$, we have $0 + n = n$.*

Proof. We prove that $(\forall n \in \omega)(0 + n = n)$ by induction. Let

$$I = \{n \in \omega : 0 + n = n\}.$$

We prove that I is inductive. Since (A1) implies that $0 + 0 = 0$, we have that $0 \in I$. Let $n \in I$. So $0 + n = n$. We prove that $0 + n^+ = n^+$ as follows:

$$0 + n^+ = (0 + n)^+ \quad \text{by (A2)}$$
$$= n^+ \qquad \text{because } n \in I.$$

Therefore, $n^+ \in I$, and this completes the proof. $\qquad \square$

Item (A2) states that $m + n^+ = (m + n)^+$ for natural numbers m and n. We will now prove that we also have that $m^+ + n = (m + n)^+$. Then we will be able to prove that addition is commutative.

Lemma 4.3.9. *For all $m \in \omega$ and $n \in \omega$, we have $m^+ + n = (m + n)^+$.*

Proof. Let $m \in \omega$ be arbitrary. We prove that $(\forall n \in \omega)(m^+ + n = (m + n)^+)$ by induction. Let

$$I = \{n \in \omega : m^+ + n = (m + n)^+\}.$$

To prove the lemma, we just need to show that I is inductive. First, we verify that $0 \in I$ by showing that $m^+ + 0 = (m + 0)^+$ as follows:

$$m^+ + 0 = m^+ \qquad \text{by (A1)}$$
$$= (m + 0)^+ \quad \text{by (A1)}.$$

Thus, $0 \in I$. Suppose that $n \in I$. Then

$$m^+ + n^+ = (m^+ + n)^+ \qquad \text{by (A2)}$$
$$= ((m + n)^+)^+ \quad \text{because } n \in I$$
$$= (m + n^+)^+ \qquad \text{by (A2)}$$

and hence, $m^+ + n^+ = (m + n^+)^+$. Therefore, $n^+ \in I$ and I is inductive. $\qquad \square$

Lemmas 4.3.8 and 4.3.9 will be used to prove that addition is commutative.

Theorem 4.3.10 (Commutative Law for Addition). *For all m and n in ω,*

$$m + n = n + m.$$

Proof. Let n be any element in ω and let

$$I = \{m \in \omega : m + n = n + m\}.$$

We prove that I is an inductive set. Observe that $0 + n = n + 0$ by Lemma 4.3.8 and (A1). So $0 \in I$. Assume that $m \in I$. Then

$$m^+ + n = (m + n)^+ \qquad \text{by Lemma 4.3.9}$$
$$= (n + m)^+ \quad \text{since } m \in I$$
$$= n + m^+ \qquad \text{by (A2)}.$$

Hence, $m^+ + n = n + m^+$, and therefore, $m^+ \in I$. $\qquad \square$

Using proof by induction, we will now show that the familiar distributive law of arithmetic holds. We will thus have additional evidence to support the assertion "Mathematics can be embedded in set theory."

Theorem 4.3.11 (Distributive Law). *For all natural numbers m, n, and p,*

$$m \cdot (n + p) = m \cdot n + m \cdot p.$$

Proof. Let n and m be in ω. Now let

$$I = \{p \in \omega : m \cdot (n + p) = m \cdot n + m \cdot p\}.$$

We will show that I is an inductive set. To verify that $0 \in I$, we have

$$
\begin{aligned}
m \cdot (n + 0) &= m \cdot n && \text{by (A1)}\\
&= m \cdot n + 0 && \text{by (A1)}\\
&= m \cdot n + m \cdot 0 && \text{by (M1).}
\end{aligned}
$$

Thus, $m \cdot (n + 0) = m \cdot n + m \cdot 0$, and so $0 \in I$. Now, suppose that $p \in I$. Then

$$
\begin{aligned}
m \cdot (n + p^+) &= m \cdot (n + p)^+ && \text{by (A2)}\\
&= m \cdot (n + p) + m && \text{by (M2)}\\
&= (m \cdot n + m \cdot p) + m && \text{as } p \in I\\
&= m \cdot n + (m \cdot p + m) && \text{by Theorem 4.3.7}\\
&= m \cdot n + m \cdot p^+ && \text{by (M2).}
\end{aligned}
$$

Hence, $m^+ + n = n + m^+$, and thus, $m^+ \in I$. $\qquad\square$

Our next two theorems verify two fundamental properties of multiplication.

Theorem 4.3.12 (Associative Law for Multiplication). *For all m, n, and p in ω,*

$$m \cdot (n \cdot p) = (m \cdot n) \cdot p.$$

Proof. Let m and n be elements in ω and let

$$I = \{p \in \omega : m \cdot (n \cdot p) = (m \cdot n) \cdot p\}.$$

We shall show that I is an inductive set. To show that $0 \in I$, observe that

$$
\begin{aligned}
m \cdot (n \cdot 0) &= m \cdot 0 && \text{by (M1)}\\
&= 0 && \text{by (M1)}\\
&= (m \cdot n) \cdot 0 && \text{by (M1)}
\end{aligned}
$$

and thus, $m \cdot (n \cdot 0) = (m \cdot n) \cdot 0$, and so $0 \in I$. Assume that $p \in I$. Then

$$
\begin{aligned}
m \cdot (n \cdot p^+) &= m \cdot (n \cdot p + n) && \text{by (M2)} \\
&= m \cdot (n \cdot p) + m \cdot n && \text{by Theorem 4.3.11} \\
&= (m \cdot n) \cdot p + m \cdot n && \text{since } p \in I \\
&= (m \cdot n) \cdot p^+ && \text{by (M2).}
\end{aligned}
$$

Therefore, $m \cdot (n \cdot p^+) = (m \cdot n) \cdot p^+$, and hence, $p^+ \in I$. $\qquad\square$

Theorem 4.3.13 (Commutative Law for Multiplication). *For all m and n in ω,*

$$m \cdot n = n \cdot m.$$

Proof. See Exercise 6. $\qquad\square$

We will now apply Theorem 4.3.1 to define exponentiation on the natural numbers, using the functions $g(m) = 1$ and $f(k, m) = k \cdot m$.

Definition 4.3.14. Let $E \colon \omega \times \omega \to \omega$ be the unique function satisfying

(1) $E(m, 0) = 1$,
(2) $E(m, n^+) = E(m, n) \cdot m$

for all natural numbers m and n. The function E is called **exponentiation**. We define the binary operation, denoted by m^n, to be

$$m^n = E(m, n)$$

for all m and n in ω.

Theorem 4.3.15. *For all natural numbers m and n,*

(E1) $m^0 = 1$,
(E2) $m^{n^+} = m^n \cdot m$.

Exercises 4.3

1. Let $m \in \omega$ and $n \in \omega$. Suppose $m + n = 0$. Prove that $m = n = 0$.

2. Let $m \in \omega$ and $n \in \omega$. Show that if $m \cdot n = 0$, then $m = 0$ or $n = 0$.

3. Let $m \in \omega$ and $n \in \omega$. Prove that for all $p \in \omega$, if $m + p = n + p$, then $m = n$.

4. Prove that for all $n \in \omega$, the equality $0 \cdot n = 0$ holds.

5. Prove for all $m \in \omega$ and $n \in \omega$, we have that $m^+ \cdot n = m \cdot n + n$.

*6. Let $n \in \omega$. Using the previous two exercises, prove Theorem 4.3.13 by showing that the set $I = \{m \in \omega : m \cdot n = n \cdot m\}$ is inductive.

7. A natural number n is *even* if $n = 2 \cdot k$ for some $k \in \omega$. If $n = 2 \cdot i + 1$ for some $i \in \omega$, then n is *odd*. Prove that for all $n \in \omega$, either n is even or n is odd.

8. Let $I = \{n \in \omega : \neg(n \text{ is even and } n \text{ is odd})\}$. Prove that I is an inductive set. Conclude that no natural number is both even and odd.

9. Prove for all m, n, and k in ω, that $m^{n+k} = m^n \cdot m^k$.

10. Prove for all m, n, and k in ω, that $(m \cdot n)^k = m^k \cdot n^k$.

11. Prove for all m, n, and k in ω, that $(m^n)^k = m^{n \cdot k}$.

Exercise Notes: For Exercise 1, suppose $n \neq 0$ and then use Theorem 4.1.6. For Exercise 2, assume that $n \neq 0$ and use Exercise 1 to prove that $m = 0$. For Exercise 3, Theorem 4.1.12 can be used in the inductive step.

4.4 Order on ω

We now want to develop a theory of order on the natural numbers. First, we define the "less than" relation on the natural numbers. Then we prove that this relation satisfies the trichotomy law (see Theorem 4.4.9) as well as the standard connections with addition and multiplication. How can we get such a relation in set theory?

Recall that for each $n \in \omega$, we defined the successor of n to be the natural number $n^+ = n \cup \{n\}$, and thus, $n \in n^+$. As a result, we have the following infinite membership list:

$$0 \in 1 \in 2 \in 3 \in 4 \in \cdots,$$

which is very much like the usual relation $<$ on the natural numbers; that is,

$$0 < 1 < 2 < 3 < 4 \cdots.$$

It turns out that the order relation that we need is just "∈," and, in fact, we will prove that this is the "less than" relation that we were looking for.

Definition 4.4.1. We say that *m* is **less than** *n* if and only if $m \in n$, whenever *m* and *n* are in ω.

In this section, we will view "∈" as the "less than" relation on the natural numbers, and we shall no longer use the symbol < in our initial discussion of ordering on ω. We will show that the relation ∈ on ω satisfies the trichotomy law, and we will show that ∈ is preserved under the operations of addition and nonzero multiplication.

Theorem 4.1.11 and Theorem 4.1.13 assert, respectively, that each natural number *n* is a transitive set and that ω is a transitive set. Hence, we have the following result.

Proposition 4.4.2. *Let n be a natural number. Then*

(1) *if $k \in n$, then $k \subseteq n$;*
(2) *if $a \in k$ and $k \in n$, then $a \in n$;*
(3) *if $k \in \omega$, then $k \subseteq \omega$;*
(4) *if $a \in k$ and $k \in \omega$, then $a \in \omega$.*

Item (2) proclaims that ∈ is a transitive relation on ω. One can also prove, without using the regularity axiom, that no natural number is "less than" itself.

Lemma 4.4.3. *Let $n \in \omega$. Then $n \notin n$.*

Proof. See Exercise 10 on page 88. □

Using Definition 4.4.1, we can now easily define the "less than or equal" relation on the natural numbers.

Definition 4.4.4. We write $m \subseteq n$ if and only if $m \in n$ or $m = n$, whenever *m* and *n* are in ω.

Since ∈ is a transitive relation on ω, it thus follows that ⊆ is also a transitive relation on ω. Clearly, ⊆ is reflexive. We will now show that it is antisymmetric (see Definition 3.4.1).

Lemma 4.4.5. *Let $m \in \omega$ and $n \in \omega$. If $m \subseteq n$ and $n \subseteq m$, then $m = n$.*

Proof. Let $m \in \omega$ and $n \in \omega$ satisfy $m \subseteq n$ and $n \subseteq m$. Thus, $m \subseteq n$ and $n \subseteq m$ by Proposition 4.4.2(1). Therefore, $m = n$. □

Hence, $\subseteq$ is a partial order on ω. Is the relation $\subseteq$ a total order on ω? To show that it is, we shall prove that the relation $\in$ satisfies the trichotomy law on ω. First, we establish two lemmas.

Lemma 4.4.6. *For all $m \in \omega$, we have that $0 \subseteq m$.*

Proof. We shall prove that the set

$$I = \{m \in \omega : 0 \subseteq m\}$$

is inductive. Clearly, $0 \in I$. Assume $m \in I$. Hence, $0 \subseteq m$. So either $0 = m$ or $0 \in m$. If $0 = m$, then, as $m \in m^+$, we see that $0 \in m^+$. If $0 \in m$, then we have $0 \in m$ and $m \in m^+$. By transitivity (see Proposition 4.4.2(2)), we conclude that $0 \in m^+$. Thus, in either case, $0 \subseteq m^+$ and so, $m^+ \in I$. Therefore, I is inductive and $I = \omega$. □

Lemma 4.4.7. *For all natural numbers m and n, if $m \in n$, then $m^+ \in n^+$.*

Proof. Let $m \in \omega$. We prove that $(\forall n \in \omega)(m \in n \rightarrow m^+ \in n^+)$ by induction. Let

$$I = \{n \in \omega : (m \in n \rightarrow m^+ \in n^+)\}.$$

Clearly, $0 \in I$ as $(m \in 0 \rightarrow m^+ \in 0^+)$ is true vacuously. Let $n \in I$. To prove that $n^+ \in I$, assume that $m \in n^+$. We shall show that $m^+ \in (n^+)^+$. Observe that (▲) $n^+ \in (n^+)^+$. Because $m \in n^+$, either $m \in n$ or $m = n$. If $m \in n$, then $m^+ \in n^+$ as $n \in I$. Since $m^+ \in n^+$, we see that (▲) and transitivity imply that $m^+ \in (n^+)^+$. If $m = n$, then $m^+ = n^+$. Thus, (▲) implies that $m^+ \in (n^+)^+$. Therefore, $n^+ \in I$. □

Corollary 4.4.8. *If m and n are in ω, then $m \in n$ if and only if $m^+ \in n^+$.*

Proof. Assume that $m \in \omega$ and $n \in \omega$. By Lemma 4.4.7, we just need to prove if $m^+ \in n^+$, then $m \in n$. So let $m^+ \in n^+$. Recall that (▼) $m \in m^+$. Since

$m^+ \in n^+$, we have either $m^+ \in n$ or $m^+ = n$. If $m^+ \in n$, then $m \in n$ by ($\blacktriangledown$) and transitivity. Also, if $m^+ = n$, then ($\blacktriangledown$) implies that $m \in n$. □

The following theorem shows that $\in$ is a total order on ω; that is, for all m and n in ω, either $m \in n$ or $n \in m$.

Theorem 4.4.9 (Trichotomy Law on ω). *For all m and n in ω, exactly one of the three relationships $m \in n$, $m = n$, $n \in m$ holds.*

Proof. First, we show that any two of the relations $m \in n$, $m = n$, $n \in m$ cannot hold at the same time, whenever m and n are natural numbers. To do this, we just have to show that the following two conjunctions do not hold:

$$m \in n \text{ and } m = n, \tag{a}$$

$$m \in n \text{ and } n \in m, \tag{b}$$

when $m \in \omega$ and $n \in \omega$. Conjunction (a) implies that $m \in m$, which contradicts Lemma 4.4.3. Conjunction (b), by transitivity, also implies that $m \in m$. Hence, both (a) and (b) do not hold.

We now prove that $(\forall m \in \omega)(\forall n \in \omega)(m \in n \text{ or } m = n \text{ or } n \in m)$. To prove this statement, let $m \in \omega$ and let

$$I = \{n \in \omega : (m \in n \text{ or } m = n \text{ or } n \in m)\}.$$

We will show that I is inductive. Lemma 4.4.6 implies that $0 \in I$. Let $n \in I$, that is, assume that

$$m \in n \text{ or } m = n \text{ or } n \in m. \tag{IH}$$

We must show that

$$m \in n^+ \text{ or } m = n^+ \text{ or } n^+ \in m. \tag{4.6}$$

To do this, we consider each of the three cases in (IH) separately. If $m \in n$, then $m \in n^+$ by the fact that $n \in n^+$ and transitivity. If $m = n$, then $m \in n^+$ since $n \in n^+$. If $n \in m$, then Lemma 4.4.7 implies that $n^+ \in m^+$. Thus, either $n^+ \in m$ or $n^+ = m$. So in each of these three cases in (IH), we obtain one of the conjuncts in (4.6). Therefore, $n^+ \in I$, and the proof is complete. □

The trichotomy law will now allow us to do the following:

(1) identify an interesting connection between the relations $\in$ and $\subset$ on the natural numbers;

(2) prove that addition and multiplication on the natural numbers preserve the "less than" relation;

(3) prove that ω, the set of natural numbers, satisfies the crucial well-ordering principle.

Corollary 4.4.10. *If m and n are in ω, then $m \in n$ if and only if $m \subset n$.*

Proof. Let $m \in \omega$ and $n \in \omega$. Assume that $m \in n$. Because n is transitive, we have that $m \subseteq n$. Since $m \in n$, Theorem 4.4.9 implies that $m \neq n$. So $m \subset n$. To prove the converse, assume $m \subset n$. Thus, $m \neq n$. From Theorem 4.4.9, we have either $m \in n$ or $n \in m$. If $n \in m$, then we would have that $n \in n$, as $m \subset n$. This contradicts Theorem 4.4.3. Thus, $m \in n$. □

Corollary 4.4.10 implies that for all natural numbers m and n,

$m \in n$ iff $m \subseteq n$.

When m and n are natural numbers, we can now define $\max(m, n) = m \cup n$, which is the larger of the two natural numbers m and n (see Exercise 5).

Our next theorem shows that the expected properties of "inequality" hold for the natural numbers.

Theorem 4.4.11 (Properties of Inequality). *For all natural numbers m, n, and p, the following hold:*

(1) *$m \in n$ if and only if $m + p \in n + p$.*
(2) *If $p \neq 0$, then $m \in n$ if and only if $m \cdot p \in n \cdot p$.*

Proof. Item (1) follows from Exercises 6 and 7 of this section. Similarly, (2) is established via Exercises 8 and 9. □

Corollary 4.4.12 (Cancellation Laws). *For all natural numbers m, n, and p, the following hold:*

(1) *If $m + p = n + p$, then $m = n$.*
(2) *If $p \neq 0$, then $m \cdot p = n \cdot p$ implies $m = n$.*

Proof. See Exercise 10. □

Corollary 4.1.5 (page 85) confirms the fact that the principle of mathematical induction is provable from the first seven axioms of Zermelo–Fraenkel (ZF) set theory. Virtually every undergraduate text in mathematics (e.g., number theory, real analysis, and abstract algebra) just cites this principle and does not attempt to prove it directly. Our next theorem shows that these seven axioms also imply the well-ordering principle: Every nonempty subset of ω has a least element.

Theorem 4.4.13 (Well-Ordering Principle). *Let A be a nonempty subset of ω. Then A has a least element; that is, there is an $\ell \in A$ such that $\ell \subseteq a$ for all $a \in A$.*

Proof. Let A be a nonempty subset of ω. Suppose, for a contradiction, that A has no least element. Let $I = \{n \in \omega : n \subseteq a \text{ for all } a \in A\}$. So I is the set of $n \in \omega$ that are less than or equal to every element in A. Note that ($\blacktriangle$) $A \cap I = \varnothing$ because if $n \in A \cap I$, then $n \in A$ and n would be a least element in A.

We will now show that I is inductive. Clearly, $0 \in I$, by Lemma 4.4.6. Let $n \in I$; that is, suppose that $n \subseteq a$ for all $a \in A$. Since A has no least element, it follows that $n \notin A$. Therefore, $n \in a$ for all $a \in A$. Hence, $n^+ \subseteq a$ for all $a \in A$ by Exercise 3. So $n^+ \in I$ and I is inductive. Thus, $I = \omega$ and ($\blacktriangle$) now implies that $A \cap \omega = \varnothing$. Nevertheless, because $A \subseteq \omega$ and A is nonempty, we also conclude that $A \cap \omega \neq \varnothing$. This contradiction completes the proof. $\square$

We end this section with a proof of the **strong induction principle on** ω.

Theorem 4.4.14 (Strong Induction Principle on ω). *Let A be a subset of ω and suppose that*

$$\text{for all } n \in \omega, \text{ if } n \subseteq A, \text{ then } n \in A. \tag{4.7}$$

Then $A = \omega$.

Proof. Let $A \subseteq \omega$ satisfy (4.7). Suppose, to the contrary, that there is a natural number that is not in A. Thus, the set $\omega \setminus A$ is nonempty. Theorem 4.4.13 now implies that $\omega \setminus A$ has a least element n. Hence, $n \notin A$ and if $m \in n$, then $m \in A$. Therefore, $n \subseteq A$. As A satisfies (4.7), we infer that $n \in A$, a contradiction. $\square$

We can now define a special relation symbol with which we are all familiar. Let m and n be in ω. We define $m < n$ if and only if $m \in n$, and define $m \leq n$ if and only if $m \subseteq m$. So in the future there may times when we shall use the symbols $<$ and $\leq$, rather than $\in$ and $\subseteq$, if it seems appropriate.

Remark 4.4.15. Using the system of natural numbers $(\omega, +, \cdot, \in)$, one can give set-theoretic constructions of the set of integers, the set of rational numbers, and then the set of real numbers. A method for constructing the set of integers is suggested in Exercise 9 on page 82. In [5], Enderton gives a more detailed description of this method and constructs the set of rational numbers and the set of real numbers.

Exercises 4.4

1. Let $n \in \omega$. Show that $1 \in n^+$.

*2. Let m and n be in ω. Show that if $m \in n^+$, then $m \subseteq n$.

*3. Let n and a be in ω. Show that if $n \in a$, then $n^+ \in a$.

4. Let I be inductive and let $a \in \omega$. Prove that $\{n \in I : n \in a \text{ or } a \subseteq n\}$ is inductive.

*5. Let m and n be in ω. Suppose that $m \subseteq n$. Prove that $\max(m, n) = n$.

*6. Let m and n be natural numbers. Prove that for all $p \in \omega$, if $m \in n$, then $m + p \in n + p$.

*7. Let $m \in \omega$ and $n \in \omega$. Prove that for all $p \in \omega$, if $m + p \in n + p$, then $m \in n$.

*8. Let m and n be natural numbers. Prove that for all $p \in \omega$, if $m \in n$, then $m \cdot p^+ \in n \cdot p^+$.

*9. Let $m \in \omega$ and $n \in \omega$. Prove that for all $p \in \omega$, if $m \cdot p^+ \in n \cdot p^+$, then $m \in n$.

*10. Use Theorem 4.4.11 and Theorem 4.4.9 to prove Corollary 4.4.12.

11. Let $m \in \omega$. Prove that $m \in m + p^+$, for all $p \in \omega$.

*12. Let $m \in \omega$. Prove that for all $n \in \omega$, if $m \in n$, then $m + p^+ = n$ for some $p \in \omega$.

13. Let m and n be in ω. Prove that $m \in n$ if and only if $m + p^+ = n$ for some $p \in \omega$.

*14. Suppose that $F : \omega \to \omega$ satisfies $F(n) \in F(n^+)$ for all $n \in \omega$.
    (a) Prove that F is an increasing function; that is, show that for all $m \in \omega$ and all $n \in \omega$, if $m \in n$, then $F(m) \in F(n)$.
    (b) Prove that F is one-to-one.

15. Prove that there is no function $F : \omega \to \omega$ satisfying $F(n^+) \in F(n)$ for all $n \in \omega$.

16. Let $n \in \omega$ and let $E \subseteq n^+$ be nonempty. Prove that E has a largest element; that is, there is an $m \in E$ such that $e \underline{\in} m$ for all $e \in E$.

17. Let $F : \omega \to \omega$. Prove that for all $n \in \omega$, the set $F[n^+]$ has a largest element.

18. Prove for all $n \in \omega$, if $f : n \to \omega$, then $\mathrm{ran}(f) \subseteq k$, for some $k \in \omega$.

19. Let d be in ω. Suppose that $0 \in d$. Let

$$I = \{n \in \omega : n = q \cdot d + r \text{ for some } q, r \in \omega \text{ where } r \in d\}.$$

Prove that I is an inductive set.

20. Let n, d, q, r, q', r' be in ω with $0 \in d$. Assume that $n = q \cdot d + r$ where $r \in d$, and $n = q' \cdot d + r'$ where $r' \in d$. Prove that $q = q'$ and $r = r'$. To prove that $q = q'$, suppose to the contrary that $q \in q'$ (without loss of generality). By Exercise 12, there is a $p \in \omega$ so that $q + p^+ = q'$. Complete the following:
    (a) Show that $r = dp^+ + r'$.
    (b) Show that $d \underline{\in} dp^+$ and $dp^+ \underline{\in} dp^+ + r'$. Then show that $d \underline{\in} r$, which is a contradiction. Conclude that $q = q'$.
    (c) Now show that $r = r'$.

Exercise Notes: For Exercise 3, apply Theorem 4.4.9. For Exercise 4, apply Exercise 3. For Exercise 6, use induction. For Exercise 7, apply Exercise 6 and Theorem 4.4.9. For Exercise 8, use induction. For Exercise 9, use Exercise 8 and Theorem 4.4.9. For Exercise 13, use Exercises 11–12. For Exercise 14, let $m \in \omega$ and prove that $I = \{n \in \omega : m \in n \to F(m) \in F(n)\}$ is inductive. For Exercise 15, one can apply Theorem 4.4.13 and proof by contradiction: Assume that there is such function F. Let $A = \{F(n) : n \in \omega\} \neq \varnothing$. Let ℓ be the least element in A. Thus, $\ell = F(k)$ for a $k \in \omega$. For Exercise 16, use the well-ordering principle and Exercise 2. For Exercise 17, use induction and review 3.3.8(3) on page 60. For Exercise 18, prove that

$$I = \{n \in \omega : \text{for all } f, \text{if } f : n \to \omega, \text{ then } \mathrm{ran}(f) \subseteq k \text{ for some } k \in \omega\}$$

is an inductive set. For Exercise 19, in the inductive step one can use Exercise 3. Exercises 19–20 yield a proof the Division Algorithm: *Let n and d be natural numbers, where $d \geq 1$. Then there exist unique natural numbers q and r such that $n = dq + r$ and $0 \leq r < d$.*

5 On the Size of Sets

The size of a finite set can be measured simply by counting; for example, the size of the set $X = \{1, 2, 3, 4, \ldots, 10\}$ is 10 because it has 10 elements, and the size of the sets $Y = \{\pi, 2, 30\}$ and $Z = \{9, 11, -1\}$ is 3. Clearly, the size of X is bigger than the size of Y. Moreover, the sets Y and Z have the same size. Can the concept of "size" be extended to infinite sets? Is there a way, which does not involve counting, of showing that two sets have the same size? Georg Cantor was the first mathematician to seriously address and answer these questions. Cantor found a way to measure the size of any infinite set. He first observed that two sets A and B have the same size if there is a *one-to-one correspondence* between A and B, that is, there is a way of evenly matching the elements in A with the elements in B. In other words, Cantor observed that A and B have the same size if there is a bijection $f : A \rightarrow B$.

The arrow diagram in Figure 5.1 represents a bijection. As a result, we can use this bijection to construct the following one-to-one correspondence (5.1) between the sets A and B:

$$
\begin{array}{llllll}
A: & a & b & c & d & e \\
& \updownarrow & \updownarrow & \updownarrow & \updownarrow & \updownarrow \\
B: & 5 & 8 & 1 & 3 & 6
\end{array}
\tag{5.1}
$$

Hence, the function f allows us to set up a pairing between the elements in A and the elements in B such that each element in A is matched with exactly one element in B, and each element in B is thereby matched with exactly one element in A. Cantor observed that we can now conclude, *without counting*, that the sets A and B have the same size.

For another illustration, let $E = \{k \in \omega : k \text{ is even}\}$, and let $f : \omega \rightarrow E$ be defined by $f(n) = 2n$. As f is a bijection, we obtain the following one-to-one correspondence between the set ω of natural numbers and the set E of even natural numbers:

$$
\begin{array}{llllllllllll}
\omega: & 0 & 1 & 2 & 3 & 4 & 5 & 6 & \cdots & n & \cdots \\
& \updownarrow & \updownarrow & \updownarrow & \updownarrow & \updownarrow & \updownarrow & \updownarrow & \cdots & \updownarrow & \cdots \\
E: & 0 & 2 & 4 & 6 & 8 & 10 & 12 & \cdots & 2n & \cdots
\end{array}
$$

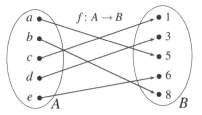

Figure 5.1. Arrow diagram of a bijection

Therefore, each natural number n corresponds to the even number $2n$, and each even natural number $2i$ is thereby matched with $i \in \omega$. The bijection $f : \omega \rightarrow E$ specifies a one-to-one match-up between the elements in ω and the elements in E. Cantor concluded that the sets ω and E have the same size.

After discovering how to determine if two infinite sets have the same size, Cantor proved that the set $\mathbb{Q}$ of rational numbers has the same size as the set ω of natural numbers. As a result, Cantor then conjectured that the set of real numbers also has the same size as ω. It came as a complete surprise to Georg Cantor when, in 1874, he discovered that these two infinite sets have different sizes. In fact, Cantor showed that the set of real numbers is much larger than the set of natural numbers. This unexpected result would have an enormous impact on the future of mathematics.

When Cantor first presented his new research on the size of infinite sets, some of his contemporaries actually refused to acknowledge his discoveries. Moreover, Henri Poincaré referred to Cantor's work as a "grave disease" that would infect mathematics. Nevertheless, Cantor and his important ideas would eventually be recognized. In 1904, the Royal Society presented Cantor with its prestigious Sylvester Medal for his "brilliant" mathematical research. David Hilbert, a very influential mathematician, described Cantor's work as

> *the finest product of mathematical genius and one of the supreme achievements of purely intellectual human activity.*

In this chapter, we shall investigate Cantor's early work in set theory. First, we will formally define and examine the notion of a finite set, and then we will classify sets into two categories: sets whose elements can be enumerated in a sequence indexed by the natural numbers (countable sets) and those sets for which it is impossible to so enumerate all of their elements (uncountable sets). We will also examine Cantor's fundamental definitions and theorems on the cardinality (size) of sets.

5.1 | Finite Sets

A set is *infinite*, of course, if it is not finite. Moreover, a set is finite if it has at most n many elements for some natural number n. The next definition is just a mathematical rewording of this notion.

Definition 5.1.1. A set X is **finite** if and only if there is a one-to-one function $f: X \to n$ for some natural number n.

It follows, vacuously, that the empty set is finite. Now that we have a formal definition of a finite set, we can now use the Well-Ordering Principle 4.4.13 to precisely define the notion of the "number of elements" in a finite set.

Definition 5.1.2. Suppose that X is a finite set. Let n be the least natural number such that there is a one-to-one function $f: X \to n$. Then n is the **number of elements** in the set X, and we write $|X| = n$.

In particular, we have that $|\varnothing| = 0$ (see Remark 3.3.7 and the paragraph just prior to Definition 3.3.17).

Theorem 5.1.3. *Let A be finite and let $n = |A|$. If $f: A \to n$ is one-to-one, then f is onto n.*

Proof. Let A be a finite set. If $A = \varnothing$, then $|A| = 0$ and any function $f: A \to 0$ is vacuously onto 0. So now assume that $A \neq \varnothing$. Let $n = |A|$ and let $f: A \to n$ be one-to-one. Suppose, to the contrary, that f is not onto n. Let $k \in n$ be so that $k \notin \mathrm{ran}(f)$. Because $n \neq 0$, there is an $m \in \omega$ such that $n = m^+$. Since $k \in n = m^+$, we have either $k = m$ or $k \in m$. If $k = m$, then $f: A \to m$ and f is one-to-one. As $m \in n$, we conclude that $|A| \in n$, which is a contradiction. If $k \in m$, then define the function $g: A \to m$ by

$$g(a) = \begin{cases} k, & \text{if } f(a) = m; \\ f(a), & \text{if } f(a) \neq m. \end{cases}$$

Since f is one-to-one, it follows that $g: A \to m$ is also one-to-one. Moreover, because $m \in n$, we conclude that $|A| \in n$, which is again a contradiction. $\qquad\square$

Corollary 5.1.4. *If A is finite and $n = |A|$, then there is a bijection $f : A \to n$.*

Proof. Suppose that A is finite and let $n = |A|$. By Definition 5.1.2 there exists a one-to-one function $f : A \to n$. Theorem 5.1.3 implies that f is a bijection. $\square$

Theorem 5.1.5. *Let $f : A \to A$ be one-to-one where A is a finite set. Then f is onto A.*

Proof. Let $f : A \to A$ be one-to-one where A is finite. Let $n \in \omega$ be such that $|A| = n$. By Definition 5.1.2, let $h : A \to n$ be one-to-one. Theorems 3.3.19 and 5.1.3 imply that $(h \circ f) : A \to n$ is a bijection. To prove that f is onto A, let $y \in A$. Since $h(y) \in n$, there exists an $x \in A$ such that $(h \circ f)(x) = h(y)$. So $h(f(x)) = h(y)$, and thus, $f(x) = y$ as h is one-to-one. Hence, f is onto A. $\square$

Corollary 5.1.6. *Let $f : A \to A$ be one-to-one and not onto A. Then A is infinite.*

The following theorem confirms that two finite sets have the same number of elements if and only if there exists a one-to-one correspondence between the two sets.

Theorem 5.1.7. *Let A and B be finite sets. Then there is a bijection $h : A \to B$ if and only if $|A| = |B|$.*

Proof. See Exercise 4. $\square$

Recall that $n^+ = \{0, 1, 2, 3, \ldots, n\}$. Since the identity function from n^+ to n^+ is one-to-one, we see from Definition 5.1.2 that $\left|n^+\right| \in n^+$.

Theorem 5.1.8. *For all natural numbers n, we have that $|n| = n$.*

Proof. We prove that $I = \{n \in \omega : |n| = n\}$ is an inductive set. As $|\varnothing| = 0$, we see that $|0| = 0$ and hence, $0 \in I$. Let $n \in I$. Thus, $|n| = n$. Let $m = \left|n^+\right|$ and let $F : n^+ \to m$ be one-to-one. So $m \in n^+$. To prove that $m = n^+$, suppose to the contrary that $m \in n^+$. Therefore, ($\blacktriangle$) $m \in n$. Since $F : n^+ \to m$ is one-to-one, it clearly follows that $(F \restriction n) : n \to m$ is one-to-one. As $|n| = n$, we have that $n \subseteq m$. It now follows from (▲) that $m = n$. Because $|n| = n$ and $(F \restriction n)$ is one-to-one, Theorem 5.1.3 implies that $F \restriction n$ is onto m. Since $F(n) \in m$, it follows that $F(n) = F(i)$ for some $i \in n$. This contradicts the fact that F is one-to-one. Therefore, $m = n^+$ and so $n^+ \in I$.   □

Corollary 5.1.9 (Pigeonhole Principle). *Let m and n be natural numbers and let $f : n \to m$ be a function. If $m \in n$, then f is not one-to-one.*

Theorem 5.1.10. *The set ω is an infinite set.*

Proof. Suppose ω is finite. Thus, there is a one-to-one function $F : \omega \to n$ for some $n \in \omega$. Since F is one-to-one, the restriction $(F \restriction n^+) : n^+ \to n$ is also one-to-one, and this contradicts Corollary 5.1.9.   □

Corollary 5.1.11. *If $G : \omega \to A$ is one-to-one, then A is an infinite set.*

Proof. See Exercise 21.   □

The pigeonhole principle implies a "dual" version involving surjections.

Lemma 5.1.12. *If $m \in n$ and $f : m \to n$ where $m \in \omega$ and $n \in \omega$, then f is not onto n.*

Proof. Let $m \in n$ where n and m are natural numbers. Suppose, to the contrary, that $f : m \to n$ is onto n. We can therefore define the function $g : n \to m$ by

$$g(x) = \text{the least } i \in m \text{ such that } f(i) = x. \tag{5.2}$$

We now show that g is one-to-one. Let $x \in n$ and $y \in n$ be such that $g(x) = g(y)$. Hence, $f(g(x)) = f(g(y))$. By the definition (5.2) of g, we have that $x = y$. So $g : n \to m$ is one-to-one, which contradicts Corollary 5.1.9, because $m \in n$.   □

Theorem 5.1.13. *If N is finite and $f : N \to N$ is onto N, then f is one-to-one.*

Proof. Let N be finite and $f : N \to N$ be onto N. Also, let $n \in \omega$ be such that $|N| = n$. Assume $f(a) = f(b)$ where $a \in N$, $b \in N$, and $a \neq b$. So $n \neq 0$ and

$n = k^+$ for some $k \in \omega$. Let $K = N \setminus \{a\}$. Thus, $|K| = k$ (see Exercise 13) and $(f \restriction K): K \to N$ is onto N. Since $|K| = k$ and $|N| = n$, there is a function $g: k \to n$ that is onto n (see Exercise 14), contradicting Lemma 5.1.12, as $k \in n$. Therefore, f is one-to-one. ◻

Theorem 5.1.14. *If A is a finite set, then $\mathcal{P}(A)$ is finite.*

Proof. We will use proof by induction. Let

$I = \{n \in \omega : \text{for all } X, \text{ if } |X| = n, \text{ then } \mathcal{P}(X) \text{ is finite}\}$.

If $|X| = 0$, then $X = \varnothing$ and $\mathcal{P}(X) = \{\varnothing\}$ is finite. Hence, $0 \in I$. Let $n \in I$. To prove that $n^+ \in I$, let X be such that $|X| = n^+$. So X is nonempty. Let $a \in X$. Thus, $|X \setminus \{a\}| = n$ (see Exercise 13). As $n \in I$, we conclude that $\mathcal{P}(X \setminus \{a\})$ is finite. By Corollary 5.1.4, there exists a bijection $f: \mathcal{P}(X \setminus \{a\}) \to m$ where $m = |\mathcal{P}(X \setminus \{a\})|$. Define $h: \mathcal{P}(X) \to m + m$ by

$$h(U) = \begin{cases} f(U), & \text{if } a \notin U; \\ m + f(U \setminus \{a\}), & \text{if } a \in U. \end{cases} \qquad (5.3)$$

To prove that h is one-to-one, suppose that $h(U) = h(V)$ with $U \in \mathcal{P}(X)$ and $V \in \mathcal{P}(X)$. Thus, $h(U) = h(V) = i$ for some $i \in 2m$. There are two cases to consider: $i \in m$ and $m \subseteq i$. If $i \in m$, then $a \notin U$ and $a \notin V$, by (5.3). Because $h(U) = h(V)$, we conclude that $f(U) = f(V)$. So $U = V$ since f is one-to-one. If $m \subseteq i$, then (5.3) implies that $a \in U$ and $a \in V$. As $h(U) = h(V)$, (5.3) asserts that $m + f(U \setminus \{a\}) = m + f(V \setminus \{a\})$. So $f(U \setminus \{a\}) = f(V \setminus \{a\})$. Hence, $U \setminus \{a\} = V \setminus \{a\}$ because f is one-to-one. It therefore follows that $U = V$. Since $h: \mathcal{P}(X) \to 2m$ is one-to-one and $2m \in \omega$, we have that $\mathcal{P}(X)$ is finite. Thus, $n^+ \in I$ and the proof is complete. ◻

Exercises 5.1

1. In our proof of Theorem 5.1.3, under the case that $k \in m$, we did not show that the function g is one-to-one. Complete the proof by proving that g is one-to-one.

2. Suppose that the set B is finite and $A \subseteq B$. Prove that A is finite.

3. Let A and B be two finite sets, and let $f: A \to B$ be one-to-one. Prove that $|A| \leq |B|$.

*4. Using Exercise 3 and Corollary 5.1.4, prove Theorem 5.1.7.

*5. Let $n \in \omega$ and let $f: n \to A$. Prove that $f[n]$ is finite.

6. Let A be finite and let $f: A \to \omega$. Prove that $\operatorname{ran}(f) \subseteq k$, for some $k \in \omega$.

7. Suppose that $f: A \to n$ is a bijection, where $n \in \omega$. Prove that $|A| = n$.

8. Let $f: \omega \times \omega \to \omega$ be defined by $f(i, j) = 2^i \cdot 3^j$. Prove that f is one-to-one. Prove that if $i < m$ and $j < n$, then $f(i, j) < f(m, n)$.

*9. Prove that $p: \omega \times \omega \to \omega$ defined by $p(i, j) = 2^i(2j + 1) - 1$ is one-to-one and is onto ω. Prove that if $i \in m$ and $j \in n$, then $p(i, j) \in p(m, n)$.

10. Prove there exists a one-to-one function $f: \omega \to \omega \times \omega$ that is onto $\omega \times \omega$.

11. Suppose that A and B are disjoint finite sets. Prove that $A \cup B$ is finite.

*12. Let A be an infinite set and B be a finite set. Prove that $A \setminus B$ is infinite.

*13. Let $k \in \omega$. Suppose that $|A| = k^+$ and $a \in A$. Prove that $|A \setminus \{a\}| = k$.

*14. Suppose that $|K| = k$, $|N| = n$, and $f: K \to N$ is onto N where $k \in \omega$ and $n \in \omega$. Show that there is a function $g: k \to n$ that is onto n.

*15. Suppose that A and B are finite sets. Prove that $A \cup B$ is finite.

16. Let $n \in \omega$. Suppose that $f: n \to A$ is onto A. Prove that A is finite.

*17. Let A be a finite set. Prove that there exists an $f: n \to A$ that is onto A for some $n \in \omega$.

18. Suppose that $f: A \to B$ is onto B where A is finite. Prove that B is finite.

19. Let $n \in \omega$ and $\{A_i : i \in n\}$ be such that A_i is finite for all $i \in n$. Prove, by induction, that $\bigcup_{i \in n} A_i$ is finite. [Use Exercise 15 in the inductive step.]

20. Suppose that A and B are finite sets. Prove that $A \times B$ is finite.

*21. Prove Corollary 5.1.11.

*22. Let $\langle A_i : i \in I \rangle$ be an indexed function where I is a finite set and $A_i \neq \emptyset$ for all $i \in I$. Without using the axiom of choice, prove by induction on $|I|$ that there is an indexed function $\langle x_i : i \in I \rangle$ such that $x_i \in A_i$ for all $i \in I$.

*23. Using Exercise 7, modify the proof of Theorem 5.1.14 to show that if $|A| = n$, then $|\mathcal{P}(A)| = 2^n$.

Exercise Notes: For Exercise **6**, apply Exercise 18 on page 109. For Exercise **7**, use Theorems 5.1.7 and 5.1.8. For Exercise **9**, if $1 \in m$, then $m - 1 = \bigcup m$ (see Theorems 4.1.6 and 4.1.10). To prove p is onto ω, apply Theorem 4.4.13 to prove that for all $n \in \omega$, if $1 \in n$, then $n = 2^i(2j + 1)$ for some $i \in \omega$ and $j \in \omega$. Note: if $k \in \omega$, then $k = k^+ - 1$. For Exercise **10**, use Exercise **9**. For Exercise **12**, apply Exercise **11**. For Exercise **16**, let $I_a = \{i \in n : f(i) = a\}$ for $a \in A$. Since f is onto A, each the set I_a is nonempty and has a least element. For Exercise **17**, use Corollary 5.1.4. For Exercise **18**, apply Exercises **16–17**. For Exercise **20**, apply Exercise **9**. For Exercise **21**, recall Theorem 3.3.19. For Exercise **22**, the result is true vacuously if $|I| = 0$. For Exercise **23**, note that if $m \in i \in m + m$, then Exercise 12 on page 108 and Theorem 4.4.11(1) imply that $i = m + k$ for a $k \in m$.

5.2   Countable Sets

Recall that $\omega = \{0, 1, 2, 3, \ldots\}$ is the set of natural numbers. A set is countable if it has the same size as some subset of ω. In other words, a set is *countable* if there is a one-to-one correspondence between the set and a subset of ω. Our next definition captures this concept in mathematical terms.

Definition 5.2.1. A set X is **countable** if and only if there exists a one-to-one function $f : X \to \omega$.

So by Definition 5.1.1, every finite set is countable (as $n \subseteq \omega$, for all $n \in \omega$). Every subset S of ω is also countable, since the identity function $i : S \to \omega$ is one-to-one, where $i(x) = x$ for all $x \in S$. Thus, in particular, the set of natural numbers ω is countable. One can also show that the set of integers and the set of rational numbers are countable (see Exercises 4 and 5). We will prove in Section 5.3 that the set of real numbers is not countable.

Definition 5.2.2. A set X is **countably infinite** if it is countable and infinite.

For example, the set ω is countably infinite as it is countable and infinite by Theorem 5.1.10. When one can prove that a set is countable and it is clear that it is infinite, then we can conclude that the set is countably infinite.

Theorem 5.2.3. *Suppose that A and B are sets where B is countable. If there is an injection g: A → B, then A is countable.*

Proof. Let A and B be sets where B is countable. Let $g: A \to B$ be one-to-one. We shall prove that A is countable. Since B is countable, there is a one-to-one function $f: B \to \omega$. Thus, by Theorem 3.3.19, $(f \circ g): A \to \omega$ is an injection. We conclude that A is countable. $\qquad\qquad\square$

Theorem 5.2.4. *If B is a countable set and A ⊆ B, then A is countable.*

Proof. Assume that B is a countable set and $A \subseteq B$. Let $i: A \to B$ be the identity function, that is, $i(x) = x$ for all $x \in A$. Since i is one-to-one, Theorem 5.2.3 implies that A is countable. $\qquad\qquad\square$

Theorem 5.2.5. *If A and B are countable sets, then A ∪ B is countable.*

Proof. Let A and B be countable. So there are one-to-one functions $f: A \to \omega$ and $g: B \to \omega$. Now define the function $h: A \cup B \to \omega$ by

$$h(x) = \begin{cases} 2 \cdot f(x), & \text{if } x \in A; \\ 2 \cdot g(x) + 1, & \text{if } x \in B \setminus A, \end{cases} \tag{5.4}$$

for each $x \in A \cup B$. We prove that $h: A \cup B \to \omega$ is one-to-one. Let x and y be in $A \cup B$ and assume $h(x) = h(y)$. We shall prove that $x = y$. First, because $h(x) = h(y)$, we cannot have $x \in A$ and $y \in B \setminus A$. To see this, suppose that $x \in A$ and $y \in B \setminus A$. So, as $h(x) = h(y)$, we infer that $2 \cdot f(x) = 2 \cdot g(x) + 1$ from (5.4); but this is impossible because a natural number cannot be both even and odd. Similarly, we cannot have $y \in A$ and $x \in B \setminus A$. Hence, either x and y are both in A, or they are both in $B \setminus A$. If x and y are in A, then $2 \cdot f(x) = 2 \cdot f(y)$. Thus, $f(x) = f(y)$ and so $x = y$, because f is one-to-one. If x and y are in $B \setminus A$, then $2 \cdot g(x) + 1 = 2 \cdot g(y) + 1$. We conclude that $g(x) = g(y)$. Since g is one-to-one, we have $x = y$. Therefore, h is one-to-one and $A \cup B$ is countable. $\qquad\square$

One can prove by mathematical induction, using Theorem 5.2.5, that a finite union of countable sets is also countable. We will not do this here, because we will soon prove a more general result (see Corollary 5.2.10).

Our next result demonstrates that every countably infinite set can be put into a one-to-one correspondence with the set of natural numbers.

Theorem 5.2.6. *If A is countably infinite, then there is a bijection $g\colon \omega \to A$.*

Proof. Let A be a countably infinite set. Thus, there exists a one-to-one function $f\colon A \to \omega$. Since A is not finite, the range of f must be an infinite subset of ω. Let $Y = \operatorname{ran}(f)$. So $f\colon A \to Y$ is a bijection. Hence, $f^{-1}\colon Y \to A$ is also a bijection, via Theorem 3.3.18. Because Y is an infinite set of natural numbers, there is a bijection $h\colon \omega \to Y$ by Exercise 13. Let $g = f^{-1} \circ h$. Theorems 3.3.19 and 3.3.20 now imply that $g\colon \omega \to A$ is a bijection. □

Corollary 5.2.7. *A set A is countably infinite if and only if there is a bijection $g\colon \omega \to A$.*

Proof. If A is countably infinite, then Theorem 5.2.6 implies that there exists a bijection $g\colon \omega \to A$. Conversely, suppose that $g\colon \omega \to A$ is a bijection. Since g is one-to-one, A is infinite by Corollary 5.1.11. As g is onto A, Theorem 3.3.18 asserts the existence of the inverse function $g^{-1}\colon A \to \omega$ that is one-to-one. Hence, A is countable and infinite. □

Theorem 5.2.8. *If A is a countably infinite set, then there exists an enumeration $a_0, a_1, a_2, \ldots, a_n, \ldots$ of all the elements in A such that each element in A appears in this enumeration exactly once.*

Proof. Assume that A is countably infinite. Thus, by Theorem 5.2.6, there is a bijection $g\colon \omega \to A$. For each $n \in \omega$, let $a_n = g(n)$. Since g is one-to-one and onto A, it follows that the enumeration $a_0, a_1, a_2, \ldots, a_n, \ldots$ lists every element in A exactly once. □

Theorem 5.2.8 implies that each countably infinite set is also *denumerable*; that is, we can list the elements of a denumerable set in the same way that we list the natural numbers, namely, $0, 1, 2, 3, 4, 5, \ldots$. We will soon show that it is impossible to list all of the real numbers in such a manner. In other words, $\mathbb{R}$ is not denumerable.

Theorem 5.2.9 (AC). *If $\{A_i : i \in \omega\}$ is such that A_i is countable for each $i \in \omega$, then $\bigcup_{i \in \omega} A_i$ is countable.*

Proof. Suppose that $\{A_i : i \in \omega\}$ is a set of countable sets. Because each A_i is countable, the axiom of choice implies that for every $i \in \omega$ there is a one-to-one function $f_i : A_i \to \omega$. In other words, by the axiom of choice, there is a choice function $\langle f_i : i \in \omega \rangle$ for the indexed function $\langle F_i : i \in \omega \rangle$ where each F_i is the set of all the one-to-one functions of the form $h : A_i \to \omega$.

Now let $p : \omega \times \omega \to \omega$ be a one-to-one function (see Exercise 9, page 116). Define the function $g : \bigcup_{i \in \omega} A_i \to \omega$ by

$$g(x) = p(i, f_i(x)) \text{ where } i \text{ is the least } i \in \omega \text{ such that } x \in A_i,$$

for all $x \in \bigcup_{i \in \omega} A_i$. To prove that g is one-to-one, assume that $g(x) = g(y)$ where x and y are in $\bigcup_{i \in \omega} A_i$. Let $i \in \omega$ be the least natural number such that $x \in A_i$, and also let $j \in \omega$ be the least such that $y \in A_j$. Since $g(x) = g(y)$, we have $p(i, f_i(x)) = p(j, f_j(x))$. Because p is one-to-one, it follows that $i = j$ and $f_i(x) = f_j(y)$. Thus, $f_i(x) = f_i(y)$ and so $x = y$, as f_i is one-to-one. Therefore, g is one-to-one and $\bigcup_{i \in \omega} A_i$ is countable. $\qquad\square$

We can now prove that a countable union of countable sets is also countable. Therefore, countable sets can be used to construct many more countable sets.

Corollary 5.2.10 (AC). *Let $\mathcal{A}$ be countable and suppose that each $B \in \mathcal{A}$ is also countable. Then $\bigcup \mathcal{A}$ is countable.*

Proof. Let $\mathcal{A}$ be as in the statement of the corollary. Thus, there is a one-to-one function $h : \mathcal{A} \to \omega$. If $\mathcal{A} = \varnothing$, then $\bigcup \mathcal{A} = \varnothing$ is countable. So assume $\mathcal{A} \neq \varnothing$ and let $B_0 \in \mathcal{A}$ be fixed. Consider the indexed set $\{A_i : i \in \omega\}$ defined by

$$A_i = \begin{cases} B, & \text{if } i \in \text{ran}(h) \text{ and } h(B) = i; \\ B_0, & \text{if } i \notin \text{ran}(h), \end{cases}$$

for each $i \in \omega$. Therefore, A_i is countable for every $i \in \omega$, and $\bigcup \mathcal{A} = \bigcup_{i \in \omega} A_i$. Theorem 5.2.9 now implies that $\bigcup \mathcal{A}$ is countable. $\qquad\square$

Since a finite set is countable, Corollary 5.2.10 implies that a finite union of countable sets is also countable.

Corollary 5.2.11 (AC). *If the indexed set $\{A_i : i \in I\}$ is such that I is countable and A_i is countable for each $i \in I$, then $\bigcup_{i \in I} A_i$ is countable.*

Let A be a set and let n be a natural number. Recalling Definition 3.3.6, we have that $^nA = \{f : f$ is a function from n to $A\}$. The set ^nA can be viewed as the set of all sequences of elements in A of length n.

Theorem 5.2.12. *Let A be a countable set. Then ^nA is countable, for all $n \in \omega$.*

Proof. Let A be countable. Thus, there is a one-to-one function $\ell\colon A \to \omega$. Let $I = \{n \in \omega : {}^nA$ is countable$\}$. We shall prove that I is inductive. Since $^0A = \{\varnothing\}$ (see Remark 3.3.7), the set ^0A is countable. Thus, $0 \in I$. Let $n \in I$. Therefore, ^nA is countable. So let $g\colon {}^nA \to \omega$ be one-to-one. By Exercise 9 on page 116, there is a one-to-one function $p\colon \omega \times \omega \to \omega$. Define $h\colon {}^{n^+}A \to \omega$ by

$$h(f) = p(g(f \restriction n), \ell(f(n)))$$

for all $f \in {}^{n^+}A$. We will now prove that h is one-to-one, and thus conclude that $^{n^+}A$ is countable. Let $f \in {}^{n^+}A$ and $q \in {}^{n^+}A$. Assume that $h(f) = h(q)$. Therefore,

$$p(g(f \restriction n), \ell(f(n))) = p(g(q \restriction n), \ell(q(n))).$$

Because p is one-to-one, we conclude that

$$g(f \restriction n) = g(q \restriction n) \text{ and } \ell(f(n)) = \ell(q(n)).$$

Since g and ℓ are one-to-one, we infer that $f \restriction n = q \restriction n$ and $f(n) = q(n)$. Thus, by Lemma 3.3.5, $f = q$. So h is one-to-one. Therefore, $n^+ \in I$. □

By applying Theorem 2.1.3, one can show that the following class is a set:

$$^{\in\omega}A = \{f : \text{for some } n \in \omega, f \text{ is a function from } n \text{ to } A\}.$$

So for every function $f\colon n \to A$, where $n \in \omega$, we have that $f \in {}^{\in\omega}A$. Clearly,

$$^{\in\omega}A = \bigcup_{n\in\omega} {}^nA. \tag{5.5}$$

Theorem 5.2.13 (AC). *Let A be a countable set. Then $^{\in\omega}A$ is countable.*

Proof. If A is countable, Theorem 5.2.12 asserts that ^nA is countable, for every $n \in \omega$. Equation (5.5) and Theorem 5.2.9 thus imply that $^{\in\omega}A$ is countable. □

Corollary 5.2.14 (AC). *If A is countable, then the set of all finite subsets of A is also countable.*

Proof. Let F_A be set of all finite subsets of A. Define the function $G: {}^{\in\omega}A \to F_A$ by $G(f) = \operatorname{ran}(f)$. Exercise 17 on page 116 implies that G is onto F_A. Hence, F_A is countable by Theorem 5.2.13 and Exercise 10. □

Remark 5.2.15. With a little more work, one can prove Theorem 5.2.13 (and Corollary 5.2.14) without appealing to the axiom of choice. We outline two such proofs below. The first proof is inspired by the proof of Theorem 5.2.12, and the second proof applies the fundamental theorem of arithmetic (see [1, Theorem 4.7.7]).

1. Suppose that $\ell: A \to \omega$ and $p: \omega \times \omega \to \omega$ are one-to-one functions. By the Recursion Theorem 4.2.1, one obtains the indexed function $\langle h_n : n \in \omega \rangle$, where $h_n: {}^nA \to \omega$ for all $n \in \omega$, such that

   (1) $h_0(\varnothing) = 0$ (recall that ${}^0A = \{\varnothing\}$);
   (2) $h_{n^+}(f) = p(h_n(f \restriction n), \ell(f(n)))$ for all $f \in {}^{n^+}A$.

   One can prove that each $h_n: {}^nA \to \omega$ is one-to-one by induction. Clearly, h_0 is one-to-one. In the inductive step, one applies the argument used in the proof of Theorem 5.2.12. One now "glues the functions h_n together" by defining $G: {}^{\in\omega}A \to \omega$ by

   $$G(f) = p(n, h_n(f)) \text{ where } n = \operatorname{dom}(f).$$

   One can now prove that G is one-to-one and thus, ${}^{\in\omega}A$ is countable.
2. Let $\ell: A \to \omega$ be one-to-one. One must first prove that ${}^{\in\omega}\omega$ is countable. To do this, define $H: {}^{\in\omega}\omega \to \omega$ by

   $$H(f) = \begin{cases} 1, & \text{if } \operatorname{dom}(f) = 0; \\ 2^{f(0)+1} \cdot 3^{f(1)+1} \cdot 5^{f(2)+1} \cdots p_k^{f(k)+1}, & \text{if } \operatorname{dom}(f) = k^+, \end{cases}$$

   where p_i is the i-th prime. By the fundamental theorem of arithmetic, the function H is one-to-one. Hence, ${}^{\in\omega}\omega$ is countable. Since $\ell: A \to \omega$, define $G: {}^{\in\omega}A \to {}^{\in\omega}\omega$ by $G(f) = \ell \circ f$. Therefore, if $f: n \to A$, then $G(f): n \to \omega$ (see Theorem 3.3.10). As ℓ is one-to-one, it follows that G is one-to-one. Therefore, because ${}^{\in\omega}\omega$ is countable, Theorem 5.2.3 implies that ${}^{\in\omega}A$ is countable.

Exercises 5.2

1. Let $A = \{4, 8, 12, 16, \ldots\}$ and let $B = \{n \in \mathbb{Z} : n < -25\}$. Define a bijection $f: A \to \omega$, and define a bijection $g: B \to \omega$.

2. Let A and B be as in Exercise 1. Define an injection $h: A \cup B \to \omega$.

3. Show that the set of negative integers $\mathbb{Z}^-$ is countable.

*4. Conclude from Exercise 3 that set of integers $\mathbb{Z}$ is countable.

*5. Let $p: \omega \times \omega \to \omega$ be one-to-one (see Exercise 9 on page 116). Using p, show that the set of positive rational numbers $\mathbb{Q}^+$ is countable. Conclude that the set of negative rational numbers $\mathbb{Q}^-$ is countable and the set $\mathbb{Q}$ is countable.

6. Let A and B be two countably infinite sets. Prove that there exists a bijection $f: A \to B$.

7. Let A and B be countable sets. Prove that $A \times B$ is countable.

8. Let A be a set. Suppose that $f: \omega \to A$ is onto A. Prove that A is countable.

9. Let A be a nonempty countable set. Prove that there exists an $f: \omega \to A$ that is onto A.

*10. Suppose that $f: A \to B$ is onto B where A is countable. Prove that B is also countable.

*11. (AC) Let A be an infinite set and let $\mathcal{C} = \mathcal{P}(A) \setminus \{\varnothing\}$. By Theorem 3.3.24, there exists a function $g: \mathcal{C} \to A$ such that $g(B) \in B$ for all $B \in \mathcal{C}$. One can show (see Theorem 6.2.3) that there is a function $h: \omega \to A$ such that

$$h(n) = g(A \setminus h[n]), \text{ for all } n \in \omega.$$

Prove that $h: \omega \to A$ is one-to-one.

12. Let A be a finite set. Prove that ^nA is finite, for all $n \in \omega$.

*13. Let $Y \subseteq \omega$ be infinite and a be the least element in Y. Define $f: Y \to Y$ by

$$f(n) = \text{ the least element in } \{k \in Y : n \in k\}.$$

By the recursion theorem, there is a function $h: \omega \to Y$ such that

(1) $h(0) = a$,
(2) $h(n^+) = f(h(n))$, for all $n \in \omega$.

Show that $h(n) \in h(n^+)$ for each $n \in \omega$. Conclude, from Exercise 14 on page 108, that h is an increasing, one-to-one function. Prove that for each $k \in \omega$ there is an $i \in \omega$ so that $k \in h(i)$. Prove that $\text{ran}(h) = Y$ and that $h: \omega \to Y$ is a bijection.

14. Let $\mathbb{Z}[x]$ be the set of polynomials in x with integer coefficients, that is,

$$\mathbb{Z}[x] = \{a_0 + a_1 x + \cdots + a_k x^k : k \in \omega \text{ and } a_i \in \mathbb{Z} \text{ for all } i \in k^+\}.$$

Define a surjection $f: {}^{\in\omega}\mathbb{Z} \to \mathbb{Z}[x]$. Now show that $\mathbb{Z}[x]$ is countable.

15. A real number a is said to be *algebraic* if $p(a) = 0$ for some $p(x) \in \mathbb{Z}[x]$ (see Exercise 14). For $p(x) \in \mathbb{Z}[x]$, let $R_{p(x)} = \{a \in \mathbb{R} : p(a) = 0\}$. Since a polynomial in $\mathbb{Z}[x]$ has only finitely many roots, each set $R_{p(x)}$ is finite. Let $\mathbb{A}$ be the set of all *algebraic numbers*, that is, $\mathbb{A} = \bigcup_{p(x)\in\mathbb{Z}[x]} R_{p(x)}$. Prove that $\mathbb{A}$ is countable.

Exercise Notes: For Exercise **5**, first assume each $x \in \mathbb{Q}^+$ is in reduced form. For Exercise **6**, first review the three Theorems 3.3.18–3.3.20. For Exercise **8**, let $I_a = \{n \in \omega : f(n) = a\}$ for each $a \in A$. Because f is onto A, each I_a is nonempty and has a least element. For Exercise **10**, use Exercises **8–9**. For Exercise **12**, modify the proof of Theorem 5.2.12. For Exercise **13**, show that for every $i \in Y$ and $j \in Y$, if $i \in j$, then $f(i) \in f(j)$. Now use induction. To prove that $\mathrm{ran}(h) = Y$, one can apply Theorem 4.4.13. For Exercise **14**, apply Theorem 5.2.13 with Exercises **4** and **10**. For Exercise **15**, use Exercise **14** and Corollary 5.2.11.

5.3 Uncountable Sets

In Section 5.2, we established the existence of many countable sets. One might begin to believe that all sets are countable. Are there sets that are not countable? We will soon prove that such sets exist. First, we identify a slightly easier way to say that a set is "not countable."

Definition 5.3.1. A set X is **uncountable** if it is not countable, that is, if there is no one-to-one function $f: X \to \omega$.

Cantor was the first mathematician to prove that an uncountable set exists. In his proof, Cantor introduced a new and powerful proof technique that is often identified as a Cantor's *diagonal argument*. This argument has had a profound influence on mathematics ever since its introduction. In particular, diagonal arguments now frequently appear in mathematical logic and in computability theory. The proof of our next theorem illustrates Cantor's diagonal argument.

Theorem 5.3.2. *Let* $\mathcal{F} = {}^{\omega}\{0, 1\}$ *be the set of all the functions* $f: \omega \to \{0, 1\}$. *Then* $\mathcal{F}$ *is uncountable.*

Proof. Let $\mathcal{F}$ be as stated in the theorem. We will prove that $\mathcal{F}$ is uncountable. Suppose, for a contradiction, that the set $\mathcal{F}$ is countable. Because $\mathcal{F}$ is infinite (see Exercise 8), Theorem 5.2.8 implies that there is an enumeration

$$f_0, f_1, f_2, \ldots, f_n, \ldots \tag{5.6}$$

of all the functions in $\mathcal{F}$; that is, every function in $\mathcal{F}$ appears in the list (5.6). Define the function $g: \omega \to \{0, 1\}$ by

$$g(i) = \begin{cases} 1, & \text{if } f_i(i) = 0; \\ 0, & \text{if } f_i(i) = 1, \end{cases} \tag{5.7}$$

for each $i \in \omega$. As $g: \omega \to \{0, 1\}$, we see that $g \in \mathcal{F}$. Since every function in $\mathcal{F}$ is in the list (5.6), we conclude that the function g appears in this list. Hence, there is an $n \in \omega$ such that $g = f_n$. Therefore, $g(i) = f_n(i)$ for all $i \in \omega$. So, in particular, (▲) $g(n) = f_n(n)$. Since $f_n \in \mathcal{F}$, either $f_n(n) = 0$ or $f_n(n) = 1$. If $f_n(n) = 0$, then $g(n) = 1$ by (5.7). Thus, (▲) implies that $1 = 0$, a contradiction. If $f_n(n) = 1$, then $g(n) = 0$ by (5.7). Again, (▲) implies that $0 = 1$, which is a contradiction. Therefore, $\mathcal{F}$ is uncountable. ☐

Where is the diagonal?
Are you wondering why the technique used in the proof of Theorem 5.3.2 is referred to as a diagonal argument? To answer this inquiry, we shall now revisit this proof. Given a function $f: \omega \to \{0, 1\}$, there is a way of writing the values of f as an infinite sequence of terms from the set $\{0, 1\}$. For example, suppose $f(i) = 0$ if i is even and $f(i) = 1$ if i is odd. So

$$f(0) = 0, \ f(1) = 1, \ f(2) = 0, \ f(3) = 1, \ f(4) = 0, \ \ldots,$$

and we can represent f as follows: $f = \langle 0, 1, 0, 1, 0, 1, 0, 1, \ldots \rangle$. Furthermore, if $h: \omega \to \{0, 1\}$ is represented by

$$h = \langle 1, 1, 0, 1, 0, 1, 1, \ldots \rangle,$$

then $h(0) = 1, h(1) = 1, h(2) = 0, h(3) = 1, h(4) = 0, h(5) = 1, \ldots$. Consider the list (5.6) of all the functions in $\mathcal{F}$. Let us represent each f_i in this list as an infinite sequence; that is, let

$$f_i = \langle f_i(0), \ f_i(1), \ f_i(2), \ f_i(3), \ f_i(4), \ f_i(5), \ \ldots \rangle.$$

Using this notation, we can rewrite the list of functions (5.6) in the following vertical form:

$$f_0 = \langle f_0(0),\ f_0(1),\ f_0(2),\ f_0(3),\ f_0(4),\ f_0(5),\ \ldots \rangle$$
$$f_1 = \langle f_1(0),\ f_1(1),\ f_1(2),\ f_1(3),\ f_1(4),\ f_1(5),\ \ldots \rangle$$
$$f_2 = \langle f_2(0),\ f_2(1),\ f_2(2),\ f_2(3),\ f_2(4),\ f_2(5),\ \ldots \rangle \qquad (5.8)$$
$$f_3 = \langle f_3(0),\ f_3(1),\ f_3(2),\ f_3(3),\ f_3(4),\ f_3(5),\ \ldots \rangle$$
$$f_4 = \langle f_4(0),\ f_4(1),\ f_4(2),\ f_4(3),\ f_4(4),\ f_4(5),\ \ldots \rangle$$
$$f_5 = \langle f_5(0),\ f_5(1),\ f_5(2),\ f_5(3),\ f_5(4),\ f_5(5),\ \ldots \rangle$$

$$\vdots$$

In the proof of Theorem 5.3.2, a function $g\colon \omega \to \{0, 1\}$ was defined so that g is not equal to any function in the list (5.8). This is done by going down this list and assigning a value to $g(i)$ that is different from the diagonal value $f_i(i)$ for each f_i appearing in (5.8). To illustrate this idea, let us give some specific values to the entries that can appear in the diagonal of (5.8). Suppose $f_0(0) = 0$, $f_1(1) = 1$, $f_2(2) = 0$, $f_3(3) = 0$, $f_4(4) = 1$, and $f_5(5) = 1$. Thus, (5.8) becomes

$$f_0 = \langle\ \mathbf{0}\ ,\ f_0(1),\ f_0(2),\ f_0(3),\ f_0(4),\ f_0(5),\ \ldots \rangle$$
$$f_1 = \langle f_1(0),\ \ \mathbf{1}\ ,\ f_1(2),\ f_1(3),\ f_1(4),\ f_1(5),\ \ldots \rangle$$
$$f_2 = \langle f_2(0),\ f_2(1),\ \ \mathbf{0}\ ,\ f_2(3),\ f_2(4),\ f_2(5),\ \ldots \rangle \qquad (5.9)$$
$$f_3 = \langle f_3(0),\ f_3(1),\ f_3(2),\ \ \mathbf{0}\ ,\ f_3(4),\ f_3(5),\ \ldots \rangle$$
$$f_4 = \langle f_4(0),\ f_4(1),\ f_4(2),\ f_4(3),\ \ \mathbf{1}\ ,\ f_4(5),\ \ldots \rangle$$
$$f_5 = \langle f_5(0),\ f_5(1),\ f_5(2),\ f_5(3),\ f_5(4),\ \ \mathbf{1}\ ,\ \ldots \rangle$$

$$\vdots$$

$$g = \langle\ \mathbf{1}\ ,\ \ \mathbf{0}\ ,\ \ \mathbf{1}\ ,\ \ \mathbf{1}\ ,\ \ \mathbf{0}\ ,\ \ \mathbf{0}\ ,\ \ldots \rangle$$

We have put the function g below the infinite list (5.9) where the values of g are determined by applying definition (5.7), in the proof of Theorem 5.3.2. For example, to evaluate $g(0)$ we see that $f_0(0) = 0$, and so $g(0) = 1$ by (5.7). Thus, $g(0) \neq f_0(0)$, and we are thereby assured that $g \neq f_0$. Now we evaluate $g(1)$. Since $f_1(1) = 1$, we obtain the value $g(1) = 0$. Hence, $g(1) \neq f_1(1)$ and $g \neq f_1$. Again, because $f_2(2) = 0$, we conclude that $g(2) = 1$ and $g \neq f_2$. Continuing in this manner we construct a function $g\colon \omega \to \{0, 1\}$ that is different from every function in the list (5.9). This is the clever diagonal argument that Cantor introduced to mathematics.

Theorem 5.3.3. *Let A and B be sets. If A is uncountable and $g: A \rightarrow B$ is a one-to-one function, then B is uncountable.*

Proof. Let A be an uncountable set, and let $g: A \rightarrow B$ be a one-to-one function. Suppose, for a contradiction, that B is countable. Since $g: A \rightarrow B$ is one-to-one, Theorem 5.2.3 thus implies that A is a countable set, contradicting the fact that A is uncountable. Therefore, B is uncountable. ☐

Before proving our next lemma, we make a simple observation. Whenever $f: \omega \rightarrow \{0, 1\}$, we have that $f(n) \in \{0, 1\}$ for all $n \in \omega$. We can thus use f to define a real number by means of an infinite decimal expansion. Let $f_n = f(n)$ for each $n \in \omega$. Then we have the real number given by $0.f_0 f_1 f_2 f_3 \cdots f_n \cdots$. For example, suppose that $f(0) = 1$, $f(1) = 0$, $f(2) = 1$, $f(3) = 1, \ldots$, then

$$0.f_0 f_1 f_2 f_3 \cdots f_n \cdots = 0.1011 \cdots f_n \cdots.$$

Decimal expansions that possess only the digits 0 and/or 1 are, in fact, unique; that is, if a real number has a decimal expansion consisting of 0's and/or 1's, then it has only one such expansion.

Lemma 5.3.4. *There is a one-to-one function $G: {}^{\omega}\{0, 1\} \rightarrow \mathbb{R}$.*

Proof. Let $\mathcal{F} = {}^{\omega}\{0, 1\}$. For each $f \in \mathcal{F}$ let us define (▲) $f_n = f(n)$, for all $n \in \omega$. So $f_n \in \{0, 1\}$ for every $n \in \omega$. Define the function $G: \mathcal{F} \rightarrow \mathbb{R}$ by

$$G(f) = 0.f_0 f_1 f_2 \cdots \tag{5.10}$$

for each $f \in \mathcal{F}$. Let f and h be functions in $\mathcal{F}$. Assume $G(f) = G(h)$. We shall prove that $f = h$. Because $G(f) = G(h)$, we conclude from (5.10) that

$$0.f_0 f_1 f_2 \cdots = 0.h_0 h_1 h_2 \cdots.$$

Since such decimal expansions are unique, we have that $f_n = h_n$ for all $n \in \omega$. Hence, $f(n) = h(n)$ for all $n \in \omega$, by (▲). Thus, $f = h$ and G is one-to-one. ☐

We can now present and prove Cantor's classic theorem.

Theorem 5.3.5 (Cantor). *The set of real numbers $\mathbb{R}$ is uncountable.*

Proof. By Lemma 5.3.4, let $G\colon {}^{\omega}\{0, 1\} \to \mathbb{R}$ be one-to-one. By Theorem 5.3.2, ${}^{\omega}\{0, 1\}$ is uncountable. So Theorem 5.3.3 implies that $\mathbb{R}$ is uncountable. $\qquad\square$

Exercises 5.3

1. Let A be uncountable. Prove that $A \times B$ is uncountable for any nonempty set B.

2. Let A be uncountable and B be countable. Prove that $A \setminus B$ is uncountable.

3. Prove that the set of irrational numbers $\mathbb{R} \setminus \mathbb{Q}$ is uncountable.

4. Suppose that A is uncountable and $A \subseteq B$. Prove that B is uncountable.

5. Prove that ${}^{\omega}\omega$ is uncountable.

6. Suppose that A is uncountable. Prove that $\mathcal{P}(A)$ is uncountable.

7. Let $A \neq \varnothing$. Prove that if B is uncountable, then ${}^{A}B$ is uncountable.

*8. Let $\mathcal{F}$ be as in Theorem 5.3.2, and let $(\star)$ $f_1, f_2, \ldots, f_n$ be a finite list of functions in $\mathcal{F}$. Using the argument in the proof of Theorem 5.3.2, define a new function $g \in \mathcal{F}$ that is not in the list $(\star)$. Conclude that $\mathcal{F}$ is infinite.

9. Suppose that someone claims that the set of real numbers in the interval $(0, 1)$ is countable and that all of these real numbers can be enumerated as in (5.11), where each such real number is represented by an infinite decimal expansion:

$$x_1 = .12345689234\cdots$$

$$x_2 = .68729958219\cdots$$

$$x_3 = .05050506620\cdots$$

$$x_4 = .57591884622\cdots \qquad\qquad (5.11)$$

$$\vdots$$

$$x_i = .x_{i1}x_{i2}x_{i3}x_{i4}x_{i5}\cdots$$

$$\vdots$$

You must show that this claim is false. Using Cantor's diagonal argument, define a decimal expansion for a real number b in $(0, 1)$ that does not appear in the list (5.11). In order to ensure uniqueness, your expansion $b = .b_1 b_2 b_3 \ldots b_i \ldots$ should not contain the digits 0 or 9. Identify the first four digits in the decimal expansion of b. Prove that $b \neq x_n$ for all $n \in \omega$.

10. Let $S = \{q \in \mathbb{Q} : 0 < q < 1\}$. Theorem 5.2.4 and Exercise 5 on page 123 imply that S is countable. We can thus enumerate all of the elements in S in a list $(\star)$ $q_1, q_2, q_3, \ldots$, by Theorem 5.2.8. Since each of these rational numbers has an infinite decimal expansion $q_i = .q_{i1} q_{i2} q_{i3} q_{i4} q_{i5} \cdots$, one can define a real number $b \in (0, 1)$ that is not in the list $(\star)$ just as in Exercise 9. Is b a rational number? Justify your answer.

11. A real number that is not algebraic is said to be a *transcendental number* (see page 124, Exercise 15). Let $\mathbb{T}$ be the set of all transcendental numbers. Show that $\mathbb{T}$ is uncountable.

Exercise Notes: For Exercise 3, apply Exercise 5 on page 123 and Exercise 2. For Exercise 9, if the decimal expansion of b does not contain a 0 or 9, then b will have a unique decimal representation. Thus, if the decimal expansions of b and x in $(0, 1)$ have different digits in at least one decimal place, then $b \neq x$. For Exercise 11, use Exercise 15 on page 124 and Exercise 2.

5.4 Cardinality

In set theory, the cardinality of a set is a measure of how many elements are in the set. The set $A = \{1, 2, 3, \ldots, 11\}$ has 11 elements, and so the cardinality of A is 11. Thus, $|A| = 11$ (see Definition 5.1.2). The cardinality of an infinite set X will also be denoted by $|X|$. In this section, we shall present Cantor's method for measuring the size of an infinite set without the use of numbers. There are infinite sets, as we will see, where one of these sets has cardinality much larger than that of some other infinite set. Therefore, it is possible for one infinite set to have "many more" elements than another infinite set.

What does it mean to say that two sets have the same cardinality, that is, the same size? Cantor discovered a mathematically precise and simple answer to this question.

Definition 5.4.1. For sets A and B, we say that A has the **same cardinality** as B, denoted by $|A| =_c |B|$, if there is a bijection $f : A \to B$.

Remark 5.4.2. The expression $|A| =_c |B|$ looks like an equation; however, the assertion $|A| =_c |B|$ should be viewed only as an abbreviation for the statement "A has the same cardinality as B." In other words, $|A| =_c |B|$ means that "there is a function $f: A \to B$ that is one-to-one and onto B." When A and B are finite sets, Theorem 5.1.7 implies that $|A| = |B|$ if and only if $|A| =_c |B|$.

The relationship $=_c$, given in Definition 5.4.1, is reflexive, symmetric, and transitive (see Exercises 4–5).

Theorem 5.4.3. *Let A be a set. Then $\left|{}^A\{0, 1\}\right| =_c |\mathcal{P}(A)|$.*

Proof. For each $f \in {}^A\{0, 1\}$, define $X_f \subseteq A$ by $X_f = \{x \in A : f(x) = 1\}$. Define the function $G: {}^A\{0, 1\} \to \mathcal{P}(A)$ by $G(f) = X_f$ for each $f \in {}^A\{0, 1\}$. Then G is one-to-one and onto $\mathcal{P}(A)$ (see Exercise 10).   □

Corollary 5.4.4. $|{}^\omega\{0, 1\}| =_c |\mathcal{P}(\omega)|$.

Theorem 5.4.5. *If $|A| =_c |B|$, then $|\mathcal{P}(A)| =_c |\mathcal{P}(B)|$.*

Proof. Suppose that $|A| =_c |B|$. Thus, there is a bijection $f: A \to B$. Consider the function $G: \mathcal{P}(A) \to \mathcal{P}(B)$ defined by $G(X) = f[X]$ for every $X \in \mathcal{P}(A)$. Exercise 15 on page 68 shows that G is one-to-one. To show that G is onto $\mathcal{P}(B)$, assume that $Y \in \mathcal{P}(A)$. Let $X = f^{-1}[Y]$. Since f is onto Y, Exercise 13 on page 68 shows that $G(X) = Y$. Therefore, G is a bijection.   □

Theorem 5.4.5 shows that the power set operation "preserves" cardinality. Our next theorem identifies three other such operations.

Theorem 5.4.6. *Suppose that $|A| =_c |K|$ and $|B| =_c |L|$.*

(1) *If $A \cap B = \varnothing$ and $K \cap L = \varnothing$, then $|A \cup B| =_c |K \cup L|$.*
(2) $|A \times B| =_c |K \times L|$.
(3) $\left|{}^A B\right| =_c \left|{}^K L\right|$.

Proof. Assume that $|A| =_c |K|$ and $|B| =_c |L|$. Let $f: A \to K$ and $g: B \to L$ be bijections.

(1) Assume $A \cap B = \varnothing$ and $K \cap L = \varnothing$. Let $H: A \cup B \to K \cup L$ be defined by

$$H(x) = \begin{cases} f(x), & \text{if } x \in A; \\ g(x), & \text{if } x \in B. \end{cases}$$

Since $A \cap B = \varnothing$, it follows that H is a function; that is, H is single-valued. As f and g are bijections, H is a bijection because $K \cap L = \varnothing$.

(2) Define $h \colon A \times B \to K \times L$ by $h(\langle a, b \rangle) = \langle f(a), g(b) \rangle$. Then h is a bijection from $A \times B$ to $K \times L$ (see Exercise 19).

(3) We have that $f \colon A \to K$ and $g \colon B \to L$ are bijections. Let $\ell \in {}^{A}B$. Thus, $\ell \colon A \to B$. Observe that $(g \circ \ell \circ f^{-1}) \colon K \to L$. Define $G \colon {}^{A}B \to {}^{K}L$ by

$$G(\ell) = g \circ \ell \circ f^{-1}$$

for each $\ell \in {}^{A}B$. To prove that G is one-to-one, let $\ell \in {}^{A}B$ and $\ell^{*} \in {}^{A}B$. Assume that $G(\ell) = G(\ell^{*})$. Thus, $(g \circ \ell \circ f^{-1})(x) = (g \circ \ell^{*} \circ f^{-1})(x)$ for all $x \in K$. Hence,

$$g((\ell \circ f^{-1})(x)) = g((\ell^{*} \circ f^{-1})(x)) \text{ for all } x \in K$$

(see Remark 3.3.11 on page 61). Since g is one-to-one, we conclude that

$$(\ell \circ f^{-1})(x) = (\ell^{*} \circ f^{-1})(x) \text{ for all } x \in K. \tag{5.12}$$

We now show that $\ell(a) = \ell^{*}(a)$ for all $a \in A$. Let $a \in A$. Thus, $f(a) \in K$. Letting $x = f(a)$ in (5.12), we obtain $\ell(a) = \ell^{*}(a)$ since $f^{-1}(f(a)) = a$ by Theorem 3.3.18. So $\ell = \ell^{*}$, and hence, G is one-to-one. To prove that G is onto ${}^{K}L$, let $h \in {}^{K}L$. Then $g^{-1} \circ h \circ f \in {}^{A}B$ and $G(g^{-1} \circ h \circ f) = h$ (see Exercise 20). Therefore, the function $G \colon {}^{A}B \to {}^{K}L$ is a bijection. □

In a similar manner, one can prove the next theorem (see Exercises 23–25).

Theorem 5.4.7. *For all sets A, B, K, and L, we have the following:*

(1) *If $A \cap B = \varnothing$, then $\left| {}^{A \cup B}K \right| =_{c} \left| {}^{A}K \times {}^{B}K \right|$.*
(2) $\left| {}^{A}(K \times L) \right| =_{c} \left| {}^{A}K \times {}^{A}L \right|$.
(3) $\left| {}^{A}({}^{B}K) \right| =_{c} \left| {}^{A \times B}K \right|$.

The proofs of our next two corollaries apply Theorems 5.4.6 and 5.4.7.

Corollary 5.4.8. $\left| {}^{\omega}\omega \times \omega \right| =_{c} \left| {}^{\omega}\omega \right|.$

Proof. Let $x \notin \omega$. It is easy to show that ($\blacktriangledown$) $\left| {}^{\{x\}}\omega \right| =_{c} |\omega|$. Thus,

$$\left| {}^{\omega}\omega \times \omega \right| =_{c} \left| {}^{\omega}\omega \times {}^{\{x\}}\omega \right| \quad \text{by ($\blacktriangledown$) and Theorem 5.4.6(2)}$$

$$=_{c} \left| {}^{\omega \cup \{x\}}\omega \right| \quad \text{by Theorem 5.4.7(1) as } \omega \cap \{x\} = \varnothing$$

$$=_{c} \left| {}^{\omega}\omega \right| \quad \text{by Theorem 5.4.6(3) as } |\omega \cup \{x\}| =_{c} |\omega|.$$

Therefore, $\left| {}^{\omega}\omega \times \omega \right| =_{c} \left| {}^{\omega}\omega \right|$ (see Exercise 5). □

Corollary 5.4.9. $|\mathcal{P}(\omega) \times \mathcal{P}(\omega)| =_c |\mathcal{P}(\omega)|.$

Proof. Theorems 5.2.5 and 5.2.6 imply that (▲) $|\omega \cup (\omega \times \{0\})| =_c |\omega|$. Note that $\omega \cap (\omega \times \{0\}) = \varnothing$. We prove that $|\mathcal{P}(\omega) \times \mathcal{P}(\omega)| =_c |\mathcal{P}(\omega)|$ as follows:

$$
\begin{aligned}
|\mathcal{P}(\omega) \times \mathcal{P}(\omega)| &=_c |{}^{\omega}\{0, 1\} \times {}^{\omega}\{0, 1\}| && \text{by Cor. 5.4.4 and Theorem 5.4.6(2)} \\
&=_c |{}^{\omega}\{0, 1\} \times {}^{\omega \times \{0\}}\{0, 1\}| && \text{by Thm. 5.4.6(3) as } |\omega| =_c |\omega \times \{0\}| \\
&=_c |{}^{\omega \cup (\omega \times \{0\})}\{0, 1\}| && \text{by Theorem 5.4.7(1)} \\
&=_c |{}^{\omega}\{0, 1\}| && \text{by Theorem 5.4.6(3) and (▲)} \\
&=_c \mathcal{P}(\omega) && \text{by Corollary 5.4.4.}
\end{aligned}
$$

Therefore, $|\mathcal{P}(\omega) \times \mathcal{P}(\omega)| =_c |\mathcal{P}(\omega)|.$ □

Using our cardinality notation, we will now restate some of the results that were previously established about two countably infinite sets.

Theorem 5.4.10 (Cantor). $|\omega| =_c |\mathbb{Z}|$ *and* $|\omega| =_c |\mathbb{Q}|.$

Proof. Exercises 4 and 5 on page 123 imply that $\mathbb{Z}$ and $\mathbb{Q}$ are countably infinite. By Theorem 5.2.6, there are bijections $f: \omega \to \mathbb{Z}$ and $g: \omega \to \mathbb{Q}$. Therefore, $|\omega| =_c |\mathbb{Z}|$ and $|\omega| =_c |\mathbb{Q}|.$ □

What does it mean to say that one set has smaller cardinality than another set? Cantor found a simple answer to this question as well.

Definition 5.4.11. A set A has **cardinality strictly less** than that of a set B, denoted by $|A| <_c |B|$, if there is a one-to-one function $f: A \to B$ and there is no function $g: A \to B$ that is both one-to-one and onto B.

Theorem 5.4.12 (Cantor). $|\omega| <_c |\mathbb{R}|.$

Proof. Let $f: \omega \to \mathbb{R}$ be defined by $f(n) = n$. This function is one-to-one. We now show that there is no function $g: \omega \to \mathbb{R}$ that is one-to-one and onto $\mathbb{R}$. Suppose, for a contradiction, that there is such a bijection g. Hence, $g^{-1}: \mathbb{R} \to \omega$ is one-to-one by Theorem 3.3.18, and therefore, $\mathbb{R}$ is countable. This contradicts Theorem 5.3.5. Thus, $|\omega| <_c |\mathbb{R}|.$ □

After proving Theorem 5.4.12, Cantor asked the following question: Is there a set of real numbers whose cardinality is *strictly* between the cardinality of ω and the cardinality of $\mathbb{R}$? In 1878, Georg Cantor announced a conjecture that is now called the *continuum hypothesis*.

Continuum Hypothesis. There is no set $A \subseteq \mathbb{R}$ such that $|\omega| <_c |A| <_c |\mathbb{R}|$.

The continuum hypothesis proclaims that every set of real numbers is either countable or has the same cardinality as the set $\mathbb{R}$. Cantor struggled to resolve this conjecture, without success, for much of his career. The problem persisted and became one of the most important unsolved mathematical problems of the twentieth century. It was only after Cantor's death that it was shown to be an unsolvable problem. The contributions of Kurt Gödel in 1940 and Paul Cohen in 1963 showed that the continuum hypothesis cannot be proven or refuted using the axioms in ZFC.

Recall that for every set A, the power set $\mathcal{P}(A) = \{X : X \subseteq A\}$ is the set of all subsets of A. For example, if $A = \{a, b\}$, then $\mathcal{P}(A) = \{\varnothing, \{a\}, \{b\}, \{a, b\}\}$ and observe that $|A| < |\mathcal{P}(A)|$ (see Definition 5.1.2). In fact, one can prove that if A is a finite set with n many elements, then $\mathcal{P}(A)$ has 2^n many elements (see Exercise 23, page 116). Thus, $|A| < |\mathcal{P}(A)|$ whenever A is a finite set; however, what happens when the set A is infinite? Georg Cantor answered this intriguing question as well, using his diagonal argument.

Theorem 5.4.13 (Cantor). *Let A be any set. Then $|A| <_c |\mathcal{P}(A)|$.*

Proof. Let A be a set. Let $f : A \to \mathcal{P}(A)$ be defined by $f(a) = \{a\}$ for all $a \in A$. It is easy to show that f is one-to-one. We must show that there is no one-to-one function $g : A \to \mathcal{P}(A)$ that is also onto $\mathcal{P}(A)$. Suppose, for a contradiction, that there exists a bijection $g : A \to \mathcal{P}(A)$. Observe that $g(x) \subseteq A$ for all $x \in A$. Let $X = \{x \in A : x \notin g(x)\}$. Clearly, $X \subseteq A$ and thus, $X \in \mathcal{P}(A)$. As g is onto $\mathcal{P}(A)$, there is an $a \in A$ such that $g(a) = X$. There are two cases to consider, namely, either $a \in X$ or $a \notin X$. If $a \in X$, then the definition of X implies that $a \notin g(a)$. Since $g(a) = X$, we have that $a \notin X$, which is a contradiction. If $a \notin X$, then the definition of X implies that $a \in g(a)$. Because $g(a) = X$, we conclude that $a \in X$, which is again a contradiction. Thus, there is no bijection $g : A \to \mathcal{P}(A)$. Therefore, $|A| <_c |\mathcal{P}(A)|$. $\qquad\square$

Definition 5.4.14. Let A and B be sets. Then A has **cardinality less than or equal** to B, denoted by $|A| \leq_c |B|$, if there is an injection $f : A \to B$.

Theorem 5.4.15. *If $|A| \leq_c |B|$, then $|\mathcal{P}(A)| \leq_c |\mathcal{P}(B)|$.*

Proof. Assume $|A| \leq_c |B|$. So there is a one-to-one function $f: A \to B$. Define $G: \mathcal{P}(A) \to \mathcal{P}(B)$ by $G(X) = f[X]$ for all $X \in \mathcal{P}(A)$. Exercise 15 on page 68 shows that G is one-to-one. Therefore, $|\mathcal{P}(A)| \leq_c |\mathcal{P}(B)|$. $\qquad\square$

Theorem 5.4.15 shows that the power set operation "preserves inequality," and its proof involves only a slight modification of the proof of Theorem 5.4.5. The following theorem identifies other set operations that also "preserve" the notion of inequality given by Definition 5.4.14. The proof can be obtained by appropriately modifying the proof of Theorem 5.4.6 (see Exercises 28–30).

Theorem 5.4.16. *Suppose that $|A| \leq_c |K|$ and $|B| \leq_c |L|$.*

(1) *If $A \cap B = \emptyset$ and $K \cap L = \emptyset$, then $|A \cup B| \leq_c |K \cup L|$.*
(2) $|A \times B| \leq_c |K \times L|$.
(3) $\left|{}^A B\right| \leq_c \left|{}^K L\right|$ *if A is nonempty.*

We will soon present the Schröder–Bernstein Theorem, which is very useful for proving many results about cardinality. The theorem states that if there are functions $f: A \to B$ and $g: B \to A$ that are both one-to-one, then there exists a function $h: A \to B$ that is one-to-one *and* onto B. The theorem sounds very reasonable; however, a proof that does not depend on the axiom of choice was seen as quite challenging and even eluded the brilliant Georg Cantor.

The following lemma, illustrated in Figure 5.2, will be used in our proof of the Schröder–Bernstein Theorem.

Lemma 5.4.17. *Let $f: A \to B$ and $g: B \to A$ be functions. Then there exists an $X \subseteq A$ that satisfies the identity*

$$g[B \setminus f[X]] = A \setminus X. \tag{5.13}$$

Proof. Let $f: A \to B$ and $g: B \to A$. Define the function $H: \mathcal{P}(A) \to \mathcal{P}(A)$ by

$$H(C) = A \setminus g[B \setminus f[C]].$$

Therefore, by the definition of H, for each $C \subseteq A$ we have that $H(C) \subseteq A$ and ($\blacktriangle$) $A \setminus H(C) = g[B \setminus f[C]]$. So, if a set $X \subseteq A$ satisfies $H(X) = X$, then ($\blacktriangle$) would imply that the set X satisfies $A \setminus X = g[B \setminus f[X]]$, and thus, (5.13) holds.

Claim 1. Let $C \subseteq A$ and $D \subseteq A$. If $C \subseteq D$, then $H(C) \subseteq H(D)$.

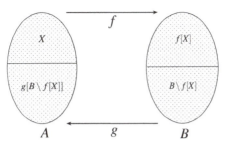

Figure 5.2. Illustration of Lemma 5.4.17 where X satisfies
$$g[B \setminus f[X]] = A \setminus X$$

Proof. See Exercise 32.   (Claim 1) $\square$

Consider the set $\mathcal{S} = \{D : D \subseteq A \text{ and } H(D) \subseteq D\}$. Since $H(A) \subseteq A$, we see that $A \in \mathcal{S}$, and so $\mathcal{S}$ is nonempty. Let $X = \bigcap \mathcal{S}$. Clearly, $X \subseteq A$ and $H(X) \subseteq A$.

Claim 2. $H(X) = X$.

Proof. Let $D \in \mathcal{S}$. Because $X = \bigcap \mathcal{S}$, we see that $X \subseteq D$. So, $H(X) \subseteq H(D)$ by Claim 1. Since $H(D) \subseteq D$, we infer that $H(X) \subseteq D$. Thus, $H(X) \subseteq D$ for every $D \in \mathcal{S}$. Hence, $H(X) \subseteq \bigcap \mathcal{S}$; that is, $H(X) \subseteq X$. We now prove that $X \subseteq H(X)$. As $H(X) \subseteq X$, Claim 1 implies that $H(H(X)) \subseteq H(X)$. So $H(X) \in \mathcal{S}$. Thus, $\bigcap \mathcal{S} \subseteq H(X)$; that is, $X \subseteq H(X)$. Therefore, $H(X) = X$.

(Claim 2) $\square$

Since $H(X) = X$, we conclude that X satisfies (5.13).   (Lemma) $\square$

Figure 5.2 depicts the set X in Lemma 5.4.17 and the identity (5.13). This figure may also help the reader to better understand the proof of the following theorem, which is named for Ernst Schröder and Felix Bernstein. The proof depends only on (at most) the first seven Zermelo–Fraenkel axioms.

Theorem 5.4.18 (Schröder–Bernstein). *Let A and B be any two sets. Suppose that $|A| \leq_c |B|$ and $|B| \leq_c |A|$. Then $|A| =_c |B|$.*

Proof. Assume that $|A| \leq_c |B|$ and $|B| \leq_c |A|$. So, there are one-to-one functions $f : A \to B$ and $g : B \to A$. By Lemma 5.4.17, there is an $X \subseteq A$ (see Figure 5.2) satisfying

$$g[B \setminus f[X]] = A \setminus X. \tag{5.14}$$

Since g is one-to-one, Theorems 3.2.7 and 3.3.15 imply that g^{-1} is a one-to-one function such that dom(g^{-1}) = ran(g) $\subseteq A$. As dom(g^{-1}) = ran(g), Equation (5.14) implies that $A \setminus X \subseteq$ dom(g^{-1}). Define the function $h: A \to B$ by

$$h(x) = \begin{cases} f(x), & \text{if } x \in X; \\ g^{-1}(x), & \text{if } x \in A \setminus X. \end{cases} \tag{5.15}$$

We will soon prove that h is one-to-one and onto B.

Claim 1. $g[f[X]] \subseteq X$.

Proof. Clearly, $f[X] \cap (B \setminus f[X]) = \varnothing$. Since g is one-to-one, Exercise 5 on page 68 implies that $g[f[X]] \cap g[B \setminus f[X]] = \varnothing$. Thus, by (5.14), we have that $g[f[X]] \subseteq X$.   (Claim 1) $\square$

Claim 2. The function $h: A \to B$ is one-to-one.

Proof. Let $x \in A$ and $y \in A$. Assume that $h(x) = h(y)$. The definition (5.15) of h yields three key cases: (1) x and y are both in X, (2) x and y are both in $A \setminus X$, or (3) $x \in X$ and $y \in A \setminus X$. The argument in case (3) will cover the fourth case: $y \in A$ and $x \in A \setminus X$.

(1) If $x \in X$ and $y \in X$, then $f(x) = f(y)$ by (5.15). Therefore, $x = y$ because f is one-to-one.

(2) If $x \in A \setminus X$ and $y \in A \setminus X$, then $g^{-1}(x) = g^{-1}(y)$ by (5.15). Since the function g^{-1} is one-to-one, we have $x = y$.

(3) If $x \in X$ and $y \in A \setminus X$, then we obtain $f(x) = g^{-1}(y)$. Thus, $g(f(x)) = y$. Claim 1 implies that $y \in X$. Therefore, case (3) is impossible.   (Claim 2) $\square$

Claim 3. The function $h: A \to B$ is onto B.

Proof. Let $b \in B$. Hence, either (1) $b \in f[X]$ or (2) $b \in B \setminus f[X]$. We consider these two cases as follows:

(1) If $b \in f[X]$, then $b = f(x)$ for some $x \in X$. From the definition (5.15) of h, it follows that $h(x) = b$.

(2) If $b \in B \setminus f[X]$, then (5.14) implies that $g(b) \in A \setminus X$. Thus, (▲) $a = g(b)$ for some $a \in A \setminus X$. By applying the function g^{-1} to both sides of (▲), we obtain $g^{-1}(a) = b$. Therefore, $h(a) = b$ as $a \in A \setminus X$.   (Claim 3) $\square$

Since $h: A \to B$ is a bijection, we see that $|A| =_c |B|$.   (Theorem) $\square$

Given sets A and B, sometimes one can easily define one-to-one functions $f: A \to B$ and $g: B \to A$; yet it may be very difficult to specify a one-to-one function from A onto B. In this situation, one can apply the Schröder–Bernstein Theorem to conclude that such a function does exist and thereby conclude that A and B have the same cardinality.

Lemma 5.4.19. *Let $\mathbb{Q}$ be the set of rational numbers and let $\mathbb{R}$ be the set of real numbers. Then $|\mathbb{R}| \leq_c |\mathcal{P}(\mathbb{Q})|$.*

Proof. For $x \in \mathbb{R}$, let $A_x = \{q \in \mathbb{Q} : q < x\}$. Let $G: \mathbb{R} \to \mathcal{P}(\mathbb{Q})$ be defined by $G(x) = A_x$ for each $x \in \mathbb{R}$. To prove that G is one-to-one, let $x \in \mathbb{R}$ and $y \in \mathbb{R}$. Assume that $G(x) = G(y)$. We shall prove that $x = y$. Assume, for a contradiction, that $x \neq y$. Without loss of generality, we can assume that $x < y$. Since the rational numbers are dense in $\mathbb{R}$, there is a $q \in \mathbb{Q}$ such that $x < q < y$. Thus, $q \notin A_x$ and $q \in A_y$. Therefore, $G(x) \neq G(y)$, which is a contradiction. So $x = y$ and G is one-to-one. $\qquad\square$

Theorem 5.4.20. $|\mathbb{R}| =_c |\mathcal{P}(\omega)|$.

Proof. By Theorem 5.4.10, we have that $|\omega| =_c |\mathbb{Q}|$. Hence, by Theorem 5.4.5, we see that $|\mathcal{P}(\omega)| =_c |\mathcal{P}(\mathbb{Q})|$. Lemma 5.4.19 allows us to now conclude that $|\mathbb{R}| \leq_c |\mathcal{P}(\omega)|$. Moreover, Corollary 5.4.4 states that $|{}^{\omega}\{0, 1\}| =_c |\mathcal{P}(\omega)|$, and Lemma 5.3.4 implies that $|{}^{\omega}\{0, 1\}| \leq_c |\mathbb{R}|$. Thus, $|\mathcal{P}(\omega)| \leq_c |\mathbb{R}|$. Therefore, by the Schröder–Bernstein Theorem 5.4.18, we have that $|\mathbb{R}| =_c |\mathcal{P}(\omega)|$. $\qquad\square$

Our next corollary shows that the real line $\mathbb{R}$ has the same cardinality as the plane $\mathbb{R} \times \mathbb{R}$.

Corollary 5.4.21. $|\mathbb{R} \times \mathbb{R}| =_c |\mathbb{R}|$.

Proof. By Corollary 5.4.9, $|\mathcal{P}(\omega) \times \mathcal{P}(\omega)| =_c |\mathcal{P}(\omega)|$. Thus, $|\mathbb{R} \times \mathbb{R}| =_c |\mathbb{R}|$ by Theorem 5.4.20 and Theorem 5.4.6(2). $\qquad\square$

Theorem 5.4.20 allows us to reformulate the continuum hypothesis:

There is no set $A \subseteq \mathcal{P}(\omega)$ such that $|\omega| <_c |A| <_c |\mathcal{P}(\omega)|$. $\qquad$ **(CH)**

For every infinite set X, Theorem 5.4.13 asserts that $|X| <_c |\mathcal{P}(X)|$. So CH can be viewed is a special case of the Generalized Continuum Hypothesis (GCH):

If X is infinite, then there is no $A \subseteq \mathcal{P}(X)$ such that $|X| <_c |A| <_c |\mathcal{P}(X)|$.

The combined work of Kurt Gödel and Paul Cohen also shows that the GCH cannot be proven or refuted using the axioms in ZFC.

Exercises 5.4

1. Let $A = \{x \in \mathbb{R} : 0 < x < 1\}$ and let $B = \{x \in \mathbb{R} : 2 < x < 5\}$. Prove that $|A| =_c |B|$.

2. Prove that $\left| {}^{[0,1]}A \right| =_c |A \times A|$.

3. Let A and B be countably infinite sets. Prove that $|A| =_c |B|$. Conclude that $|\omega \times \omega| =_c |\omega|$.

*4. Let A and B be sets. Prove that $|A| =_c |A|$, and prove that if $|A| =_c |B|$, then $|B| =_c |A|$.

*5. Prove that if $|A| =_c |B|$ and $|B| =_c |C|$, then $|A| =_c |C|$.

6. Prove that if $|A| <_c |B|$ and $|B| =_c |C|$, then $|A| <_c |C|$.

7. Let B be countable. Prove that if $|A| \leq_c |B|$, then A is countable.

8. Let A be uncountable. Prove that if $|A| \leq_c |B|$, then B is uncountable.

9. Prove that if $|A| <_c |B|$ and $|B| <_c |C|$, then $|A| <_c |C|$.

*10. Prove that the function G in the proof of Theorem 5.4.3 is a bijection.

11. Prove that the set $\mathcal{P}(\omega)$ is uncountable.

12. Using Theorem 5.4.18, prove that $|A| =_c |\mathbb{R}|$, where $A = [-\pi, \pi)$.

13. Let $A = \{a, b, c, d, w, y, z\}$, and let $g: A \to \mathcal{P}(A)$ be the function given by

$$g(a) = \{b, c, d\},$$
$$g(b) = \{a, b, c, w, z\},$$
$$g(c) = \{b, c\},$$
$$g(d) = \{d\},$$
$$g(w) = A,$$
$$g(y) = \{a, b, c, d, w\},$$
$$g(z) = \varnothing.$$

The function g is one-to-one. The proof of Theorem 5.4.13 shows that g is not onto $\mathcal{P}(A)$ because the subset of A defined by $X = \{x \in A : x \notin g(x)\}$ is not in the range of g. Evaluate the set X.

14. Let $f : A \to B$ be onto B, and let $C = \{x \in A : x \notin f(x)\}$. Show that $C \notin B$.

15. Suppose $A \subseteq B$ and $|B| \leq_c |A|$. Prove that $|A| =_c |B|$.

16. Suppose $A \subseteq B \subseteq C$ and $|A| =_c |C|$. Prove that $|B| =_c |C|$.

*17. Prove that $|A \times B| =_c |B \times A|$.

18. Suppose that $|A| \leq_c |B|$, $|B| \leq_c |C|$, and $|C| \leq_c |A|$. Using Theorem 5.4.18, prove that $|A| =_c |B|$, $|B| =_c |C|$, and $|A| =_c |C|$.

*19. Let $f : A \to K$ and $g : B \to L$ be bijections. Let $h : A \times B \to K \times L$ be defined by $h(\langle a, b \rangle) = \langle f(a), g(b) \rangle$ for all $\langle a, b \rangle \in A \times B$. Prove that h is a bijection.

*20. Prove that $G(g^{-1} \circ h \circ f) = h$ where the functions G, g^{-1}, h, and f are as in the proof of Theorem 5.4.6(3).

21. Recall that $^A B$ is the set of all functions from A to B (see Definition 3.3.6). Suppose A is nonempty. Let $c \in B$ and $d \in B$ be distinct. Using a diagonal argument, show that there is no function $F : A \to {}^A B$ that is onto $^A B$. Now, find a one-to-one function $G : A \to {}^A B$. Conclude that $|A| <_c |^A B|$.

22. Prove that $|^\omega \omega| =_c |\mathcal{P}(\omega)|$. Conclude that $|^\omega \omega| =_c |\mathbb{R}|$.

*23. Prove Theorem 5.4.7(1).

*24. Prove Theorem 5.4.7(2).

*25. Prove Theorem 5.4.7(3).

26. Using Theorem 5.4.7 and Exercise 3, prove that $|^\omega(^\omega \omega)| =_c |^\omega \omega|$. Using Exercise 22, show that $|^\omega \mathbb{R}| =_c |^\mathbb{Q} \mathbb{R}| =_c |\mathbb{R}|$ and $|^\omega \mathcal{P}(\omega)| =_c |\mathcal{P}(\omega)|$.

27. Using Theorem 5.4.7 and Corollary 5.4.8, prove that

$$\left|^{(^\omega \omega)}(^\omega \omega)\right| =_c \left|^{(^\omega \omega)} \omega\right| \text{ and } \left|^{(^\omega \omega)}(^\omega\{0, 1\})\right| =_c \left|^{(^\omega \omega)}\{0, 1\}\right|.$$

Conclude, using Exercise 22, that $\left|^\mathbb{R} \mathbb{R}\right| =_c \left|^\mathbb{R} \omega\right| =_c \left|^\mathbb{R}\{0, 1\}\right|$.

*28. Prove Theorem 5.4.16(1).

*29. Prove Theorem 5.4.16(2).

*30. Prove Theorem 5.4.16(3). Hint: Let $c \in A$ and let $f: A \to K$ be one-to-one. Define $h: K \to A$ by

$$h(x) = \begin{cases} f^{-1}(x), & \text{if } x \in \operatorname{ran} f; \\ c, & \text{if } x \in K \setminus \operatorname{ran} f. \end{cases}$$

Show that h is a "left inverse for f"; that is, show that $h(f(a)) = a$ for all $a \in A$. Now replace the use of f^{-1} in the proof of Theorem 5.4.6(3) with the function h and modify the proof appropriately.

31. Let A and B be nonempty sets. Using the axiom of choice, prove that there exists a one-to-one $f: A \to B$ if and only if there exists a $g: B \to A$ that is onto A.

*32. Let $f: A \to B$ and $g: B \to A$ be functions, and let $C \subseteq A$ and $D \subseteq A$. Prove that if $C \subseteq D$, then $A \setminus g[B \setminus f[C]] \subseteq A \setminus g[B \setminus f[D]]$.

33. Let $f: A \to B, g: B \to A$, and $X \subseteq A$ be as in Lemma 5.4.17. Show that if $X = \varnothing$, then g is onto A and if $X = A$, then f is onto B.

34. Let $A = (0, 1)$. Using Theorem 5.4.18, show that $|A \times A| =_c |A|$.

35. Let $A = (0, 1]$ and $B = (0, 1)$. The functions $f: A \to B$ defined by $f(x) = \frac{1}{2}x$ and $g: B \to A$ defined by $g(x) = x$ are both one-to-one. Let $X \subseteq A$ be $X = \{1, \frac{1}{2}, \frac{1}{4}, \frac{1}{8}, \frac{1}{16}, \ldots\}$. Show that $g[B \setminus f[X]] = A \setminus X$. Now, using (5.15) as a guide, explicitly define a bijection $h: A \to B$.

36. Let $f: A \to B$ and $g: B \to A$ be functions. Consider the set $\{C_n : n \in \omega\}$ whose elements are defined by the following recursion:
(a) $C_0 = A \setminus \operatorname{ran}(g)$,
(b) $C_{n^+} = g[f[C_n]]$.
Let $Y = \bigcup_{n \in \omega} C_n$. Prove that if g is one-to-one, then $g[B \setminus f[Y]] = A \setminus Y$.

37. Let $f: A \to B$ and $g: B \to A$. Let Y be as defined in Exercise **36**. Assume g is one-to-one and $X \subseteq A$ satisfies $g[B \setminus f[X]] = A \setminus X$. Prove $Y \subseteq X$.

38. Let X, f, and g be as in the proof of Lemma 5.4.17. Now assume that g is one-to-one and let Y be as in Exercise **36**. Prove that $X = Y$.

39. Let A be a set and let $H: \mathcal{P}(A) \to \mathcal{P}(A)$. Suppose for all C and D in $\mathcal{P}(A)$, we have that $H(C) \subseteq H(D)$ whenever $C \subseteq D$. Let

$$S = \{D : D \subseteq A \text{ and } H(D) \subseteq D\}$$

and let $X = \bigcap S$. Prove that $H(X) = X$. Moreover, suppose that $Y \subseteq X$ satisfies $H(Y) \subseteq Y$. Prove that $X = Y$.

Exercise Notes: For Exercise **12**, use the inverse tangent function $\tan^{-1}(x)$. For Exercise **21**, review the proofs of Theorems 5.3.2 and 5.4.13. For Exercise **22**, use Theorem 5.4.18. To show that $|{}^{\omega}\omega| \leq_c |\mathcal{P}(\omega)|$, given $f \in {}^{\omega}\omega$, let $A_f \subseteq \omega$ be the set $A_f = \{p(i, j) : \langle i, j \rangle \in f\}$ where p is the function in Exercise 9 on page 116. For Exercise **23**, if $f \in {}^{A \cup B}K$, then $\langle f \restriction A, f \restriction B \rangle$ is in ${}^{A}K \times {}^{B}K$. For Exercise **24**, let $\pi_1(\langle x, y \rangle) = x$ and $\pi_2(\langle x, y \rangle) = y$. If $f \in {}^{A}(K \times L)$, then let $f_1 \in {}^{A}K$ be defined by $f_1(a) = \pi_1(f(a))$ and let $f_2 \in {}^{A}L$ satisfy $f_2(a) = \pi_2(f(a))$. For Exercise **25**, if $f \in {}^{A}({}^{B}K)$, then let $g \colon A \times B \to K$ be defined by $g(\langle a, b \rangle) = f(a)(b)$. For Exercise **31**, in the implication ($\Rightarrow$), let g be a left inverse for f as in Exercise **30**. The implication ($\Leftarrow$) uses the axiom of choice. For Exercise **32**, see Exercise 3 on page 36 and Exercise 2 on page 68. For Exercise **34**, merge the two infinite decimal expansions, each one not ending with repeating 9's, into one expansion. Exercise **36** outlines a different proof of Lemma 5.4.17 when g is one-to-one. Use the "double subset" strategy. For Exercise **37**, since g is one-to-one, Claim 1 (in the proof of Theorem 5.4.18) implies that $g[f[X]] \subseteq X$. For Exercise **38**, from the proof of Lemma 5.4.17, it follows from Exercise **37** that $X \subseteq Y$. One can show that $C_0 \subseteq X$ and $C_n \subseteq X$ for all $n \in \omega$. For Exercise **39**, review the proof of Lemma 5.4.17.

REMARK: Exercises **36**, **37**, and **39** imply that the set X given in the proof of Lemma 5.4.17 equals the set Y in Exercise **36** when g is one-to-one.

6 Transfinite Recursion

In Definition 3.4.2, we introduced the notion of a partial order on a set. The concept of a total order was then presented in Definition 3.4.3. We also proved that there exists a total ordering on the natural numbers so that every nonempty set of natural numbers has a least element with respect to this ordering (see Theorem 4.4.13). As a result, we were able to prove the important recursion theorem. In this chapter, we will generalize these ideas.

6.1 Well-Ordering

Definition 6.1.1. (Well-Ordering) A **well-ordering** is a total ordering on a set A for which every nonempty subset of A has a smallest (least) element.

Suppose that $\preccurlyeq$ is a well-ordering on a set A. Whenever $S \subseteq A$ is nonempty, then the set S has a least element ℓ; that is, $\ell \in S$ and $\ell \preccurlyeq x$ for all $x \in S$.

Let $\le$ be the usual order on set $\mathbb{Z}$ of integers. Then $\le$ is a total order on $\mathbb{Z}$. Since $S = \{n \in \mathbb{Z} : n < 0\}$ has no least element, $\le$ is not a well-ordering on $\mathbb{Z}$.

Theorem 6.1.2. *Let $\preccurlyeq$ be a well-ordering on A. Then there exists no function $f: \omega \to A$ satisfying $f(n^+) \prec f(n)$ for all $n \in \omega$.*

Proof. Suppose there is a function such that $f(n^+) \prec f(n)$ for all $n \in \omega$, where $\preccurlyeq$ is a well-ordering on A. Let $S = \{f(n) : n \in \omega\}$. Clearly, S is a nonempty subset of A. Thus, S has a least element $f(k)$. So $f(k) \preccurlyeq f(n)$ for all $n \in \omega$; however, by assumption, we have that $f(k^+) \prec f(k)$, a contradiction. $\square$

Let A be a set with well-ordering $\preccurlyeq$ and let $S \subseteq A$ be nonempty. When $\ell \in S$ is the smallest element in S, we shall write

$\ell = $ the $\preccurlyeq$-least element in S.

Lemma 3.4.12 implies that ℓ is unique.

Theorem 6.1.3. *If A is an infinite set with the well-ordering $\preccurlyeq$, then there is a function $h\colon \omega \to A$ such that $h(n) \prec h(n^+)$ for all $n \in \omega$.*

Proof. For each $x \in A$, let $G_x = \{y \in A : x \prec y\}$ be the set of elements in A that are "greater" than x, and let $\underline{L}_x = \{y \in A : y \preccurlyeq x\}$ be the set of elements in A that are "less than or equal" to x. As $\preccurlyeq$ is a total order, $A = \underline{L}_x \cup G_x$ for each $x \in A$. Moreover, since A is infinite, it follows (see Exercise 15, page 116) that for any $x \in A$,

if $\underline{L}_x$ is finite, then G_x is infinite. $\qquad\qquad (6.1)$

Let p be the $\preccurlyeq$-least element in A, and define $f\colon A \to A$ by

$$f(x) = \begin{cases} \text{the } \preccurlyeq\text{-least element in } G_x, & \text{if } G_x \neq \varnothing; \\ p, & \text{if } G_x = \varnothing. \end{cases} \qquad (6.2)$$

So ($\blacktriangle$) if $x \in A$ and $G_x \neq \varnothing$, then $x \prec f(x)$. By the recursion theorem, there is a function $h\colon \omega \to A$ such that

(1) $h(0) = p$,
(2) $h(n^+) = f(h(n))$, for all $n \in \omega$.

We now prove that $\underline{L}_{h(n)}$ is finite for all $n \in \omega$. Let $I = \{n \in \omega : \underline{L}_{h(n)} \text{ is finite}\}$. Because $h(0) = p$ is the smallest element in A, we see that $\underline{L}_{h(0)} = \{p\}$, which is finite. So $0 \in I$. Let $n \in I$. Thus, $\underline{L}_{h(n)}$ is finite, and $G_{h(n)} \neq \varnothing$ by (6.1). As $h(n^+) = f(h(n))$ and $G_{h(n)} \neq \varnothing$, we infer from (6.2) that $h(n^+)$ is the $\preccurlyeq$-least element in $G_{h(n)}$. It follows that $\underline{L}_{h(n^+)} = \underline{L}_{h(n)} \cup \{h(n^+)\}$, which is a finite set. Hence, $n^+ \in I$. Therefore, $I = \omega$ and $\underline{L}_{h(n)}$ is finite for all $n \in \omega$. Thus, by (6.1), $G_{h(n)} \neq \varnothing$ for all $n \in \omega$. Let $n \in \omega$. Since $f(h(n)) = h(n^+)$ and $G_{h(n)} \neq \varnothing$, the above ($\blacktriangle$) implies that $h(n) \prec h(n^+)$. $\qquad\square$

Theorem 6.1.4. *Let $\preccurlyeq'$ be a well-ordering on the set B. Suppose that $h\colon A \to B$ is one-to-one. Define the relation $\preccurlyeq$ on A by $x \preccurlyeq y$ if and only if $h(x) \preccurlyeq' h(y)$, for all x and y in A. Then $\preccurlyeq$ is a well-ordering on A.*

Proof. See Exercise 6. $\qquad\square$

Let $\preccurlyeq$ be a well-ordering on a set W. For each $u \in W$, we define the following **initial segments** up to u:

$$s^\prec(u) = \{x \in W : x \prec u\}, \qquad (6.3)$$
$$s^\preccurlyeq(u) = \{x \in W : x \preccurlyeq u\}. \qquad (6.4)$$

The set $s^\prec(u)$ consists of the elements in W that are **less than** (smaller than) u, whereas $s^{\preceq}(u)$ is the set of elements that are **less than or equal to** u. In other words, $u \in s^{\preceq}(u)$ and $u \notin s^\prec(u)$. Furthermore, for u and v in W, if $u \prec v$, then

$$s^\prec(u) \subset s^{\preceq}(u) \subseteq s^\prec(v) \subset s^{\preceq}(v).$$

If the well-ordering $\preceq$ is understood, we shall write $s(u)$ for $s^\prec(u)$ as well as $\underline{s}(u)$ in place of $s^{\preceq}(u)$. For example, because ω is well-ordered by $\in$ and it is transitive, if $n \in \omega$, then

$$s(n) = \{i \in \omega : i \in n\} = n, \tag{6.5}$$

$$\underline{s}(n) = \{i \in \omega : i \underline{\in} n\} = n^+. \tag{6.6}$$

Theorem 4.4.14, the strong induction principle, applies to the finite elements in ω. The term *transfinite* applies to elements that may not be finite.

Theorem 6.1.5 (Transfinite Induction Principle). *Let $\preceq$ be a well-ordering on W. Let A be a subset of W, and suppose that*

for all $u \in W$, if $s(u) \subseteq A$, then $u \in A$. $\qquad$ (6.7)

Then $A = W$.

Proof. Suppose that $A \subseteq W$ satisfies (6.7) and assume, for a contradiction, that there exists a $u \in W \setminus A$. Thus, the set $W \setminus A$ is nonempty. Since $\preceq$ is a well-ordering, $W \setminus A$ has a least element u. So $u \notin A$, and if $v \prec u$, then $v \in A$. Hence, $s(u) \subseteq A$. As A satisfies (6.7), we have that $u \in A$, a contradiction. $\qquad \square$

Exercises 6.1

1. Let $\mathbb{R}$ be the set of real numbers and let $\leq$ be the standard order on $\mathbb{R}$. Suppose $f : \omega \to \mathbb{R}$ is such that for all n and m in ω, if $n \in m$, then $f(n) < f(m)$. Let $A = \{f(i) : i \in \omega\}$.
   (a) Let i and j be in ω. Prove that if $f(i) < f(j)$, then $i \in j$.
   (b) Prove that $\leq$ is a well-ordering on A.

2. Let $A = \omega \cup \{\omega\}$, and for all x and y in A, define the relation $\underline{\in}$ on A by $x \underline{\in} y$ if and only if $x \in y$ or $x = y$. Show that $\underline{\in}$ is a well-ordering on A.

3. Suppose that $\preccurlyeq$ is a well-ordering on A and that $f: A \to A$ is such that if $x \prec y$, then $f(x) \prec f(y)$ whenever x and y are in A.
   (a) Show that f is one-to-one.
   (b) Prove that $x \preccurlyeq f(x)$ for all $x \in A$.

4. Let A be an infinite set with the well-ordering $\preccurlyeq$. Now let $h: \omega \to A$ be the function defined in the proof of Theorem 6.1.3. Let $y \in A$. Prove that for all $n \in \omega$, if $y \prec h(n)$, then $y = h(i)$ for some $i \in n$.

*5. Suppose that $\preccurlyeq$ is a well-ordering on W and let $C \subseteq W$. Show that $\preccurlyeq_C$ is a well-ordering on C.

*6. Prove Theorem 6.1.4.

7. Prove that every finite set has a well-ordering.

8. Prove that every countably infinite set has a well-ordering.

Exercise Notes: For Exercise **5**, apply the result of Exercise 14 on page 76 and review Definition 3.4.15. For Exercise **6**, Exercise 12 on page 76 implies that $\preccurlyeq$ is a total order on A. For Exercises **7–8**, use Theorem 6.1.4.

6.2 Transfinite Recursion Theorem

A powerful tool in set theory is definition by transfinite recursion. If one wants to define a function on a set using transfinite recursion, then one must first have a well-ordering on the set. Many of the important functions that we shall introduce later in the book are defined by means of transfinite recursion. This recursion principle plays a crucial role in set theory, and its importance cannot be overstated.

The concept of transfinite recursion is grounded in recursion on ω. Let us recall the "finite" Recursion Theorem 4.2.1: *Let $f: A \to A$ be a function and let $a \in A$. Then there exists a unique function $h: \omega \to A$ such that*

(1) $h(0) = a$ *and* (2) $h(n^+) = f(h(n))$, *for all $n \in \omega$.*

The value $h(n^+)$ depends on the "previous" value $h(n)$ and on the function f. Our proof of the finite recursion theorem used the fact that f was a set.

We would like to extend Theorem 4.2.1 and prove that there exists a function $h: \omega \to A$ so that $h(n^+)$ depends on all of the prior values $h(0), h(1), \ldots, h(n)$. We would prefer, however, to prove that there is a function h where each value $h(n^+)$ depends on the restricted function $h \restriction n^+$, that is, on $h \restriction \{0, 1, 2, \ldots, n\}$. Can we prove such a result?

As noted on page 121, Theorem 2.1.3 implies that the following is a set:

$${}^{\in\omega}A = \{p : \text{for some } n \in \omega, p \text{ is a function from } n \text{ to } A\}.$$

Thus, for every function $p: n \to A$, where $n \in \omega$, we have that $p \in {}^{\in\omega}A$. Now let $f: {}^{\in\omega}A \to A$ be a function and let $a \in A$. To positively address the above question, we would need to prove that there exists a function $h: \omega \to A$ that satisfies the following two conditions:

(i) $h(0) = a$ and (ii) $h(n^+) = f(h \upharpoonright n^+)$, for all $n \in \omega$.

One can combine conditions (i) and (ii) into one condition. Since the empty function $\varnothing$ is in ${}^{\in\omega}A$, one can let the function $f: {}^{\in\omega}A \to A$ satisfy $f(\varnothing) = a$ and then require that h meet the following single condition:

$$h(n) = f(h \upharpoonright n), \text{ for all } n \in \omega. \tag{6.8}$$

If h satisfies (6.8), then it will satisfy (i) and (ii); namely, we will have

$$h(0) = f(h \upharpoonright 0) = f(\varnothing) = a,$$
$$h(1) = f(h \upharpoonright 1) = f(\{\langle 0, a \rangle\}),$$
$$h(2) = f(h \upharpoonright 2) = f(\{\langle 0, a \rangle, \langle 1, h(1) \rangle\}),$$
$$h(3) = f(h \upharpoonright 3) = f(\{\langle 0, a \rangle, \langle 1, h(1) \rangle, \langle 2, h(2) \rangle\}).$$

Our proof of the Recursion Theorem 4.2.1 can be adapted to show that such a function h actually does exist; however, rather than do this, we will pursue a much more general result. In fact, we will establish two such generalizations, the first of which will imply that there exists a function h satisfying (6.8) (see Corollary 6.2.2).

6.2.1   Using a Set Function

Let $\preccurlyeq$ be a well-ordering on W and let A be a set. Recall the initial segment functions s and $\underline{s}$ that were defined on page 143. Theorem 2.1.3 implies that the following sets exist:

$${}^{\prec W}A = \{g : \text{for some } u \in W, g \text{ is a function from } s(u) \text{ to } A\},$$

$${}^{\preccurlyeq W}A = \{g : \text{for some } u \in W, g \text{ is a function from } \underline{s}(u) \text{ to } A\}.$$

Our next theorem is an extension of the finite recursion theorem on ω. The proof of this theorem uses a method that will be applied again in Section 6.2.2. We shall be using the expression *set function* only to emphasize that a given function is a set.

Transfinite Recursion Theorem 6.2.1. *Let $\preccurlyeq$ be a well-ordering on a set W. If $F: {}^{\prec W}\!A \to A$ is a set function, then there exists a unique function $H: W \to A$ such that*

$$H(u) = F(H \upharpoonright s(u)), \ \textit{for all } u \in W. \tag{6.9}$$

Proof. Let $\preccurlyeq$ be a well-ordering on W and let $F: {}^{\prec W}\!A \to A$ be a function. When $u \in W$, we will say that $g \colon \underline{s}(u) \to A$ is *F-recursive up to u* if and only if for all $x \in \underline{s}(u)$, we have that $g(x) = F(g \upharpoonright s(x))$.

Claim 1. Let $g \colon \underline{s}(u) \to A$ and $h \colon \underline{s}(u) \to A$ be F-recursive up to u. Then $g = h$.

Proof. Suppose that $g \neq h$, and let $v \in \underline{s}(u)$ be the $\preccurlyeq$-least element such that $g(v) \neq h(v)$. Thus, $g(x) = h(x)$ for all $x \prec v$. So ($\blacktriangle$) $g \upharpoonright s(v) = h \upharpoonright s(v)$. Hence,

$$g(v) = F(g \upharpoonright s(v)) \quad \text{as } g \text{ is } F\text{-recursive up to } u \text{ and } v \in \underline{s}(u)$$
$$ = F(h \upharpoonright s(v)) \quad \text{by } (\blacktriangle)$$
$$ = h(v) \quad \text{as } h \text{ is } F\text{-recursive up to } u \text{ and } v \in \underline{s}(u).$$

We conclude that $g(v) = h(v)$, a contradiction. $\qquad\qquad$ (Claim 1) $\square$

Claim 2. Let $u \in W$ and let $g \colon \underline{s}(u) \to A$ be F-recursive up to u. Then for all $v \prec u$, the function $g \upharpoonright \underline{s}(v)$ is F-recursive up to v.

Proof. Let $v \prec u$ and let $h = g \upharpoonright \underline{s}(v)$. Hence, $h \colon \underline{s}(v) \to A$ and

$$h(y) = g(y) \text{ for all } y \in \underline{s}(v). \tag{6.10}$$

For each $x \in \underline{s}(v)$, we have that $x \preccurlyeq v \prec u$, and therefore,

$$h(x) = g(x) \quad \text{by (6.10), as } x \in \underline{s}(v)$$
$$ = F(g \upharpoonright s(x)) \quad \text{as } g \text{ is } F\text{-recursive up to } u \text{ and } x \in \underline{s}(u)$$
$$ = F(h \upharpoonright s(x)) \quad \text{by (6.10), as } s(x) \subseteq \underline{s}(v).$$

So $h(x) = F(h \upharpoonright s(x))$, and thus, $g \upharpoonright \underline{s}(v)$ is F-recursive up to v. (Claim 2) $\square$

Consider the relation $G \subseteq W \times {}^{\preccurlyeq W}\!A$ defined by

$$\langle x, g \rangle \in G \text{ if and only if } g \colon \underline{s}(x) \to A \text{ is } F\text{-recursive up to } x. \tag{6.11}$$

Claim 1 implies that G is a function. For each $u \in \mathrm{dom}(G)$, let $g_u = G(u)$. Therefore, we have that the function $g_u \colon \underline{s}(u) \to A$ is F-recursive up to u whenever $u \in \mathrm{dom}(G)$. In particular, $g_u(x) \in A$ for all $x \in \underline{s}(u)$.

Claim 3. Let $v \in \mathrm{dom}(G)$. If $x \in s(v)$, then $x \in \mathrm{dom}(G)$ and $g_v(x) = g_x(x)$.

Proof. Let $v \in \mathrm{dom}(G)$ and $x \prec v$. Hence, $g_v : \underline{s}(v) \to A$ is F-recursive up to v. Claim 2 implies that $g_v \upharpoonright \underline{s}(x)$ is F-recursive up to x. Therefore, $x \in \mathrm{dom}(G)$ and $g_v \upharpoonright \underline{s}(x) = g_x$ by Claim 1. So $g_v(x) = g_x(x)$ as $x \in \underline{s}(x)$.   (Claim 3) $\square$

Claim 4. $\mathrm{dom}(G) = W$.

Proof. Clearly, $\mathrm{dom}(G) \subseteq W$. Suppose, to the contrary, that $W \setminus \mathrm{dom}(G) \neq \varnothing$. Let $u \in W$ be the $\prec$-least such that $u \notin \mathrm{dom}(G)$. So $v \in \mathrm{dom}(G)$ for all $v \prec u$. Let $h : s(u) \to A$ be the function defined by ($\blacklozenge$) $h(x) = g_x(x)$ for all $x \in s(u)$, and let $v \prec u$. Claim 3 implies that $g_v(x) = g_x(x)$ for all $x \in s(v)$. Thus, $g_v(x) = h(x)$ for all $x \in s(v)$, and therefore, ($\blacktriangledown$) $h \upharpoonright s(v) = g_v \upharpoonright s(v)$. Define $\underline{h} : \underline{s}(u) \to A$ by

$$\underline{h}(x) = \begin{cases} F(h), & \text{if } x = u; \\ h(x), & \text{if } x \prec u. \end{cases} \tag{6.12}$$

We will show that $\underline{h}$ is F-recursive up to u. Let $v \in \underline{s}(u)$. There are two cases to consider: $v \prec u$ and $v = u$. If $v \prec u$, we can prove that $\underline{h}(v) = F(\underline{h} \upharpoonright s(v))$ as follows:

$$\underline{h}(v) = h(v) \qquad \text{by (6.12), as } v \prec u$$
$$= g_v(v) \qquad \text{by the definition } (\blacklozenge) \text{ of } h$$
$$= F(g_v \upharpoonright s(v)) \qquad \text{as } g_v \text{ is } F\text{-recursive up to } v$$
$$= F(h \upharpoonright s(v)) \qquad \text{by } (\blacktriangledown)$$
$$= F(\underline{h} \upharpoonright s(v)) \qquad \text{by definition (6.12) of } \underline{h}, \text{ because } v \prec u.$$

When $v = u$, we need to show that $\underline{h}(u) = F(\underline{h} \upharpoonright s(u))$. The definition of $\underline{h}$ in (6.12) implies that $\underline{h} \upharpoonright s(u) = h$ and hence,

$$\underline{h}(u) = F(h) \qquad \text{by (6.12), as } u = u$$
$$= F(\underline{h} \upharpoonright s(u)) \qquad \text{since } \underline{h} \upharpoonright s(u) = h.$$

Thus, $\underline{h}$ is F-recursive up to u. So $u \in \mathrm{dom}(G)$, a contradiction.   (Claim 4) $\square$

Define $H : W \to A$ by $H(x) = g_x(x)$. The argument used to prove Claim 4, involving h, shows that $H(u) = F(H \upharpoonright s(u))$ for all $u \in W$. Moreover, the proof of Claim 1 adapts to show that H is unique. (See Exercise 1.)

    (Theorem) $\square$

The relation $\in$ is a well-ordering on ω, as established by Theorem 4.4.13. Theorem 6.2.5 and (6.5) thereby imply the following corollary.

Corollary 6.2.2. *Let $f: {}^{\in\omega}A \to A$ be a function. There exists a unique function $h: \omega \to A$ such that*

$$h(n) = f(h \restriction n), \text{ for all } n \in \omega. \tag{6.13}$$

Corollary 6.2.2 allows us to prove the converse of Corollary 5.1.11.

Theorem 6.2.3 (AC). *Suppose that A is an infinite set. Then there is one-to-one function $h: \omega \to A$.*

Proof. Let A be an infinite set. Recall that (see page 146)

$${}^{\in\omega}A = \{p : \text{ for some } n \in \omega, p \text{ is a function from } n \text{ to } A\}.$$

Whenever $p \in {}^{\in\omega}A$ and $n = \text{dom}(p)$, we observe that $\text{ran}(p) = p[n]$ is a finite subset of A, and therefore, $A \setminus \text{ran}(p)$ is infinite by Exercise 12 on page 116. Let $C = \mathcal{P}(A) \setminus \{\varnothing\}$. By Theorem 3.3.24, there is a function $g: C \to A$ such that $g(B) \in B$ for all $B \in C$. Define $f: {}^{\in\omega}A \to A$ by

$$f(p) = g(A \setminus \text{ran}(p)).$$

By Corollary 6.2.2, there exists a function h with domain ω such that

$$h(n) = f(h \restriction n) = g(A \setminus h[n]), \text{ for all } n \in \omega. \tag{6.14}$$

Thus, $h: \omega \to A$, and Exercise 11 on page 123 shows that h is one-to-one. $\square$

Richard Dedekind [3] defined a set to be infinite if and only if it has the same cardinality as a proper subset of itself. Such a set is said to be *Dedekind-infinite*. For example, the function $f: \omega \to \omega$ defined by $f(n) = n^+$ is one-to-one and shows that ω and $\omega \setminus \{0\}$ have the same cardinality. So, according to Dedekind, ω is an infinite set. Our next result confirms Dedekind's definition in general.

Corollary 6.2.4 (AC). *A set A is infinite if and only if there exists a one-to-one function $f: A \to A$ that is not onto A.*

Proof. Let A be set. Assume that A is infinite. Theorem 6.2.3 asserts that there is a one-to-one function $h: \omega \to A$. Define $f: A \to A$ by

$$f(x) = \begin{cases} h(n^+), & \text{if } x = h(n) \text{ for some } n \in \omega; \\ x, & \text{if } x \notin \text{ran}(h). \end{cases} \tag{6.15}$$

Since h is one-to-one, it follows that f is a function and that f is one-to-one. As $h(0) \in A$ and $h(0)$ is not in the range of f, we see that f is not onto A.

To prove the converse, suppose $f: A \to A$ is a one-to-one function that is not onto A. Corollary 5.1.6 implies that A is infinite.   □

6.2.2   Using a Class Function

Let us say that a formula $\varphi(g, u)$ is **functional** if $\forall g \exists! u \varphi(g, u)$; that is, for all g there exists a unique u such that $\varphi(g, u)$. Let $\varphi(g, u)$ be functional formula and consider the class

$$\mathcal{F} = \{\langle g, u \rangle : \varphi(g, u)\}.$$

Because $\varphi(g, u)$ is functional, we can now view $\mathcal{F}$ as a *class function*. Let $\preccurlyeq$ be a well-ordering on a set W. If there were a set A such that for all $g \in {}^{<W}A$, there is a $v \in A$ such that $\varphi(g, v)$, then Theorem 2.1.3 would imply that the following function $F: {}^{<W}A \to A$ is a set:

$$F = \{\langle g, v \rangle : g \in {}^{<W}A, \ v \in A, \ \text{and} \ \varphi(g, v)\}.$$

Theorem 6.2.1 would then yield the existence of a function $H: W \to A$ so that

$$H(u) = F(H \restriction s(u)) \text{ for all } u \in W$$

(where the initial segment $s(u)$ is defined on page 143). Thus, the set function H would satisfy

$$\varphi(H \restriction s(u), H(u)) \text{ for each } u \in W.$$

Can one prove the existence of this function H without a set such as A? Yes, and this will be verified by our next theorem, whose proof is based on the argument used to prove Theorem 6.2.1. This theorem also validates an essential general tool in set theory, namely, *definition by transfinite recursion.*

Transfinite Recursion Theorem 6.2.5 (Class Form). *Suppose that $\varphi(x, y)$ is a functional formula (which may contain parameters). If $\preccurlyeq$ is a well-ordering on a set W, then there exists a unique function H with domain W such that*

$$\varphi(H \restriction s(u), H(u)) \text{ for all } u \in W. \tag{6.16}$$

The crucial ingredient in the proof of Theorem 6.2.1 was the existence of the set function G (see (6.11)). In our proof of Theorem 6.2.5, using the functional formula $\varphi(x, y)$ and the replacement axiom, we will be able to get a function such as G. Recall that the replacement axiom asserts that for any set A, if for each $x \in A$, there is an element y that is "uniquely connected" to x, then we can *replace* every $x \in A$ with its unique connection y, and the result is a new set.

Replacement Axiom. Let $\psi(x, y)$ be a formula. For every set A, if for each $x \in A$ there is a unique y such that $\psi(x, y)$, then there is a set S that consists of all the elements y such that $\psi(x, y)$ for some $x \in A$.

Thus, if $\forall x \exists! y \psi(x, y)$, then for each set A, we have the following set:

$$S = \{y : \exists x(x \in A \land \psi(x, y))\}.$$

Hence, $f = \{\langle x, y \rangle \in A \times S : \psi(x, y)\}$ is a set function $f : A \to S$.

We are now ready to prove Theorem 6.2.5. In a loose sense, our proof will be the result of replacing all of the expressions $F(x) = y$ and $y = F(x)$, appearing in the proof of Theorem 6.2.1, with the expression $\varphi(x, y)$. We could just ask the reader to do this as an exercise, but since the proof requires the replacement axiom, we shall provide a complete proof.

Proof (of Theorem 6.2.5). Let $\preccurlyeq$ be a well-ordering on W. Suppose that $\varphi(x, y)$ is functional. When $u \in W$, we will say that a function g is φ-*recursive up to* u if and only if $\operatorname{dom}(g) = \underline{s}(u)$ and

$$\varphi(g \restriction \underline{s}(x), g(x)), \quad \text{for all } x \in \underline{s}(u).$$

We now establish four claims.

Claim 1. Let g and h be φ-recursive up to u. Then $g = h$.

Proof. Let g and h be φ-recursive up to u. Suppose that $g \neq h$. Let $v \in \underline{s}(u)$ be the $\prec$-least element such that $g(v) \neq h(v)$. Thus, $g(x) = h(x)$ for all $x \prec v$. Hence, (▲) $g \restriction \underline{s}(v) = h \restriction \underline{s}(v)$. Since g and h are φ-recursive up to u, we have that $\varphi(g \restriction \underline{s}(v), g(v))$ and $\varphi(h \restriction \underline{s}(v), h(v))$. Since φ is functional, (▲) implies that $g(v) = h(v)$, a contradiction.   (Claim 1) □

Claim 2. If $u \in W$ and g is φ-recursive up to u, then for all $v \prec u$, the function $g \restriction \underline{s}(v)$ is φ-recursive up to v.

Proof. Let $u \in W$ and g be φ-recursive up to u. Let $v \prec u$ and $h = g \restriction \underline{s}(v)$. So

$$h(y) = g(y) \text{ for all } y \in \underline{s}(v). \tag{6.17}$$

Let $x \in \underline{s}(v)$. Therefore, $x \preccurlyeq v \prec u$ and $s(x) \subseteq \underline{s}(v)$. Hence, $h(x) = g(x)$ and $h \restriction s(x) = g \restriction s(x)$, by (6.17). As g is φ-recursive up to u and $x \in \underline{s}(u)$, we have that $\varphi(g \restriction s(x), g(x))$. Since $h \restriction s(x) = g \restriction s(x)$ and $h(x) = g(x)$, we conclude that $\varphi(h \restriction s(x), h(x))$. Thus, $g \restriction \underline{s}(v)$ is φ-recursive up to v.   (Claim 2) □

Let $\psi(x, g)$ represent the property

"g is a function that is φ-recursive up to x."

Let $A = \{x \in W : \text{ there exists a } g \text{ such that } \psi(x, g)\}$. Claim 1 implies that for all $x \in A$ there is a unique g such that $\psi(x, g)$. Thus, by the replacement axiom we have that $S = \{g : \exists x(x \in A \wedge \psi(x, g))\}$ is a set. Now define the relation $G \subseteq W \times S$ by

$$\langle x, g \rangle \in G \text{ if and only if } g \text{ is } \varphi\text{-recursive up to } x. \tag{6.18}$$

Claim 1 implies that G is a function. For each $u \in \text{dom}(G)$, let $g_u = G(u)$, and hence, g_u is φ-recursive up to u.

Claim 3. Let $v \in \text{dom}(G)$. If $x \in s(v)$, then $x \in \text{dom}(G)$ and $g_v(x) = g_x(x)$.

Proof. Let $v \in \text{dom}(G)$ and $x \prec v$. So g_v is φ-recursive up to v. Claim 2 implies that $g_v \upharpoonright s(x)$ is φ-recursive up to x. Thus, $x \in \text{dom}(G)$ and $g_v \upharpoonright s(x) = g_x$ by Claim 1. Since $x \in s(x)$, we conclude that $g_v(x) = g_x(x)$.   (Claim 3) $\square$

Claim 4. $\text{dom}(G) = W$.

Proof. Clearly, $\text{dom}(G) \subseteq W$. Suppose, to the contrary, that $W \setminus \text{dom}(G) \neq \varnothing$. Let $u \in W$ be the $\preccurlyeq$-least such that $u \notin \text{dom}(G)$. So $v \in \text{dom}(G)$ for all $v \prec u$. Let h be the function with domain $s(u)$ defined by ($\blacklozenge$) $h(x) = g_x(x)$, and let $v \prec u$. Claim 3 implies that $g_v(x) = g_x(x) = h(x)$ for all $x \in s(v)$. Therefore, ($\blacktriangledown$) $h \upharpoonright s(v) = g_v \upharpoonright s(v)$. Let y be the unique element satisfying $\varphi(h, y)$. Define the function $\underline{h}$ with domain $s(u)$ by

$$\underline{h}(x) = \begin{cases} y, & \text{if } x = u; \\ h(x), & \text{if } x \prec u. \end{cases} \tag{6.19}$$

We will show that $\underline{h}$ is φ-recursive up to u. Let $v \in s(u)$. There are two cases to consider: $v \prec u$ and $v = u$. If $v \prec u$, then we have the following equivalences:

$$\varphi(\underline{h} \upharpoonright s(v), \underline{h}(v)) \text{ iff } \varphi(h \upharpoonright s(v), h(v)) \quad \text{by (6.19), as } v \prec u$$

$$\text{iff } \varphi(g_v \upharpoonright s(v), g_v(v)) \quad \text{by } (\blacktriangledown) \text{ and the definition } (\blacklozenge) \text{ of } h.$$

Since g_v is φ-recursive up to v, we have that $\varphi(g_v \upharpoonright s(v), g_v(v))$ holds, and thus its equivalent $\varphi(\underline{h} \upharpoonright s(v), \underline{h}(v))$ also holds.

When $v = u$, we need to show that $\varphi(\underline{h} \upharpoonright s(u), \underline{h}(u))$ holds. The definition of $\underline{h}$ in (6.19) implies that $\underline{h} \upharpoonright s(u) = h$, and therefore the following are equivalent:

$$\varphi(\underline{h} \upharpoonright s(u), \underline{h}(u)) \text{ iff } \varphi(h, \underline{h}(u)) \quad \text{because } \underline{h} \upharpoonright s(u) = h$$

$$\text{iff } \varphi(h, y) \qquad \text{as } \underline{h}(u) = y, \text{ by (6.19)}.$$

Since y was chosen to satisfy $\varphi(h, y)$, we conclude that $\varphi(\underline{h} \upharpoonright s(u), \underline{h}(u))$ is true. Thus, $\underline{h}$ is φ-recursive up to u. So $u \in \text{dom}(G)$, a contradiction.   (Claim 4) $\square$

Now define the function H with domain W by $H(x) = g_x(x)$. The argument in the proof of Claim 4, involving h, shows that $\varphi(H \upharpoonright s(u), H(u))$

for all $u \in W$. Moreover, the proof of Claim 1 adapts to show that H is unique. (Theorem) $\square$

Remark 6.2.6. Let $\preceq$ be a well-ordering on W. In Theorem 6.2.5, let $\varphi(x, y)$ be a formula that contains parameters (free variables other than x and y). The function H given in Theorem 6.2.5 depends on the values that are assigned to these parameters. To illustrate this, suppose that p is a parameter appearing in $\varphi(x, y)$. Because p is a free variable in φ, we can express $\varphi(x, y)$ as $\varphi(p, x, y)$. Now let γ be a set that can be assigned to the parameter p so that $\varphi(\gamma, x, y)$ is functional. Theorem 6.2.5 implies that there is a unique function H_γ such that $\varphi(\gamma, H_\gamma \restriction s(u), H_\gamma(u))$, for all $u \in W$. So H_γ depends on γ, and Theorem 6.2.5 holds whenever the formula $\varphi(x, y)$ contains fixed sets, because these sets can be viewed as values that have been assigned to parameters (see page 25).

Since the relation $\in$ is a well-ordering of ω, Theorem 6.2.5 and (6.5) (see page 144) imply the next corollary.

Corollary 6.2.7. *Let $\varphi(x, y)$ be a formula that is functional. Then there exists a unique function F with domain ω such that*

$$\varphi(F \restriction n, F(n)), \text{ for all } n \in \omega. \tag{6.20}$$

Let F be as in Corollary 6.2.7. Because $F \restriction 0 = \varnothing$, (6.20) implies that $F(0)$ equals the unique y satisfying $\varphi(\varnothing, y)$. Since $F \restriction 1 = \{\langle 0, F(0) \rangle\}$, we see that $F(1)$ is the unique y for which $\varphi(\{\langle 0, F(0) \rangle\}, y)$ holds, and $F(2)$ equals the unique y satisfying $\varphi(\{\langle 0, F(0) \rangle, \langle 1, F(1) \rangle\}, y)$, etc. We now apply Corollary 6.2.7 to show that every set is a subset of a transitive set.

Theorem 6.2.8. *For every set A, there exists a transitive set $\overline{A}$ such that $A \subseteq \overline{A}$.*

Proof. Let A be a set. Now consider the following definition by cases:

$$y = \begin{cases} A, & \text{if } f = \varnothing; \\ \bigcup f(k), & \text{if } f \text{ is a function with domain } k^+ \text{ for some } k \in \omega; \\ \varnothing, & \text{if none of the above hold.} \end{cases} \tag{6.21}$$

Let $\varphi(f, y)$ be a formula that expresses (6.21). Clearly, for each f, there is a unique y so that $\varphi(f, y)$. By Corollary 6.2.7, there is a function F with domain ω such that $\varphi(F \restriction m, F(m))$ for all $m \in \omega$. Thus, for each $m \in \omega$, we have that

$$F(m) = \begin{cases} A, & \text{if } m = 0; \\ \bigcup F(n), & \text{if } m = n^+ \text{ for some } n \in \omega. \end{cases} \tag{6.22}$$

Let $\overline{A} = \bigcup_{n \in \omega} F(n)$. We see that $F(0) = A$, by (6.22). So $A \subseteq \bigcup_{n \in \omega} F(n)$. Thus, $A \subseteq \overline{A}$. To show that $\overline{A}$ is a transitive set, let $x \in y$ and $y \in \overline{A}$. We show that $x \in \overline{A}$. Because $y \in \overline{A}$, we have $y \in F(n)$ for some $n \in \omega$. Thus, $y \subseteq \bigcup F(n)$. Since $F(n^+) = \bigcup F(n)$, we see that $x \in F(n^+)$. Hence, $x \in \overline{A}$ and $\overline{A}$ is transitive. □

Let A, $\overline{A}$, and F be as in the statement and the proof of Theorem 6.2.8. By evaluating $F(n)$ for $n = 0, 1, 2, \ldots$, one can verify that

$$\overline{A} = A \cup \bigcup A \cup \bigcup\bigcup A \cup \bigcup\bigcup\bigcup A \cup \cdots .$$

Definition 6.2.9. Let A be a set. Then the set $\overline{A}$ constructed in the proof of Theorem 6.2.8 is called the **transitive closure** of A.

Exercise 6 shows that $\overline{A}$, the transitive closure of a set A, is the "smallest" transitive set of which A is a subset.

Exercises 6.2

*1. Complete the proof of Theorem 6.2.1 by showing that the function H satisfies $H(u) = F(H \restriction s(u))$, for all $u \in W$. Also show that H is the only such function.

2. Show that Theorem 6.2.5 implies Theorem 6.2.1.

3. Let $\preccurlyeq$ be a well-ordering on a set W and A be a set. Define $\ell \colon {}^{\prec W}A \to W$ by $\ell(g)$ equals the $\preccurlyeq$-least u in W such that $u \notin \mathrm{dom}(g)$. So if $g \in {}^{\prec W}A$ and $g \colon s(u) \to A$, then $\ell(g) = u$. Let $G \colon {}^{\prec W}A \times W \to A$ be a function. Show that Theorem 6.2.1 implies there exists a function $H \colon W \to A$ such that

$$H(u) = G(H \restriction s(u), u), \quad \text{for all } u \in W.$$

4. Suppose that $\preccurlyeq$ is a well-ordering on a set W and $F \colon {}^{\prec W}A \to A$ is a function. Let $H \colon W \to A$ be as in Theorem 6.2.1.
   (a) Let u and v be in W. Prove that if $s(u) = s(v)$, then $u = v$.
   (b) Suppose that F is one-to-one. Prove that H is also one-to-one.

5. Let $\preccurlyeq$ be a well-ordering on W and $\preccurlyeq^*$ be a well-ordering on A. Assume for each $g \in {}^{\prec W}A$, there is a $y \in A$ such that $g(x) \prec^* y$ for all $x \in \mathrm{dom}(g)$.

Show that there is an $H : W \to A$ so that for all $v \in W$ and $u \in W$, if $v \prec u$, then $H(v) \prec^* H(u)$.

*6. Let A be a set. Suppose that $A \subseteq C$ where C is a transitive set. Prove that $\overline{A} \subseteq C$. Thus, $\overline{A}$ is the "smallest" transitive set for which A is a subset.

7. Let B be a set. Conclude from Exercise **6** that $\overline{\{B\}}$ is the smallest transitive set that contains B as an element.

8. Let A be a set. Using the replacement axiom, one can show that $\{\overline{x} : x \in A\}$ is a set. Establish the following:
   (a) Prove that if $x \in A$, then $\overline{x} \subseteq \overline{A}$.
   (b) Prove that $A \cup \bigcup \{\overline{x} : x \in A\} \subseteq \overline{A}$.
   (c) Prove that $A \cup \bigcup \{\overline{x} : x \in A\}$ is a transitive set.
   (d) Prove that $A \cup \bigcup \{\overline{x} : x \in A\} = \overline{A}$.

9. Let A be a set. Using the proof of Theorem 6.2.8 as a model, prove that there is a function F with domain ω such that
   1. $F(0) = A$,
   2. $F(n^+) = \{F(n)\}$ for all $n \in \omega$.
   Let $C = F[\omega]$. Prove that $A \in C$ and for all X, if $X \in C$, then $\{X\} \in C$.

10. Let A be a set. Using the proof of Theorem 6.2.8 as a model, prove that there is a function F with domain ω such that
   1. $F(0) = A$,
   2. $F(n^+) = \mathcal{P}(F(n))$ for all $n \in \omega$.
   Let $C = F[\omega]$. Prove that $A \in C$ and for all X, if $X \in C$, then $\mathcal{P}(X) \in C$.

Exercise Notes: For Exercise **6**, let F be as in the proof of Theorem 6.2.8, and prove that $(\forall n \in \omega)(F(n) \subseteq C)$. For Exercise **8**, use Exercise **6**.

7

The Axiom of Choice (Revisited)

In Chapter 3 (see page 66), we first introduced the following principle, which is applied in many areas of mathematics.

Axiom of Choice. Let $\langle A_i : i \in I \rangle$ be an indexed function with nonempty terms. Then there is an indexed function $\langle x_i : i \in I \rangle$ such that $x_i \in A_i$, for all $i \in I$.

The above function $\langle x_i : i \in I \rangle$, where $x_i \in A_i$ for all $i \in I$, is called a **choice function** for $\langle A_i : i \in I \rangle$. The axiom of choice is used to prove many theorems in mathematics. In real analysis, for example, the axiom of choice is (tacitly) applied to prove that a real valued function $f : \mathbb{R} \to \mathbb{R}$ is continuous at a point $c \in \mathbb{R}$ if and only if $\lim_{n \to \infty} f(x_n) = f(c)$ for every sequence $\langle x_n \rangle$ that converges to c. It is thus said that continuity is equivalent to sequential continuity. The axiom of choice is also used to prove that every vector space has a basis.

In Chapter 3, it was shown that the axiom of choice implies the following theorem, which asserts the existence of another **choice function** H.

Theorem 3.3.24 (AC). *Let C be a set of nonempty sets. Then there is a function $H : C \to \bigcup C$ such that $H(A) \in A$ for all $A \in C$.*

It was then left as an exercise to show that Theorem 3.3.24 implies the axiom of choice. Hence, the theorem is actually equivalent to the axiom. Moreover, the axiom of choice is equivalent to a number of seemingly unrelated theorems. In this chapter, we will show that the axiom implies two other results that are also equivalent to it; but before we do this, we mention an application of the axiom of choice to topology, which concerns "product sets."

Let $\langle B_i : i \in I \rangle$ be an indexed function with nonempty terms. The **product set**, denoted by $\prod_{i \in I} B_i$, is defined by

$$\prod_{i \in I} B_i = \{f : f \text{ is a function, } \mathrm{dom}(f) = I, \text{ and } (\forall i \in I)(f(i) \in B_i)\}. \qquad (7.1)$$

Typically, when I is an infinite set, the only way to conclude that this product set is nonempty is to appeal to the axiom of choice as follows: Let $\langle x_i : i \in I \rangle$

be a choice function such that $x_i \in B_i$ for all $i \in I$. Define $f : I \to \bigcup_{i \in I} B_i$ by $f(i) = x_i$. Then $f \in \prod_{i \in I} B_i$, and thus, the product is nonempty.

Product sets are used to create a variety of topological spaces and to prove Tychonoff's compactness theorem, a classic result in topology. In Chapter 9 (see Exercise 19 on page 226), one is asked to complete a proof of a theorem on product sets due to Julius König.

Many important consequences of the axiom of choice can be proved from the following weaker version.

Countable Axiom of Choice. Suppose that $\langle A_i : i \in I \rangle$ is an indexed function with nonempty terms. If I is countable, then there is a function $\langle x_i : i \in I \rangle$ such that $x_i \in A_i$, for all $i \in I$.

Thus, if $\mathcal{C}$ is a countable set of nonempty sets, then the countable axiom of choice implies there is a function $H : \mathcal{C} \to \bigcup \mathcal{C}$ such that $H(A) \in A$ for all $A \in \mathcal{C}$. The axiom of choice, of course, implies the countable axiom of choice, but not conversely.

Finally, we remark that the replacement axiom shall not be applied in this chapter; however, it will be used in Chapters 8 and 9.

7.1 Zorn's Lemma

Zorn's Lemma, also known as the maximum principle, is an important theorem about partially ordered sets that is normally used to prove the existence of a mathematical object when it cannot be explicitly produced. According to Max Zorn, he first formulated his principle at Hamburg in 1933. Zorn published this maximum principle in 1935. Amil Artin then used Zorn's principle to establish theorems in algebra. Artin later proved that the principle implies the axiom of choice.

In 1935, Zorn proposed adding his maximum principle to ZF, the axioms of set theory. Although Zorn was not the first to suggest such a principle, he demonstrated how useful his formulation was in applications, particularly in topology, abstract algebra, and real analysis. Zorn also asserted (but did not prove) that his lemma and the axiom of choice are equivalent. To review the concepts of a maximal element and a chain, see Definitions 3.4.6 and 3.4.14.

Zorn's Lemma 7.1.1 (AC). *Let $(A, \preccurlyeq)$ be a partially ordered set. If every chain in A has an upper bound, then A contains a maximal element.*

Our proof of Zorn's Lemma 7.1.1 will apply "proof by contradiction"; that is, we shall let $(A, \preccurlyeq)$ be a partially ordered set in which every chain has an upper bound. We will then assume that A has no maximal elements and thereby derive a contradiction. The proof will depend on a few lemmas and definitions, which we now present.

Let $(A, \preccurlyeq)$ be a partially ordered set and $S \subseteq A$. Recalling Definition 3.4.8, an element $b \in A$ is called an *upper bound* for S when b satisfies $(\forall x \in S)(x \preccurlyeq b)$. We will say that b is a **proper** upper bound if, in addition, we have that $b \notin S$.

Definition 7.1.2. Let $(A, \preccurlyeq)$ be a partially ordered set in which C is a chain. Then C_p is the set of *proper* upper bounds for C.

If $(A, \preccurlyeq)$ is a poset and $C \subseteq A$ is a chain, then

$$C_p = \{q \in A : c \prec q, \text{ for all } c \in C\}.$$

Thus, C_p is nonempty if and only if C has proper upper bounds in A.

Definition 7.1.3. Let C be a chain in A, where $(A, \preccurlyeq)$ is a partially ordered set. We shall say that C' is **cofinal** in C when $C' \subseteq C$ and for all $c \in C$, there is a $c' \in C'$ such that $c \preccurlyeq c'$.

Lemma 7.1.4. *Suppose that $(A, \preccurlyeq)$ is a poset and let $C \subseteq A$ be a chain. If C' is cofinal in C, then $C'_p = C_p$.*

Lemma 7.1.5. *Let $(A, \preccurlyeq)$ be a partially ordered set in which every chain has an upper bound. If A has no maximal elements, then every chain C in A has a proper upper bound, and hence, C_p is nonempty.*

Proof. Assume that A has no maximal elements. Let C be a chain in A. Since every chain in A has an upper bound, let $b \in A$ be an upper bound for C. If $b \notin C$, then b is a proper upper bound for C. Suppose $b \in C$. So, because b is not a maximal element, there is a $q \in A$ such that $b \prec q$. Thus, $q \notin C$ and q is a proper upper bound for C. $\qquad\square$

We now present a proof of Zorn's Lemma.

Proof (of Zorn's Lemma 7.1.1). Let $(A, \preccurlyeq)$ be a partially ordered set in which every chain $C \subseteq A$ has an upper bound in A. We will prove that there is an $m \in A$ such that m is a maximal element. Assume, for a contradiction, that A

has no maximal elements. By Lemma 7.1.5 and the axiom of choice, there is a function f such that $f(C_p) \in C_p$ whenever C is a chain in A. As $C_p \cap C = \varnothing$, we see that

$$f(C_p) \notin C \text{ for all chains } C \text{ in } A. \tag{7.2}$$

A chain L in A shall be called a **ladder** if whenever $C \subseteq L$ and $C_p \cap L \neq \varnothing$, then $f(C_p)$ is the smallest element in $C_p \cap L$ (see Definition 3.4.11 on page 73). In other words, *if $C \subseteq L$ and C has a proper upper bound in L, then $f(C_p) \in L$ and $f(C_p)$ is the smallest element in L that is a proper upper bound for C.*

Ladders exist. To confirm this, note that $\varnothing$ is (vacuously) a ladder and that $\varnothing_p = A$. For another example, let $a = f(\varnothing_p)$. Since $\{a\}$ is a chain, we have that $a \prec f(\{a\}_p)$. One can show that the set $L = \{a, f(\{a\}_p)\}$ is also a ladder.

Claim 1. Suppose $L \subseteq K$ where L and K are ladders. Let $\ell \in L$ and $k \in K$. If $k \preccurlyeq \ell$, then $k \in L$.

Proof. Let L and K are as in the claim. Let $\ell \in L$ and $k \in K$. Assume that $k \preccurlyeq \ell$. Suppose, for a contradiction, that $k \notin L$. Thus, $k \prec \ell$. Let C be defined by

$$C = \{x \in L : x \prec k\}. \tag{7.3}$$

So, $C \subseteq L$. Since $L \subseteq K$, we also have that $C \subseteq K$. Clearly, $k \in K$ is a proper upper bound for C. Because K is a ladder, it follows that

$$f(C_p) \in K \text{ and } f(C_p) \preccurlyeq k. \tag{7.4}$$

Now, because $k \prec \ell$ and $\ell \in L$, the set C also has a proper upper bound in L. Thus, $f(C_p) \in L$. Since $k \notin L$ and $f(C_p) \in L$, (7.4) implies that $f(C_p) \prec k$. Hence, $f(C_p) \in C$ by (7.3), and this contradicts (7.2). (Claim 1) $\square$

Claim 2. Suppose that L and K are ladders. Then either $L \subseteq K$ or $K \subseteq L$.

Proof. Let L and K be ladders. Assume, for a contradiction, that $L \nsubseteq K$ and $K \nsubseteq L$. Hence, there is an $x \in L$ so that $x \notin K$ and a $y \in K$ so that $y \notin L$. Let

$$C = \{z \in L : z \prec x\} \cap \{z \in K : z \prec y\}. \tag{7.5}$$

Clearly, $C \subseteq L$ and $C \subseteq K$. Because $C \subseteq L$ and $x \in L$ is a proper upper bound for C, it follows that $f(C_p) \in L$ and $f(C_p) \preccurlyeq x$. Similarly, as $C \subseteq K$ and $y \in K$ is a proper upper bound for C, we see that $f(C_p) \in K$ and $f(C_p) \preccurlyeq y$. So, in summary, we have

(1) $x \notin K$, $f(C_p) \preccurlyeq x$, $f(C_p) \in K$; and (2) $y \notin L$, $f(C_p) \preccurlyeq y$, $f(C_p) \in L$.

So $f(C_p) \prec x$ by (1), and $f(C_p) \prec y$ by (2). Since $f(C_p) \in L$ and $f(C_p) \in K$, we conclude from (7.5) that $f(C_p) \in C$, contradicting (7.2). (Claim 2) $\square$

Now let $\mathcal{B} = \{L \in \mathcal{P}(A) : L \text{ is a ladder}\}$. By the power set and subset axioms, we see that $\mathcal{B}$ is a set. Therefore, $\mathcal{B}$ is the set of all ladders.

Claim 3. $\bigcup \mathcal{B}$ is a ladder.

Proof. To see that $\bigcup \mathcal{B}$ is a chain, let x and y be in $\bigcup \mathcal{B}$. Hence, $x \in L$ for some $L \in \mathcal{B}$, and $y \in K$ for some $K \in \mathcal{B}$. By Claim 2, either $L \subseteq K$ or $K \subseteq L$. So, we can assume that x and y are in L. As L is a chain, either $x \preccurlyeq y$ or $y \preccurlyeq x$. Thus, $\bigcup \mathcal{B}$ is a chain. Suppose $C \subseteq \bigcup \mathcal{B}$ has a proper upper bound in $\bigcup \mathcal{B}$, say, b. We shall show that $f(C_p) \in \bigcup \mathcal{B}$ and $f(C_p) \not\preccurlyeq b$. Since $b \in \bigcup \mathcal{B}$, we have that $b \in L$ for some $L \in \mathcal{B}$. We will now show that $C \subseteq L$. Let $c \in C$. So, $c \prec b$. Because $C \subseteq \bigcup \mathcal{B}$, it follows that $c \in K$ for some $K \in \mathcal{B}$. By Claim 2, either $L \subseteq K$ or $K \subseteq L$. If $K \subseteq L$, then $c \in L$. If $L \subseteq K$, then (as $c \prec b$ and $b \in L$) Claim 1 implies that $c \in L$. Hence, $C \subseteq L$, and $b \in L$ is a proper upper bound for C. Since L is a ladder, $f(C_p) \in L$ and $f(C_p) \not\preccurlyeq b$. Consequently, $f(C_p) \in \bigcup \mathcal{B}$ and $f(C_p) \not\preccurlyeq b$. Therefore, $\bigcup \mathcal{B}$ is a ladder.   (Claim 3) □

Let $\pounds = \bigcup \mathcal{B}$. As $\mathcal{B}$ is the set of all ladders, (♦) $L \subseteq \pounds$ whenever L is a ladder. Claim 3 states that $\pounds$ is a ladder. Since $\pounds$ is a chain, we have by assumption that $\pounds_p$ is nonempty. Thus, $f(\pounds_p) \in \pounds_p$. Let $\pounds^* = \pounds \cup \{f(\pounds_p)\}$.

Claim 4. $\pounds^*$ is a ladder.

Proof. Since $\pounds$ is a chain and $x \prec f(\pounds_p)$ for each $x \in \pounds$, it follows that $\pounds^*$ is a chain. Let $C \subseteq \pounds^*$ have a proper upper bound b in $\pounds^*$. Thus, $b \preccurlyeq f(\pounds_p)$ and $C \subseteq \pounds$. If $b \in \pounds$, then $f(C_p) \in \pounds \subseteq \pounds^*$ and $f(C_p) \not\preccurlyeq b$, because $\pounds$ is a ladder. If $b \notin \pounds$, then $b = f(\pounds_p)$, and there are two cases to consider.

CASE 1: C is not cofinal in $\pounds$. Thus, C has a proper upper bound u in $\pounds$. Hence, $f(C_p) \in \pounds$ and $f(C_p) \not\preccurlyeq u$ as $\pounds$ is a ladder. Therefore, $f(C_p) \in \pounds^*$ and $f(C_p) \not\preccurlyeq b$ because $u \prec f(\pounds_p) = b$.

CASE 2: C is cofinal in $\pounds$. Lemma 7.1.4 implies that $f(C_p) = f(\pounds_p) = b$. So we clearly have that $f(C_p) \in \pounds^*$ and $f(C_p) \not\preccurlyeq b$.   (Claim 4) □

Since $\pounds^*$ is a ladder, (♦) implies that $\pounds^* \subseteq \pounds$. As $f(\pounds_p) \in \pounds^*$, we conclude that $f(\pounds_p) \in \pounds$, and this contradicts (7.2).   (Zorn's Lemma) □

Hence, the axiom of choice implies Zorn's Lemma 7.1.1. Moreover, one can also prove that Zorn's Lemma implies the axiom of choice (see Exercise 10). A different proof of Zorn's Lemma is summarized in Exercise 14 on page 193; however, this alternative proof relies on ordinals and the replacement axiom.

Corollary 7.1.6 (AC). *Suppose that $(A, \preccurlyeq)$ is a partially ordered set in which every chain has an upper bound. For each $a \in A$, there exists a maximal element $m \in A$ such that $a \preccurlyeq m$.*

Proof. Let $(A, \preccurlyeq)$ be a poset in which every chain has an upper bound, and let $a \in A$. Define $B = \{x \in A : a \preccurlyeq x\}$ and $\preccurlyeq_B = \preccurlyeq \cap B \times B$. Thus, $(B, \preccurlyeq_B)$ is a partially ordered set (see Exercise 14 on page 76) in which every chain has an upper bound. By Zorn's Lemma, B has a maximal element m. So $a \preccurlyeq m$ and m is a maximal element in A. $\qquad\square$

7.1.1 Two Applications of Zorn's Lemma

Zorn's Lemma is frequently applied to posets of the form $(\mathcal{F}, \subseteq)$ where $\mathcal{F}$ is a set (of sets) and $\subseteq$ is the subset relation (see Problem 1 on page 70). In such applications, one proceeds as described in Exercise 5. We will now present two theorems whose proofs apply Zorn's Lemma as outlined by Exercise 5.

The first theorem, often called the *comparability theorem*, shows that one can compare the cardinalities of any two sets (see Definition 5.4.14). First, we shall state and prove a relevant lemma.

Lemma 7.1.7. *Let $P = \{h \subseteq A \times B : h \text{ is a one-to-one function}\}$. If $f \in P$ is a maximal element in the poset $(P, \subseteq)$, then either $\mathrm{dom}(f) = A$ or $\mathrm{ran}(f) = B$.*

Proof. Let $(P, \subseteq)$ be the poset defined in the statement of the lemma, and let f be a maximal element in P. Clearly, $\mathrm{dom}(f) \subseteq A$ and $\mathrm{ran}(f) \subseteq B$. We shall prove the either $\mathrm{dom}(f) = A$ or $\mathrm{ran}(f) = B$. Suppose, for a contradiction, that $\mathrm{dom}(f) \neq A$ and $\mathrm{ran}(f) \neq B$. Let $a \in A \setminus \mathrm{dom}(f)$ and $b \in B \setminus \mathrm{ran} f$. Now let $g = f \cup \{\langle a, b \rangle\}$. Clearly, $g \subseteq A \times B$. Because $a \notin \mathrm{dom}(f)$ and $b \notin \mathrm{ran}(f)$, it follows that g is a one-to-one function. Thus, $g \in P$. Since $f \subseteq g$ and $f \neq g$, this contradicts the fact that f is maximal. Therefore, either $\mathrm{dom}(f) = A$ or $\mathrm{ran}(f) = B$. $\qquad\square$

We can now state and prove the comparability theorem, which is actually equivalent to the axiom of choice (see Exercise 23, page 218).

Theorem 7.1.8 (AC). *For any two sets A and B, either $|A| \leq_c |B|$ or $|B| \leq_c |A|$.*

Proof. Let A and B be sets, and let $P = \{h \subseteq A \times B : h \text{ is a one-to-one function}\}$. We will show that every chain in the poset $(P, \subseteq)$ has an upper bound. To do this, suppose that $\mathcal{C} \subseteq P$ is a chain. Thus, for all $f \in \mathcal{C}$ and $g \in \mathcal{C}$, we have that either $f \subseteq g$ or $g \subseteq f$. Clearly, $\bigcup \mathcal{C} \subseteq A \times B$. Exercise 11 on page 68 shows that $\bigcup \mathcal{C}$ is a one-to-one function. So $\bigcup \mathcal{C} \in P$ and $f \subseteq \bigcup \mathcal{C}$ for each $f \in C$. We conclude that $\bigcup C$ is an upper bound for C. Hence, by Zorn's Lemma, the poset $(P, \subseteq)$ has a maximal element f. Lemma 7.1.7 asserts that either $\mathrm{dom}(f) = A$ or $\mathrm{ran}(f) = B$. If $\mathrm{dom}(f) = A$, then $|A| \leq_c |B|$ because $f \colon A \to B$ is one-to-one. If $\mathrm{ran}(f) = B$, then $|B| \leq_c |A|$ as $f^{-1} \colon B \to A$ is one-to-one by Corollary 3.3.15 and Theorem 3.2.7. $\qquad\square$

Our second theorem, identified as the *order-extension principle*, shows that any partial order on a set can be extended to a total order on the same set. A proof of this theorem was first published by Edward Marczewski in 1930. Our proof will apply Zorn's Lemma to a partial order on a set of partial orders. We first state and prove a useful lemma, which shows that if a partial order is not a total order, then it can be extended to a larger partial order.

Lemma 7.1.9. *If $\preccurlyeq$ is a partial order on A that is not a total order on A, then there is a partial order $\preccurlyeq'$ on A such that $\preccurlyeq \subseteq \preccurlyeq'$ and $\preccurlyeq \neq \preccurlyeq'$.*

Proof. Let $\preccurlyeq$ is a partial order on A that is not a total order on A. Since $\preccurlyeq$ is not a total order, there exist $a \in A$ and $b \in A$ such that ($\blacktriangle$) $a \not\preccurlyeq b$ and $b \not\preccurlyeq a$. Let $R \subseteq A \times A$ be defined by

$$R = \{\langle x, y \rangle \in A \times A : x \preccurlyeq a \text{ and } b \preccurlyeq y\}.$$

Because $\preccurlyeq$ is reflexive on A, we see that $a \preccurlyeq a$ and $b \preccurlyeq b$. Thus, $\langle a, b \rangle \in R$. Let $\preccurlyeq' = \preccurlyeq \cup R$. So $\preccurlyeq \subseteq \preccurlyeq'$. Also, $\preccurlyeq \neq \preccurlyeq'$ as $\langle a, b \rangle \in R$ and $\langle a, b \rangle \notin \preccurlyeq$ by ($\blacktriangle$). We will now show that $\preccurlyeq'$ is a partial order on A. Since $\preccurlyeq' = \preccurlyeq \cup R$, we have that

$$x \preccurlyeq' y \text{ if and only if } x \preccurlyeq y \text{ or } \langle x, y \rangle \in R. \tag{7.6}$$

Let $x \in A$. Since $x \preccurlyeq x$, we see from (7.6) that $x \preccurlyeq' x$. So $\preccurlyeq'$ is reflexive on A. To show that $\preccurlyeq'$ is antisymmetric, suppose that $x \preccurlyeq' y$ and $y \preccurlyeq' x$. We shall show that $x = y$. Since $x \preccurlyeq' y$ and $y \preccurlyeq' x$, the equivalence in (7.6) yields the following four cases (three of which, as we will see, cannot hold):

1. $x \preccurlyeq y$ and $y \preccurlyeq x$. Since $\preccurlyeq$ is antisymmetric, we conclude that $x = y$.
2. $x \preccurlyeq y$ and $\langle y, x \rangle \in R$. So $(\alpha)\, x \preccurlyeq y$, $(\beta)\, y \preccurlyeq a$, $(\gamma)\, b \preccurlyeq x$. Conditions (α) and (β) imply that $x \preccurlyeq a$. Hence, (γ) implies that $b \preccurlyeq a$, which contradicts ($\blacktriangle$).
3. $\langle x, y \rangle \in R$ and $y \preccurlyeq x$. As in the previous case 2, this is impossible.
4. $\langle x, y \rangle \in R$ and $\langle y, x \rangle \in R$. Thus, $(\alpha)\, x \preccurlyeq a$, $(\beta)\, b \preccurlyeq y$, $(\gamma)\, y \preccurlyeq a$, $(\delta)\, b \preccurlyeq x$. Clearly, (β) and (γ) imply that $b \preccurlyeq a$, contradicting ($\blacktriangle$).

Because case 1 is the only possibility, we conclude that $\preccurlyeq'$ is antisymmetric. Finally, we now show that $\preccurlyeq'$ is transitive. Assume $x \preccurlyeq' y$ and $y \preccurlyeq' z$. We must show that $x \preccurlyeq' z$. As $x \preccurlyeq' y$ and $y \preccurlyeq' z$, the equivalence (7.6) again yields four cases to consider:

1. $x \preccurlyeq y$ and $y \preccurlyeq z$. Since $\preccurlyeq$ is transitive, we conclude that $x \preccurlyeq z$.
2. $x \preccurlyeq y$ and $\langle y, z \rangle \in R$. Thus, $(\alpha)\ x \preccurlyeq y$, $(\beta)\ y \preccurlyeq a$, $(\gamma)\ b \preccurlyeq z$. As (α) and (β) imply that $x \preccurlyeq a$, we see from (γ) that $x \preccurlyeq a$ and $b \preccurlyeq z$. Hence, $\langle x, z \rangle \in R$.
3. $\langle x, y \rangle \in R$ and $y \preccurlyeq z$. So $(\alpha)\ x \preccurlyeq a$, $(\beta)\ b \preccurlyeq y$, $(\gamma)\ y \preccurlyeq z$. Since (β) and (γ) imply that $b \preccurlyeq z$, we conclude from (α) that $x \preccurlyeq a$ and $b \preccurlyeq z$. Thus, $\langle x, z \rangle \in R$.
4. $\langle x, y \rangle \in R$ and $\langle y, z \rangle \in R$. Hence, $(\alpha)\ x \preccurlyeq a$, $(\beta)\ b \preccurlyeq y$, $(\gamma)\ y \preccurlyeq a$, $(\delta)\ b \preccurlyeq z$. Clearly, (β) and (γ) imply that $b \preccurlyeq a$, contradicting $(\blacktriangle)$.

In each of the valid cases, we have either $x \preccurlyeq z$ or $\langle x, z \rangle \in R$. Thus, by (7.6), $x \preccurlyeq' z$. Hence, $\preccurlyeq'$ is a partial order on A such that $\preccurlyeq \subseteq \preccurlyeq'$ and $\preccurlyeq \neq \preccurlyeq'$. □

We can now show that Zorn's Lemma implies the order-extension principle.

Theorem 7.1.10 (AC). *Let $\preccurlyeq'$ be a partial order on A. Then there exists a total order $\preccurlyeq$ on A such that $\preccurlyeq' \subseteq \preccurlyeq$.*

Proof. Let $\preccurlyeq'$ be a partial order on A, and let

$$P = \{ \unlhd \subseteq A \times A : \unlhd \text{ is a partial order on } A \}.$$

Hence, $\preccurlyeq' \in P$. We now show that every chain in the poset $(P, \subseteq)$ has an upper bound. Let $C \subseteq P$ be a chain. Thus, for all $\unlhd \in C$ and $\preccurlyeq \in C$, either $\unlhd \subseteq \preccurlyeq$ or $\preccurlyeq \subseteq \unlhd$. Clearly, $\bigcup C \subseteq A \times A$. Exercises 17 and 18 on page 76 imply that $\bigcup C$ is a partial order on A. So $\bigcup C \in P$ and $\preccurlyeq \subseteq \bigcup C$ for all $\preccurlyeq \in C$. Hence, $\bigcup C$ is an upper bound for C. Hence, by Corollary 7.1.6, there is a maximal element $\preccurlyeq \in P$ such that $\preccurlyeq' \subseteq \preccurlyeq$. Since $\preccurlyeq$ is maximal, Lemma 7.1.9 implies that $\preccurlyeq$ is a total order on A. □

Exercises 7.1

1. Let $\langle A_i : i \in I \rangle$ be an indexed function so that $A_i \neq \varnothing$, for all $i \in I$. Without appealing to the axiom of choice, prove that if $\bigcup_{i \in I} A_i$ is countably infinite, then there is a function $f : I \to \bigcup_{i \in I} A_i$ such that $f(i) \in A_i$, for all $i \in I$.

2. Let $R \subseteq A \times B$, and let $R(x, y)$ denote $\langle x, y \rangle \in R$. Now suppose that for all $x \in A$, there is a $y \in B$ such that $R(x, y)$. Using the axiom of choice, show that there is a function $f \colon A \to B$ such that $R(x, f(x))$ for all $x \in A$.

3. Let $R \subseteq A \times A$, and let $R(x, y)$ denote $\langle x, y \rangle \in R$. Suppose that $A \neq \varnothing$ and for all $x \in A$, there is a $y \in A$ such that $R(x, y)$. By Exercise **2** and the axiom of choice, there is an $f \colon A \to A$ such that $R(x, f(x))$ for all $x \in A$. Using Theorem 4.2.1, show that there is a function $h \colon \omega \to A$ such that $R(h(n), h(n^+))$ for all $n \in \omega$.

4. Prove Lemma 7.1.4.

*5. Let $(\mathcal{F}, \subseteq)$ be a partially ordered set where $\mathcal{F}$ is a set (of sets). Suppose that for every chain $C \subseteq \mathcal{F}$ we have that $\bigcup C$ is in $\mathcal{F}$.
   (a) Let $C \subseteq \mathcal{F}$ be a chain. Prove that $\bigcup C$ is an upper bound for C.
   (b) Using Zorn's Lemma, show that there exists an $M \in \mathcal{F}$ that is maximal, that is, M is not the proper subset of any $A \in \mathcal{F}$.

6. Let $f \colon A \to B$ and $\mathcal{F} = \{X \subseteq A : (f \restriction X) \colon X \to B \text{ is one-to-one}\}$. Consider the partially ordered set $(\mathcal{F}, \subseteq)$.
   (a) Using Zorn's Lemma, show that there is an $M \in \mathcal{F}$ that is maximal.
   (b) Prove that if f is onto B, then $(f \restriction M) \colon X \to B$ is a bijection.

7. Let $(A, \preccurlyeq)$ be a poset. Let $\mathcal{F} = \{D \in \mathcal{P}(A) : D \text{ is a chain in } A\}$. Consider the partially ordered set $(\mathcal{F}, \subseteq)$.
   (a) Let C be a chain in $\mathcal{F}$. Show that $\bigcup C$ is a chain in A.
   (b) Conclude, via Zorn's Lemma, that there exists a chain D in A that is maximal, that is, D is not the proper subset of any other chain in A.

8. Let P be the set of nonzero natural numbers. Then $(P, \mid)$ is a poset where $\mid$ is the divisibility relation on P (see Problem 2 on page 71). Define a partial order $\preccurlyeq$ on P such that $\mid \, \subseteq \, \preccurlyeq$ and $2 \preccurlyeq 3$.

9. Using Theorem 7.1.10, show there is a total order $\preccurlyeq$ on $\mathcal{P}(\omega)$ such that for all x and y in $\mathcal{P}(\omega)$, if $x \subseteq y$, then $x \preccurlyeq y$. Show that $\preccurlyeq \, \neq \, \subseteq$ on $\mathcal{P}(\omega)$.

*10. Let $\langle A_i : i \in I \rangle$ be an indexed function with nonempty terms, and let

$$\mathcal{F} = \{g : g \text{ is a function}, \operatorname{dom}(g) \subseteq I, \text{ and } (\forall i \in \operatorname{dom}(g))(g(i) \in A_i)\}.$$

Thus, $(\mathcal{F}, \subseteq)$ is a partially ordered set. Assuming Zorn's Lemma, prove the axiom of choice as follows:
   (a) Let $C \subseteq \mathcal{F}$ be a chain. Prove that $\bigcup C$ is in $\mathcal{F}$ and is an upper bound for C.
   (b) Conclude, via Zorn's Lemma, that $\mathcal{F}$ has a maximal element h.

(c) Show that $\mathrm{dom}(h) = I$ and $(\forall i \in I)(h(i) \in A_i)$. Thus, $\langle h(i) : i \in I \rangle$ is a choice function.

Exercise Notes: The conclusion of Exercise **3** is called the *axiom of dependent choices* (DC). For Exercise **6**(a), apply Exercise **5**. Exercise **7** shows that Zorn's Lemma implies the *Hausdorff maximal principle*, which proclaims that every partially ordered set contains a maximal chain. This principle is also equivalent to the axiom of choice. For Exercise **10**(a), see Exercise **11**(a) on page 68.

7.2 Filters and Ultrafilters

We now introduce the concept of a filter, which has a variety of applications in set theory and in topology. Filters were introduced by Henri Cartan in 1937 and subsequently used by Bourbaki in the book *Topologie générale*.

The motivation for our next definition is to give meaning to the intuitive notion of a "large" subset of X.

Definition 7.2.1. Let X be nonempty and let $F \subseteq \mathcal{P}(X)$. Then F is a **filter** on X if it satisfies the following three conditions:

(1) $\varnothing \notin F$ and $X \in F$.
(2) If $Y \in F$ and $Z \in F$, then $Y \cap Z \in F$.
(3) If $Y \in F$ and $Y \subseteq Z \subseteq X$, then $Z \in F$.

So if $Y \in F$ and F is a filter, then Y behaves like a "large" set.

If F is a filter on X and $A \in F$, then $X \setminus A \notin F$. To prove this, suppose that $X \setminus A \in F$. Thus, by Definition 7.2.1(2), $\varnothing = A \cap (X \setminus A) \in F$, which contradicts Definition 7.2.1(1).

Clearly, the singleton set $\{X\}$ is a filter on any nonempty set X. For another example, let S be a nonempty subset of X and let $F = \{A \subseteq X : S \subseteq A\}$. One can verify (see Exercise 1) that F satisfies the conditions of Definition 7.2.1. Hence, F is a filter on X, and it is called the *filter on X generated by S*. When S is a singleton, then F shall be called a **principal filter**. For example, if $a \in X$, then the principal filter generated by $\{a\}$ can be expressed by

$$F = \{A \subseteq X : a \in A\}. \tag{7.7}$$

Definition 7.2.2. Let F be a filter on X. We say that F is a **maximal filter** if for all filters G on X, if $F \subseteq G$, then $F = G$.

If $a \in X$, then the principal filter F generated by $\{a\}$ is a maximal filter. To see why it is maximal, let G be a filter on X so that $F \subseteq G$. Let $B \in G$. To prove that $B \in F$, assume $B \notin F$. Thus, $a \notin B$ and $a \in X \setminus B$. So $X \setminus B \in F$, and hence, $X \setminus B \in G$. Since $B \in G$, we conclude from Definition 7.2.1(1) that $\varnothing = B \cap (X \setminus B) \in G$, a contradiction. So $B \in F$ and $F = G$. On the other hand, for $a \in X$ and $b \in X$ where $a \neq b$, one can verify that the filter generated by $\{a, b\}$ is not maximal.

We now present an example of a *nonprincipal* filter. Let X be an infinite set. Let F be defined by

$$F = \{S \subseteq X : X \setminus S \text{ is finite}\}.$$

One can show that F is a filter on X. For example, to verify Definition 7.2.1(2), let S and T be in F. Since $X \setminus (S \cap T) = (X \setminus S) \cup (X \setminus T)$, we conclude that $X \setminus (S \cap T)$ is finite, as $X \setminus S$ and $X \setminus T$ are finite. Thus, $S \cap T$ is in F. A set S is said to be a *cofinite subset* of X, if $S \subseteq X$ and $X \setminus S$ is finite. So the filter F is called the filter of all *cofinite* subsets of X. To show that F is not a principal filter, let $A \in F$ and $a \in A$. Consider the set $B = X \setminus \{a\}$. Then $B \in F$ (because $X \setminus B = \{a\}$ is finite) and $a \notin B$. Therefore, F is not generated by any singleton.

Definition 7.2.3. Let F be a filter on X. Then F is said to be an **ultrafilter** on X if for all $Y \subseteq X$, either $Y \in F$ or $X \setminus Y \in F$.

Let $a \in X$. The principle filter F generated by $\{a\}$ (see (7.7)) is an ultrafilter. So there are principal ultrafilters. Do nonprincipal ultrafilters exist? Let X be an infinite set and let F be the filter of all cofinite subsets of X. Our goal in this section is to prove a result that implies the existence of an ultrafilter $\mathcal{U}$ such that $F \subseteq \mathcal{U}$. Exercise 9 then implies that $\mathcal{U}$ is nonprincipal. Thus, nonprincipal ultrafilters do, in fact, exist.

The following lemma identifies two important properties that an ultrafilter possesses.

Lemma 7.2.4. *Let $\mathcal{U}$ be an ultrafilter on X.*

(1) *For all $A \subseteq X$ and $B \subseteq X$, if $A \cup B \in \mathcal{U}$, then either $A \in \mathcal{U}$ or $B \in \mathcal{U}$.*
(2) *Let F be a filter on X. If $\mathcal{U} \subseteq F$, then $\mathcal{U} = F$.*

Proof. The proofs of (1) and (2) are left for Exercises 7 and 8, respectively.

□

Lemma 7.2.4(2) shows that an ultrafilter is maximal. Furthermore, if a filter is maximal, then it is an ultrafilter (see Exercise 15).

Definition 7.2.5. Let X and $G \subseteq \mathcal{P}(X)$ be nonempty. Then G has the **finite intersection property** if whenever $S \subseteq G$ is nonempty and finite, then $\bigcap S \neq \emptyset$.

Lemma 7.2.6. *Let X and $G \subseteq \mathcal{P}(X)$ be nonempty. Suppose that G has the finite intersection property. Then there is a filter F on X such that $G \subseteq F$.*

Proof. Let $G \subseteq \mathcal{P}(X)$. Assume G has the finite intersection property. Define

$$F = \left\{ A \in \mathcal{P}(X) : \bigcap S \subseteq A \text{ for some nonempty finite } S \subseteq G \right\}.$$

Clearly, $G \subseteq F$, $\emptyset \notin F$, and $X \in F$. We shall prove that F is a filter; that is, we will establish items (2) and (3) of Definition 7.2.1. Let $Y \in F$ and $Z \in F$. So $\bigcap S \subseteq Y$ and $\bigcap T \subseteq Z$ where $S \subseteq G$ and $T \subseteq G$ are nonempty and finite. Thus, $\bigcap S \cap \bigcap T \subseteq Y \cap Z$. Exercise 12 on page 40 implies that $\bigcap S \cap \bigcap T = \bigcap (S \cup T)$. Hence, $\bigcap (S \cup T) \subseteq Y \cap Z$. Since $S \cup T$ is a finite subset of G (see Exercise 15 on page 116), we have that $Y \cap Z \in F$. Thus, (2) of Definition 7.2.1 holds. To prove (3), let $Y \in F$ and $Y \subseteq Z \subseteq X$. Since $\bigcap S \subseteq Y$ for some nonempty finite $S \subseteq G$, it follows that $Z \in F$. Therefore, F is a filter. □

Definition 7.2.7. Let X and $G \subseteq \mathcal{P}(X)$ be nonempty. Then G is **closed under finite intersections** if whenever $S \subseteq G$ is nonempty and finite, then $\bigcap S \in G$.

One can prove by induction that a filter is closed under finite intersections, and therefore, a filter has the finite intersection property (see Exercise 2).

Lemma 7.2.8. *Let F be a filter on X. Suppose that $A \subseteq X$ is such that $A \notin F$ and $X \setminus A \notin F$. Then the set $F \cup \{A\}$ has the finite intersection property.*

Proof. Let F be a filter on X, and suppose that $A \subseteq X$ is such that $A \notin F$ and (▲) $X \setminus A \notin F$. Because F is a filter, (▲) implies that $A \neq \emptyset$.

Claim. For every $Y \in F$, we have that $Y \cap A \neq \emptyset$.

Proof. Assume, for a contradiction that $Y \cap A = \emptyset$ for some $Y \in F$. Since $Y \cap A = \emptyset$ and $Y \subseteq X$, it follows that $Y \subseteq X \setminus A$. Thus, because F is a filter, we have that $X \setminus A \in F$, which contradicts (▲).   (Claim) □

Let $S \subseteq F \cup \{A\}$ be finite and nonempty. So either $A \notin S$ or $A \in S$. If $A \notin S$, then $S \subseteq F$. As F has the finite intersection property, it follows that $\bigcap S \neq \emptyset$. If $A \in S$, then $S = T \cup \{A\}$ for a finite $T \subseteq F$. If $T = \emptyset$, then $\bigcap S = A$, and thus, $\bigcap S \neq \emptyset$. If $T \neq \emptyset$, then one can show that $\bigcap S = A \cap (\bigcap T)$. Since F is closed under finite intersections, $\bigcap T \in F$. Hence, $A \cap (\bigcap T) \neq \emptyset$ by the above claim. Thus, $\bigcap S \neq \emptyset$ and $F \cup \{A\}$ has the finite intersection property.   (Lemma) □

Lemmas 7.2.6 and 7.2.8, together with Zorn's Lemma, imply the following theorem, which has many applications in mathematics.

Ultrafilter Theorem 7.2.9 (AC). *If F is a filter on X, then there is an ultrafilter $\mathcal{U}$ on X such that $F \subseteq \mathcal{U}$.*

Proof. See Exercise 14.   □

7.2.1 Ideals

Filters allowed us to investigate the concept of a "large" subset of a set X. Ideals allow us to pursue the concept of a "small" subset of X. As we will see, given a filter on X, one can construct an ideal on X, and conversely.

Definition 7.2.10. Let X be nonempty and let $I \subseteq \mathcal{P}(X)$. We say that I is an **ideal** on X if it satisfies the following three conditions:

(1) $\emptyset \in I$ and $X \notin I$.
(2) If $Y \in I$ and $Z \in I$, then $Y \cup Z \in I$.
(3) If $Y \in I$ and $Z \subseteq Y$, then $Z \in I$.

The singleton $\{\varnothing\}$ is an ideal on any nonempty set X. Let $S \subseteq X$ be a proper subset and let

$$I = \{A \subseteq X : A \subseteq S\}.$$

The set I is an ideal, since it clearly satisfies the conditions of Definition 7.2.10. This ideal I is called the *ideal on X generated by S*.

The following lemma shows that there is an intimate connection between filters and ideals.

Lemma 7.2.11. *Let X be nonempty. Then the following hold:*

(1) *If F is a filter on X, the $X_F = \{B \in \mathcal{P}(X) : X \setminus B \in F\}$ is an ideal on X.*
(2) *If I is an ideal on X, the $X_I = \{A \in \mathcal{P}(X) : X \setminus A \in I\}$ is a filter on X.*

Proof. See Exercises 19 and 20. $\qquad\qquad\qquad\qquad\qquad\qquad\qquad\qquad$ □

Let F and I be as in Lemma 7.2.11. Then X_F is called the *dual ideal*, and X_I is called the *dual filter*. It is easy to show that $X_{X_F} = F$ and $X_{X_I} = I$.

Definition 7.2.12. Let I be an ideal on X. Then I is said to be a **prime ideal** on X if for all $Y \subseteq X$, either $Y \in I$ or $X \setminus Y \in I$.

Theorem 7.2.13. *Let X be a nonempty set. Then $\mathcal{U}$ is an ultrafilter on X if and only if $X_{\mathcal{U}}$ is a prime ideal on X.*

Proof. See Exercise 21. $\qquad\qquad\qquad\qquad\qquad\qquad\qquad\qquad\qquad\qquad\qquad$ □

Finally, let I be an ideal on X. Then I is said to be a **maximal ideal** if for all ideals J, if $I \subseteq J$, then $I = J$. One can prove that a prime ideal is maximal (see Exercise 22).

Exercises 7.2

*1. Let $S \subseteq X$ where S is nonempty. Let $F = \{A \subseteq X : S \subseteq A\}$. Prove that F is a filter on X.

*2. Let F be a filter on X. Prove that F is closed under finite intersections. Conclude that F has the finite intersection property.

3. Let $a \in X$ and $b \in X$ where $a \neq b$. Prove that the filter generated by $\{a, b\}$ is not maximal.

4. Let F be a filter on a nonempty set X. Let $A \in F$. Prove that $\mathcal{P}(A) \cap F$ is a filter on A.

5. Suppose that X is an infinite set, and let A be a cofinite subset of X. Prove the following two items:
   (a) If $A \subseteq B \subseteq X$, then B is a cofinite subset of X.
   (b) If C is a cofinite subset of A, then C is a cofinite subset of X.

6. Let F be the filter of cofinite subsets of ω. Show that F is not an ultrafilter.

*7. Prove Lemma 7.2.4(2).

*8. Prove Lemma 7.2.4(3).

9. Let X be an infinite set and let F be the filter of all cofinite subsets of X. Suppose G is a filter on X such that $F \subseteq G$.
   (a) Prove that every $A \in G$ is infinite.
   (b) Prove that G is nonprincipal.
   (c) (AC) Conclude that there is nonprincipal ultrafilter on X.

10. Let $\mathcal{F}$ be a nonempty set of filters on X. Prove that $\bigcap \mathcal{F}$ is a filter on X.

11. Let F be filter on X and let $a \in X$.
   (a) Prove that if $\{a\} \in F$, then $a \in A$ for all $A \in F$.
   (b) Suppose that F is an ultrafilter on X and $\{a\} \in F$. Conclude that F is the principal filter on X generated by $\{a\}$.

12. Let $\mathcal{U}$ be an ultrafilter on X.
   (a) Suppose there is a finite set in $\mathcal{U}$. Prove that $\{a\} \in \mathcal{U}$ for some $a \in X$.
   (b) Conclude that if $\mathcal{U}$ is nonprincipal, then every $B \in \mathcal{U}$ is infinite.

13. Show that $\mathcal{P}(X) \setminus \mathcal{U} = \{B \in \mathcal{P}(X) : X \setminus B \in \mathcal{U}\}$, whenever $\mathcal{U}$ is an ultrafilter on X.

*14. Let F be a filter on X. Let $\mathcal{S} = \{E \subseteq \mathcal{P}(X) : E \text{ is a filter on } X\}$. Note that $F \in \mathcal{S}$. Consider the partially ordered set $(\mathcal{S}, \subseteq)$.
   (a) Let $C \subseteq \mathcal{S}$ be a chain. Prove that $\bigcup C \in \mathcal{S}$ and that $\bigcup C$ is an upper bound for C.
   (b) Show, via Corollary 7.1.6, that $\mathcal{S}$ has a maximal element $\mathcal{U}$ such that $F \subseteq \mathcal{U}$.
   (c) Using Lemma 7.2.6 and Lemma 7.2.8, prove that $\mathcal{U}$ is an ultrafilter.

*15. Let F be a filter on X. Prove that F is an ultrafilter if and only if F is maximal.

16. Let F be a filter on A. Let $^AB = \{f : f$ is a function from A to $B\}$. Define the relation $\sim$ on ^AB by

$$f \sim g \text{ if and only if } \{x \in A : f(x) = g(x)\} \in F$$

for all f and g in ^AB. Prove that $\sim$ is an equivalence relation on ^AB.

17. Let $\mathcal{U}$ be an ultrafilter on A. Let $^AB = \{f : f$ is a function from A to $B\}$. Suppose that $\preccurlyeq$ is a total preorder on A. Define the relation $\preccurlyeq^*$ on ^AB by

$$f \preccurlyeq^* g \text{ if and only if } \{x \in A : f(x) \preccurlyeq g(x)\} \in \mathcal{U}$$

for all f and g in ^AB. Prove that $\preccurlyeq^*$ is total preorder on ^AB.

18. Let $\mathcal{U}$ be an ultrafilter on A and let $f : A \to B$. Define the set

$$\mathcal{U}^* = \{Y \in \mathcal{P}(B) : f^{-1}[Y] \in \mathcal{U}\}.$$

Prove that $\mathcal{U}^*$ is an ultrafilter on B.

*19. Prove Theorem 7.2.11(1).

*20. Prove Theorem 7.2.11(2).

*21. Prove Theorem 7.2.13.

*22. Suppose I is a prime ideal on X. Prove that I is a maximal ideal.

Exercise Notes: For Exercise **6**, find a set $A \subseteq \omega$ and a set $B \subseteq \omega$ that violate Lemma 7.2.4(1). For Exercise **12**(a), let $A \in \mathcal{U}$ be a finite set such that for all $B \in \mathcal{U}$, if B is finite then $|A| \leq |B|$. Using Lemma 7.2.4(1), deduce that A is a singleton. For Exercise **12**(b), use Exercise **11**. For Exercise **18**, see Exercise 9 on page 68.

7.3 Well-Ordering Theorem

The following theorem is due to Ernst Zermelo and states that every set can be well-ordered. Therefore, the powerful principle of transfinite recursion can be applied to any set.

Well-Ordering Theorem 7.3.1 (AC). *Each set has a well-ordering.*

Proof. Assume the axiom of choice. Let A be a set. We shall use Zorn's Lemma to prove that there is a well-ordering on A. Let W be the following set:

$$W = \{\preceq : \text{the relation } \preceq \text{ is a well-ordering on a subset of } A\}$$

(Theorem 2.1.3 implies that W is a set). Define the *continuation* relation $\trianglelefteq$ on W by

$$\preceq \trianglelefteq \preceq' \text{ iff (i) } \preceq \subseteq \preceq' \text{ and (ii) if } x \preceq' y \text{ and } y \prec y, \text{ then } x \prec y \qquad (7.8)$$

where x and y in (7.8)(ii) are arbitrary. The relationship $\preceq \trianglelefteq \preceq'$ is illustrated by the following diagram (7.9) where the order $\preceq'$ *continues* the order $\preceq$; that is, the order $\preceq'$ can only add new elements that are greater than all of the elements ordered by $\preceq$ (see Exercise 3):

$$\overbrace{a \prec b \prec c \prec \cdots \prec h}^{\preceq} \prec \cdots \underbrace{\prec' \ell \prec' x \prec' \cdots}_{} . \qquad (7.9)$$

$$\underbrace{}_{\preceq'}$$

The structure $(W, \trianglelefteq)$ is a partially ordered set (see Exercise 4). Let $\mathcal{C} \subseteq W$ be a chain. Thus, for all $\preceq \in \mathcal{C}$ and $\preceq' \in \mathcal{C}$, either $\preceq \trianglelefteq \preceq'$ or $\preceq' \trianglelefteq \preceq$. Now let $\preceq^{\mathcal{C}}$ be the relation $\bigcup \mathcal{C}$ with field $\bigcup \{\text{fld}(\preceq) : \preceq \in \mathcal{C}\}$, which is a subset of A.

Claim 1. $\preceq^{\mathcal{C}}$ is a total order on its field.

Proof. Exercise 18 on page 76 implies that $\preceq^{\mathcal{C}}$ is a partial order on its field. To show that $\preceq^{\mathcal{C}}$ is a total order, let x and y be in the field of $\preceq^{\mathcal{C}}$. Thus, $x \in \text{fld}(\preceq)$ and $y \in \text{fld}(\preceq')$ for some $\preceq$ and $\preceq'$ in $\mathcal{C}$. Since $\mathcal{C}$ is a chain, it follows from (7.8) that either $\text{fld}(\preceq) \subseteq \text{fld}(\preceq')$ or $\text{fld}(\preceq') \subseteq \text{fld}(\preceq)$. Without loss of generality, let us assume that $\text{fld}(\preceq) \subseteq \text{fld}(\preceq')$. Hence, $x \in \text{fld}(\preceq')$ and $y \in \text{fld}(\preceq')$. As $\preceq'$ is a total order, we see that either $x \preceq' y$ or $y \preceq' x$. Since $\preceq' \subseteq \preceq^{\mathcal{C}}$, we conclude that either $x \preceq^{\mathcal{C}} y$ or $y \preceq^{\mathcal{C}} x$. Therefore, $\preceq^{\mathcal{C}}$ is a total order.   (Claim 1) $\square$

Claim 2. $\preceq^{\mathcal{C}}$ is a well-ordering on its field.

Proof. Suppose that $S \subseteq \bigcup \{\text{fld}(\preceq) : \preceq \in \mathcal{C}\}$ is nonempty. So let $y \in S$. Hence, (▲) $y \in \text{fld}(\preceq)$ for some $\preceq \in \mathcal{C}$. We now show that $\{z : z \preceq^{\mathcal{C}} y\} = \{z : z \prec y\}$. Since $\preceq \subseteq \preceq^{\mathcal{C}}$, we see that $\{z : z \prec y\} \subseteq \{z : z \preceq^{\mathcal{C}} y\}$. Assume $x \in \{z : z \preceq^{\mathcal{C}} y\}$. So $x \preceq^{\mathcal{C}} y$. Thus, $x \preceq' y$ for some $\preceq' \in \mathcal{C}$. We must verify that $x \in \{z : z \prec y\}$. Since $\preceq \in \mathcal{C}$, $\preceq' \in \mathcal{C}$, and $\mathcal{C}$ is a chain, we have either $\preceq \trianglelefteq \preceq'$ or $\preceq' \trianglelefteq \preceq$.

CASE 1: $\preceq \trianglelefteq \preceq'$. So (7.8)(ii) holds. We know that $x \preceq' y$. Since $y \in \text{fld}(\preceq)$ by (▲), we have that $y \prec y$ because $\preceq$ is reflexive. Thus, $x \preceq' y$ and $y \prec y$. From (7.8)(ii), we conclude that $x \prec y$, and so $x \in \{z : z \prec y\}$.

CASE 2: $\preccurlyeq' \trianglelefteq \preccurlyeq$. Item (i) of (7.8) implies that $\preccurlyeq' \subseteq \preccurlyeq$. Because $x \preccurlyeq' y$, we infer that $x \preccurlyeq y$, and therefore, $x \in \{z : z \preccurlyeq y\}$.

So $\{z : z \preccurlyeq^{\mathcal{C}} y\} = \{z : z \preccurlyeq y\}$. Thus, ($\blacklozenge$) $\{z \in S : z \preccurlyeq^{\mathcal{C}} y\} = \{z \in S : z \preccurlyeq y\}$. Since $y \in S$, the set $\{z \in S : z \preccurlyeq y\}$ is nonempty. As $\preccurlyeq$ is a well-ordering, there is a $\preccurlyeq$-least element $\ell \in \{z \in S : z \preccurlyeq y\}$. Because $\preccurlyeq \subseteq \preccurlyeq^{\mathcal{C}}$, ($\blacklozenge$) implies that ℓ is the $\preccurlyeq^{\mathcal{C}}$-least element in S. So $\preccurlyeq^{\mathcal{C}}$ is a well-ordering on its field.   (Claim 2) $\square$

Claim 3. $\preccurlyeq^{\mathcal{C}}$ is in W and $\preccurlyeq^{\mathcal{C}}$ is an upper bound for C.

Proof. Claim 2 implies that $\preccurlyeq^{\mathcal{C}}$ is in W. Let $\preccurlyeq \in C$. We prove that $\preccurlyeq \trianglelefteq \preccurlyeq^{\mathcal{C}}$. Clearly, $\preccurlyeq \subseteq \preccurlyeq^{\mathcal{C}}$. Thus, (i) of (7.8) holds. For (ii), suppose that $x \preccurlyeq^{\mathcal{C}} y$ and $y \preccurlyeq y$. Since $x \preccurlyeq^{\mathcal{C}} y$, there is a $\preccurlyeq' \in C$ such that $x \preccurlyeq' y$. As C is a chain, we have either $\preccurlyeq \trianglelefteq \preccurlyeq'$ or $\preccurlyeq' \trianglelefteq \preccurlyeq$. The arguments used in case 1 and case 2 of the proof of Claim 2 show that $x \preccurlyeq y$. Therefore, $\preccurlyeq \trianglelefteq \preccurlyeq^{\mathcal{C}}$.   (Claim 3) $\square$

By Zorn's Lemma, there exists a maximal element $\preccurlyeq$ in W. Hence, $\preccurlyeq$ is a well-ordering on a set $X \subseteq A$ and $\preccurlyeq \trianglelefteq \preccurlyeq'$ for all $\preccurlyeq' \in W$. We now prove that $X = A$. Suppose, for a contradiction, that there is an $a \in A$ such that $a \notin X$. Let $\preccurlyeq' = \preccurlyeq \cup \{\langle x, a \rangle : x \in X \text{ or } x = a\}$. One can easily show (see Exercise 5) that $\preccurlyeq'$ is a well-ordering on $X \cup \{a\}$ and therefore, $\preccurlyeq' \in W$. In addition, one can also show that $\preccurlyeq \triangleleft \preccurlyeq'$, which contradicts the maximality of $\preccurlyeq$. Hence, $X = A$ and $\preccurlyeq$ is a well-ordering on A.

The proof of Theorem 7.3.1 now shows that the axiom of choice implies that every set can be well-ordered. Moreover, in Exercise 1 you are asked to show that the well-ordering theorem implies the axiom of choice. Another proof of Theorem 7.3.1 is outlined in Exercise 15 on page 194. This alternative proof depends on the replacement axiom.

Exercises 7.3

*1. Let $\langle A_i : i \in I \rangle$ be an indexed function such that $A_i \neq \varnothing$ for all $i \in I$. By Theorem 7.3.1, the set $\bigcup_{i \in I} A_i$ has a well-ordering $\preccurlyeq$. Define a function f such that $\text{dom}(f) = I$ and $(\forall i \in I)(f(i) \in A_i)$. Thus, $\langle f(i) : i \in I \rangle$ is a choice function.

2. Let $\preccurlyeq$ be a total order on A. For each $u \in A$, let $s(u) = \{x \in A : x \prec u\}$. Suppose that for every $u \in A$ and X, if $X \subseteq s(u)$ is nonempty, then X has a $\preccurlyeq$-least element. Prove that $\preccurlyeq$ is a well-ordering on A.

*3. Let $\preccurlyeq$ be a well-ordering on C and let $\preccurlyeq'$ be a well-ordering on D. Suppose that (i) $\preccurlyeq \subseteq \preccurlyeq'$ and (ii) whenever $x \preccurlyeq' y$ and $y \preccurlyeq y$, then $x \preccurlyeq y$.
   (a) Prove that $C \subseteq D$.
   (b) Prove that $\preccurlyeq = \preccurlyeq' \cap (C \times C)$.
   (c) Prove that if $C \neq D$, then $C = \{z : z \prec' \ell\}$ for some $\ell \in D$.

*4. Let $\trianglelefteq$ be the relation defined on W in the proof of Theorem 7.3.1. Prove that $\trianglelefteq$ is a partial ordering on W.

*5. Let $\preccurlyeq$ be a well-ordering on a set X and let $a \notin X$. Define a relation $\preccurlyeq'$ on $X \cup \{a\}$ by $\preccurlyeq' = \preccurlyeq \cup \{\langle x, a \rangle : x \in X \text{ or } x = a\}$.
   (a) Show that $\preccurlyeq'$ is well-ordering on $X \cup \{a\}$.
   (b) Prove that $\preccurlyeq \trianglelefteq \preccurlyeq'$ where $\trianglelefteq$ is the continuation relation defined by (7.8) in the proof of Theorem 7.3.1.

*6. Suppose that $\preccurlyeq$ is a well-ordering on A and that $\preccurlyeq'$ is a well-ordering on B. Let $\preccurlyeq_\ell$ be the lexicographic ordering on $A \times B$. Prove that $\preccurlyeq_\ell$ is a well-ordering.

Exercise Notes: For Exercise **6**, see Exercise 19 on page 76.

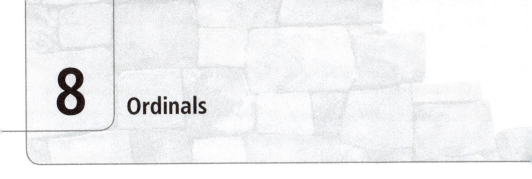

8 Ordinals

We can easily count the number of elements in a finite set; but can we count the number of elements in an infinite set? If a set A is countably infinite, then Theorem 5.2.6 suggests that A has ω many elements. Is ω a number? In order to count the number of elements in infinite sets, we need to define some new numbers that will allow us to count far beyond the natural numbers. These new numbers are called *ordinals*.

8.1 Ordinal Numbers

Each natural number $n \in \omega$ is a transitive set, and the set ω is also a transitive set (see Theorems 4.1.11 and 4.1.13). In addition, Theorem 4.4.13 shows that ω is well-ordered by the relation $\in$. By extending these concepts, we can easily define new counting numbers. For sets x and y, we shall write $x \subseteq y$ if and only if $x \in y$ or $x = y$.

Definition 8.1.1. An **ordinal** is a transitive set that is also well-ordered by the relation $\in$.

Thus, the set ω is an ordinal, and each natural number is also an ordinal. Let α be an ordinal. Then α is a transitive set, and the relation $\subseteq$ is a well-ordering on α. Therefore, in particular, $\subseteq$ is a total order on α (recall Definitions 6.1.1 and 3.4.3). Hence, Lemma 3.4.5 implies the following result.

Lemma 8.1.2. *Let α be an ordinal. Then for all x, y, z in α,*

(a) $x \notin x$,
(b) *if $x \in y$, then $y \notin x$,*
(c) *if $x \in y$ and $y \in z$, then $x \in z$,*
(d) *exactly one of the following holds: $x \in y$, or $x = y$, or $y \in x$.*

An ordinal α is a transitive set. So if $x \in y \in \alpha$, then we can conclude that $x \in \alpha$. We shall use lowercase Greek letters $\alpha, \beta, \gamma, \eta, \nu, \ldots$ to denote ordinals.

Lemma 8.1.3. *The following hold:*

(1) $\varnothing$ *is an ordinal.*
(2) *If α is an ordinal, then $\alpha \notin \alpha$.*
(3) *If α is an ordinal and $\gamma \in \alpha$, then γ is an ordinal.*
(4) *If α is an ordinal, then α^+ is also an ordinal.*

Proof. (1) This follows vacuously. (2) Suppose α is an ordinal. If $\alpha \in \alpha$, then this would contradict Lemma 8.1.2(a). Therefore, $\alpha \notin \alpha$.

(3) Assume α is an ordinal and let $\gamma \in \alpha$. We first prove that γ is a transitive set. Let ($\blacktriangle$) $\eta \in \nu$ and let ($\blacklozenge$) $\nu \in \gamma$. We prove that $\eta \in \gamma$. Since α is a transitive set and $\nu \in \gamma \in \alpha$, we have that $\nu \in \alpha$. Thus, $\eta \in \nu \in \alpha$, and we conclude that $\eta \in \alpha$. Now, since $\gamma \in \alpha$ and $\eta \in \alpha$, Lemma 8.1.2(d) implies that either $\eta \in \gamma$, $\eta = \gamma$, or $\gamma \in \eta$. If $\eta = \gamma$, then ($\blacktriangle$) implies that $\gamma \in \nu$. So Lemma 8.1.2(b) implies that $\nu \notin \gamma$, which contradicts ($\blacklozenge$). If $\gamma \in \eta$, then ($\blacklozenge$) implies that $\nu \in \eta$ by Lemma 8.1.2(c). Hence, $\eta \notin \nu$ by Lemma 8.1.2(b), contradicting ($\blacktriangle$). So we must have $\eta \in \gamma$, and therefore, γ is a transitive set. Since $\gamma \subseteq \alpha$ and α is well-ordered by $\in$, it follows that γ is also well-ordered by $\in$ (see Exercise 5 on page 145). Therefore, γ is an ordinal.

(4) Suppose that α is an ordinal, and recall that $\alpha^+ = \alpha \cup \{\alpha\}$. Exercise 2 on page 87 implies that the successor α^+ is a transitive set. Let $A \subseteq \alpha^+$ be nonempty. If $A \cap \alpha = \varnothing$, then $A = \{\alpha\}$ and A clearly has an $\in$-least element. Suppose $A \cap \alpha \neq \varnothing$. Since α is an ordinal, there is a $u \in A \cap \alpha$ so that $u \in x$ for all $x \in A \cap \alpha$. Because α is the $\in$-largest element in α^+, it follows that $u \in x$ for all $x \in A$. Consequently, α^+ is an ordinal. $\qquad\square$

We will show how to construct an ordinal using any well-ordered set. First, for each set x, let $\mathrm{ran}(x) = \{d : \exists c(\langle c, d \rangle \in x)\}$, which is a set (see page 45). Now let $\preccurlyeq$ be a well-ordering on a set A. Let $\varphi(x, y)$ be the formula $y = \mathrm{ran}(x)$. Clearly, for each x, there exists a unique y satisfying $\varphi(x, y)$. By Theorem 6.2.5, there is a unique function H with domain A such that

$$H(u) = \mathrm{ran}(H \restriction s(u)), \quad \text{for all } u \in A. \tag{8.1}$$

Recall that $s(u) = \{x \in A : x \prec u\}$ is the "less than" initial segment up to u (see (6.3) on page 143). Since $\mathrm{ran}(H \restriction s(u)) = \{H(x) : x \prec u\}$, we can rewrite (8.1) as

$$H(u) = \{H(x) : x \prec u\}, \quad \text{for all } u \in A. \tag{8.2}$$

Suppose that u is the $\preceq$-least element in A, that v is the $\preceq$-least element in A such that $u \prec v$, and that w is the $\preceq$-least element in A so that $v \prec w$. Then

$$H(u) = \{H(x) : x \prec u\} = \varnothing = 0,$$

$$H(v) = \{H(x) : x \prec v\} = \{H(u)\} = \{\varnothing\} = 1,$$

$$H(w) = \{H(x) : x \prec w\} = \{H(u), H(v)\} = \{\varnothing, \{\varnothing\}\} = 2.$$

Since 0, 1, and 2 are ordinals, it appears that the range of H consists of ordinals. We now show that every well-ordered set can be used to construct an ordinal.

Theorem 8.1.4. *Let $\preceq$ be a well-ordering on a set A, and let H be the unique function with domain A that satisfies (8.2). If $\delta = \mathrm{ran}(H)$, then δ is an ordinal.*

Proof. Let $\preceq$, A, H, and δ be as in the statement of the theorem.

Claim 1. For all $x \in A$, $H(x) \notin H(x)$.

Proof. Suppose, for a contradiction, that the set $C = \{x \in A : H(x) \in H(x)\}$ is nonempty. Let $a \in A$ be the $\preceq$-least element in C. Hence, (▲) $H(a) \in H(a)$. Thus, by (8.2), $H(a) = H(u)$ for some $u \prec a$. So by (▲), $H(u) \in H(u)$. Thus, $u \in C$ and $u \prec a$. Therefore, a is not the $\preceq$-least element in C. (Claim 1) □

Claim 2. H is one-to-one.

Proof. Let $x \in A$ and $y \in A$. Assume $H(x) = H(y)$. If $x \neq y$, then either $x \prec y$ or $y \prec x$, as $\preceq$ is a total order. Suppose $x \prec y$. So, by (8.2), $H(x) \in H(y)$. Since $H(x) = H(y)$, we conclude that $H(x) \in H(x)$, which contradicts Claim 1. Similarly, the relationship $y \prec x$ is also impossible. So $x = y$. (Claim 2) □

Claim 3. If x and y are in A, then $x \prec y$ if and only if $H(x) \in H(y)$.

Proof. Let x and y be in A. If $x \prec y$, then $H(x) \in H(y)$ by (8.2). Conversely, if $H(x) \in H(y)$, then $H(x) = H(u)$ for some $u \prec y$, again by (8.2). Since H is one-to-one, we see that $x = u$ and consequently, $x \prec y$. (Claim 3) □

Claim 4. δ is a transitive set.

Proof. Let $\alpha \in \beta$ and $\beta \in \delta$. As $\delta = \mathrm{ran}(H)$, let x in A be such that $H(x) = \beta$. Since $\alpha \in \beta$ and $H(x) = \{H(y) : y \prec x\}$ by (8.2), we conclude that $\alpha = H(y)$ for some $y \prec x$. Therefore, α is in the range of H and so, $\alpha \in \delta$. (Claim 4) □

Claim 5. The relation $\subseteq$ is a total order on δ.

Proof. First we show that $\subseteq$ is a partial order on δ. Let α, β, and γ be in δ. As $\delta = \mathrm{ran}(H)$, let x, y, and z be in A so that $H(x) = \alpha$, $H(y) = \beta$, and $H(z) = \gamma$. Clearly, $\alpha \subseteq \alpha$. Suppose that $\alpha \subseteq \beta$ and $\beta \subseteq \alpha$. Claims 2 and 3 imply that $x \preceq y$

and $y \preccurlyeq x$. Thus, $x = y$ since $\preccurlyeq$ is a partial order. So $H(x) = H(y)$ and $\alpha = \beta$. Finally, assume $\alpha \in \beta$ and $\beta \in \gamma$. Then $x \preccurlyeq y$ and $y \preccurlyeq z$. Hence, $x \preccurlyeq y$. Thus, by Claim 3, $H(x) \in H(y)$. So $\alpha \in \gamma$. To show that $\in$ is a total order, note that since $\preccurlyeq$ is a total order on A, we have that either $x \preccurlyeq y$ or $y \preccurlyeq x$. Claim 3 implies that either $\alpha \in \beta$ or $\beta \in \alpha$. (Claim 5) $\square$

Claim 6. The relation $\in$ is a well-ordering on δ.

Proof. Since $\delta = \text{ran}(H)$, the function H is onto δ. Let $D \subseteq \delta$ be nonempty. Therefore, $H^{-1}[D]$ is a nonempty subset of A. Let a be the $\preccurlyeq$-least element in $H^{-1}[D]$. Let $\alpha = H(a)$. Clearly, $\alpha \in D$. Now let $\beta \in D$ be arbitrary. Because H is onto δ, let $b \in H^{-1}[D]$ be such that $\beta = H(b)$. Since a is the $\preccurlyeq$-least element in $H^{-1}[D]$, we see that $a \preccurlyeq b$. Claim 3 now implies that $H(a) \in H(b)$. Hence, $\alpha \in \beta$ and α is the $\in$-least element in D. (Claim 6) $\square$

Therefore, δ is an ordinal. $\square$

Definition 8.1.5. We shall say that $(A, \preccurlyeq)$ is a **well-ordered structure** when $\preccurlyeq$ is a well-ordering on A. [We shall write $(A, \preccurlyeq)$ for the ordered pair $\langle A, \preccurlyeq \rangle$.]

Definition 8.1.6. Let $(A, \preccurlyeq)$ and $(B, \preccurlyeq^*)$ be well-ordered structures. A function $f: A \to B$ is an **isomorphism** from $(A, \preccurlyeq)$ onto $(B, \preccurlyeq^*)$ if f is one-to-one, f is onto B, and f satisfies

$$x \preccurlyeq y \text{ if and only if } f(x) \preccurlyeq^* f(y) \tag{8.3}$$

for all x and y in A. If such an isomorphism exists, then $(A, \preccurlyeq)$ and $(B, \preccurlyeq^*)$ are said to be **isomorphic**, and we write $(A, \preccurlyeq) \cong (B, \preccurlyeq^*)$.

We now present a lemma about isomorphic well-ordered structures that will be used shortly.

Lemma 8.1.7. *If $f: A \to B$ is an isomorphism between the two well-ordered structures $(A, \preccurlyeq)$ and $(B, \preccurlyeq^*)$, then*

(1) $x \prec y$ *if and only if $f(x) \prec^* f(y)$, for all x and y in A;*
(2) $\{f(x) : x \prec a\} = \{y \in B : y \prec^* f(a)\}$, *for each $a \in A$.*

Proof. Let $f: A \to B$ be an isomorphism between the well-ordered structures $(A, \preccurlyeq)$ and $(B, \preccurlyeq^*)$. First, we prove (1). Let x and y be in A, and assume that $x \prec y$. Thus, $x \preccurlyeq y$. So by (8.3), $f(x) \preccurlyeq^* f(y)$. Suppose that $f(x) = f(y)$. As

f is one-to-one, we would then have $x = y$, which contradicts the fact that $x \prec y$ (see Definition 3.4.4). Therefore, $f(x) \prec^* f(y)$. To prove the converse, assume that $f(x) \prec^* f(y)$. Thus, $x \preccurlyeq y$ by (8.3). If $x = y$, then $f(x) = f(y)$, which contradicts our assumption that $f(x) \prec^* f(y)$. Hence, $x \prec y$.

To prove (2), let $a \in A$ and let $z \in \{f(x) : x \prec a\}$. Thus, $z = f(x)$ for an $x \in A$ where $x \prec a$. Hence, $z = f(x) \prec^* f(a)$ by (1), and so $z \in \{y \in B : y \prec^* f(a)\}$. Let $z \in \{y \in B : y \prec^* f(a)\}$. So $z \in B$ and $z \prec^* f(a)$. Since f is onto B, there is an $x \in A$ so that $z = f(x)$. Hence $z = f(x) \prec^* f(a)$, and (1) implies that $x \prec a$. Thus, $z \in \{f(x) : x \prec a\}$. So $\{f(x) : x \prec a\} = \{y \in B : y \prec^* f(a)\}$. $\square$

Remark 8.1.8. Let $(A, \preccurlyeq)$ be a well-ordered structure. Theorem 8.1.4 and its proof show that there exists an ordinal δ and a one-to-one function $H : A \to \delta$ that is onto δ. The proof also shows that

$$x \preccurlyeq y \text{ if and only if } H(x) \subseteq H(y)$$

for all x and y in A. Therefore, H is an isomorphism from $(A, \preccurlyeq)$ onto $(\delta, \subseteq)$.

Definition 8.1.9. If $(A, \preccurlyeq)$ is a well-ordered structure, then the **order type** of $(A, \preccurlyeq)$ is the unique ordinal δ given by Theorem 8.1.4.

Our next theorem shows that if two well-orderings are isomorphic, then they have the same order type.

Theorem 8.1.10. *If $(A, \preccurlyeq)$ and $(B, \preccurlyeq^*)$ are isomorphic well-ordered structures, then $(A, \preccurlyeq)$ and $(B, \preccurlyeq^*)$ have the same order type.*

Proof. Let $(A, \preccurlyeq)$ and $(B, \preccurlyeq^*)$ be well-ordered structures. Let $f : A \to B$ be an isomorphism from $(A, \preccurlyeq)$ onto $(B, \preccurlyeq^*)$. Now let $H : A \to \delta$ and $H^* : B \to \delta^*$ be as given in the proof of Theorem 8.1.4 where $\operatorname{ran}(H) = \delta$ and $\operatorname{ran}(H^*) = \delta^*$. We shall prove that the ordinals δ and δ^* are equal. By (8.2), we have that

$$H(u) = \{H(x) : x \prec u\}, \text{ for all } u \in A. \tag{8.4}$$

$$H^*(v) = \{H^*(y) : y \prec^* v\}, \text{ for all } v \in B. \tag{8.5}$$

Claim 1. $H(u) = H^*(f(u))$ for all $u \in A$.

Proof. Assume, to the contrary, that $C = \{u \in A : H(u) \neq H^*(f(u))\} \neq \varnothing$. Let $a \in A$ be the $\preccurlyeq$-least element in C. So ($\blacktriangle$) $H(a) \neq H^*(f(a))$, and

$$H(x) = H^*(f(x)) \text{ for all } x \prec a. \tag{8.6}$$

Hence,

$$H(a) = \{H(x) : x \prec a\} \qquad \text{by (8.4)}$$
$$= \{H^*(f(x)) : x \prec a\} \quad \text{by (8.6)}$$
$$= \{H^*(y) : y \prec^* f(a)\} \quad \text{by Lemma 8.1.7(2)}$$
$$= H^*(f(a)) \qquad \text{by (8.5)}.$$

Therefore, $H(a) = H^*(f(a))$, which contradicts (▲).   (Claim 1) □

Since f is onto B, Claim 1 implies that $\text{ran}(H) = \text{ran}(H^*)$. So $\delta = \delta^*$.   □

Lemma 8.1.11. *If α is an ordinal, then α is the order type of the well-ordered structure $(\alpha, \in)$.*

Proof. See Exercise 2.   □

Theorem 8.1.12. *Let α and β be ordinals. If $(\alpha, \in) \cong (\beta, \in)$, then $\alpha = \beta$.*

Proof. This follows immediately from Theorem 8.1.10 and Lemma 8.1.11.

□

Theorems 8.1.10 and 8.1.12 imply that every well-ordered set is isomorphic to a unique ordinal number, namely, its order type (see Exercise 1). Therefore, our next theorem implies that the ordinals can now be used to measure and compare the "lengths" of any two well-orderings.

Theorem 8.1.13. *Let α, β, and γ be ordinals.*
(1) *If $\alpha \in \beta$ and $\beta \in \gamma$, then $\alpha \in \gamma$.*
(2) *If $\alpha \subseteq \beta$, then $\alpha \in \beta$ or $\alpha = \beta$.*
(3) *Exactly one of the following holds: (i) $\alpha \in \beta$, (ii) $\alpha = \beta$, or (iii) $\beta \in \alpha$.*

Proof. Let α, β, and γ be ordinals. We first prove (1). Suppose that $\alpha \in \beta$ and $\beta \in \gamma$. Since γ is transitive, we see that $\alpha \in \gamma$.

To prove (2), suppose $\alpha \subseteq \beta$. If $\alpha \neq \beta$, then $X = \beta \setminus \alpha$ is a nonempty subset of β. Let η be the $\in$-least element in X. Because (▲) $\eta \in X$, we have $\eta \in \beta$ and (♦) $\eta \notin \alpha$. We will now show that $\eta = \alpha$. First, we prove that $\eta \subseteq \alpha$. Let $\nu \in \eta$. Consequently, $\nu \in \beta$, as $\eta \in \beta$ and β is a transitive set. Since $\nu \in \eta$ and η is the $\in$-least element in X, we conclude that $\nu \notin X$. Thus, $\nu \in \alpha$ and so, $\eta \subseteq \alpha$. To prove that $\alpha \subseteq \eta$, let (★) $\nu \in \alpha$. Since $\alpha \subseteq \beta$, we see that $\nu \in \beta$.

So, as $v \in \alpha$, we conclude that ($\blacklozenge$) $v \notin X$. Because v and η are in β, we have by Lemma 8.1.2(d) that either $v \in \eta$, or $v = \eta$, or $\eta \in v$. If $v = \eta$, then $v \in X$ by ($\blacktriangle$), contradicting ($\blacklozenge$). If $\eta \in v$, then ($\bigstar$) implies that $\eta \in v \in \alpha$ and thus, $\eta \in \alpha$ because α is a transitive set; but this contradicts ($\blacklozenge$). Hence, we must have that $v \in \eta$. Therefore, $\eta = \alpha$, and so $\alpha \in \beta$.

For (3), let $\delta = \alpha \cap \beta$. We show that δ is an ordinal. Let $\eta \in \delta$. So, $\eta \in \alpha$ and $\eta \in \beta$. As α and β are transitive, $\eta \subseteq \alpha$ and $\eta \subseteq \beta$. Thus, $\eta \subseteq \alpha \cap \beta = \delta$. Hence, δ is transitive and, in addition, δ is well-ordered by $\in$ because $\delta \subseteq \alpha$ and α is well-ordered by $\in$. Therefore, δ is an ordinal. Since $\delta \subseteq \alpha$ and $\delta \subseteq \beta$, item (2) and Lemma 8.1.3(2) imply that

(a) either $\delta \in \alpha$ or $\delta = \alpha$, but not both;
(b) either $\delta \in \beta$ or $\delta = \beta$, but not both.

We will now show that the conjunction "$\delta \in \alpha$ and $\delta \in \beta$" is impossible, and thus, we must have either (i) $\alpha \in \beta$, (ii) $\alpha = \beta$, or (iii) $\beta \in \alpha$. If both $\delta \in \alpha$ and $\delta \in \beta$ hold, then $\delta \in \alpha \cap \beta = \delta$, and this contradicts Lemma 8.1.3(2). One can now easily show that exactly one of (i)–(iii) holds. $\qquad\square$

Corollary 8.1.14. *Let α, β, and γ be ordinals. Then*

(1) $\alpha \subseteq \alpha$.
(2) *If $\alpha \subseteq \beta$ and $\beta \subseteq \alpha$, then $\alpha = \beta$.*
(3) *If $\alpha \subseteq \beta$ and $\beta \subseteq \gamma$, then $\alpha \subseteq \gamma$.*
(4) *Either $\alpha \subseteq \beta$ or $\beta \subseteq \alpha$.*

Proof. (1) is clear. For (2), assume that $\alpha \subseteq \beta$ and $\beta \subseteq \alpha$. To prove that $\alpha = \beta$, suppose to the contrary that $\alpha \neq \beta$. Thus, $\alpha \in \beta$ and $\beta \in \alpha$. Theorem 8.1.13(1) implies that $\alpha \in \alpha$, contradicting Lemma 8.1.3(2). Hence, we must have that $\alpha = \beta$. To establish item (3), suppose $\alpha \subseteq \beta$ and $\beta \subseteq \gamma$. Theorem 8.1.13(1) implies $\alpha \subseteq \gamma$. Finally, item (4) follows from Theorem 8.1.13(3). $\qquad\square$

Theorem 8.1.15. *Let A and C be sets of ordinals.*

(1) *If A is nonempty, then there is an $\alpha \in A$ such that $(\forall \beta \in A)(\alpha \subseteq \beta)$.*
(2) $\bigcup C$ *is an ordinal.*
(3) $\bigcap C$ *is an ordinal whenever C is nonempty.*

Proof. Let A and C be sets of ordinals. To prove item (1), assume that $A \neq \varnothing$. Let $\lambda \in A$. Corollary 8.1.14(4) implies that $\eta \subseteq \lambda$ or $\lambda \subseteq \eta$ for all $\eta \in A$. Let

$S = \{\eta \in A : \eta \underline{\in} \lambda\}$. Since $\lambda \in S$, we have that S is nonempty. In addition, $\eta \underline{\in} \lambda$ for all $\eta \in S$. Thus, $S \subseteq \lambda^+$ by Theorem 8.1.13, as $\lambda \in \lambda^+$. Because λ^+ is an ordinal, S contains an $\underline{\in}$-least element α. It now follows that α is also the $\underline{\in}$-least element in A.

We now prove item (2). Since C is a set of ordinals, Lemma 8.1.3(3) implies that $\bigcup C$ is also a set of ordinals. Corollary 8.1.14 allows us to conclude that $\underline{\in}$ is a total order on $\bigcup C$. Item (1) now implies that $\bigcup C$ is also well-ordered by $\underline{\in}$. So to show that $\bigcup C$ is an ordinal, we just need to prove that it is a transitive set. Let $\alpha \in \beta$ and let $\beta \in \bigcup C$. Thus, $\beta \in \gamma$ for an ordinal $\gamma \in C$. Since $\alpha \in \beta$ and $\beta \in \gamma$, we see that $\alpha \in \gamma$. Hence, $\alpha \in \bigcup C$. Therefore, $\bigcup C$ is an ordinal. For the proof of (3), see Exercise 14. □

Definition 8.1.16. Let C be a set of ordinals. The **supremum** of C is the ordinal $\bigcup C$ and is denoted by sup C. When C is nonempty, we denote the ordinal $\bigcap C$ by inf C, which is called the **infimum** of C.

In Definition 8.1.16, the ordinal sup C is the least upper bound for C, and inf C is the greatest lower bound for C (see Exercises 13 and 14, respectively). In particular, if α and β are ordinals, then $\max(\alpha, \beta) = \sup\{\alpha, \beta\} = \alpha \cup \beta$ is the larger of the two ordinals.

We now classify ordinals into three distinct types: A given ordinal is either 0, a successor ordinal, or a limit ordinal.

Definition 8.1.17. Let α be a nonzero ordinal. Then α is a **successor ordinal** if $\alpha = \eta^+$ for some ordinal η. We shall say that α is a **limit ordinal** when α is not a successor ordinal.

Lemma 8.1.18. *Let α be a nonzero ordinal. Then α is a limit ordinal if and only if for all $\lambda \in \alpha$, we have that $\lambda^+ \in \alpha$.*

Proof. Let $\alpha \neq 0$. Suppose α is a limit ordinal. Let $\lambda \in \alpha$. Exercise 9(b) implies that $\lambda^+ \underline{\in} \alpha$. Since α is not a successor ordinal, we conclude that $\lambda^+ \in \alpha$. For the converse, assume that

$$\lambda^+ \in \alpha \text{ whenever } \lambda \in \alpha. \tag{8.7}$$

Suppose, to the contrary, that α is a successor ordinal. Thus, $\eta^+ = \alpha$ for some ordinal η. Since $\eta \in \eta^+ = \alpha$, (8.7) implies that $\alpha \in \alpha$, a contradiction. □

Lemma 8.1.19. *If α is a limit ordinal, then $\omega \subseteq \alpha$.*

Proof. Let α be a limit ordinal. By Theorem 8.1.13(3), either $\alpha \in \omega$ or $\omega \in \alpha$. If $\alpha \in \omega$, then by Theorem 4.1.6 we conclude that $\alpha = k^+$ for some $k \in \omega$. So α would be a successor ordinal. Thus, we must have $\omega \in \alpha$. Therefore, by transitivity, $\omega \subseteq \alpha$. $\qquad\square$

Theorem 8.1.20 (Burali–Forti). *There is no set that contains all of the ordinals.*

Proof. Assume, to the contrary, that there is a set A in which each ordinal is a member. By the subset axiom there exists a set B such that

$$B = \{\alpha \in A : \alpha \text{ is an ordinal}\}. \tag{8.8}$$

Thus, B is the set of all ordinals. Corollary 8.1.14 allows us to conclude that $\in$ is a total order on B. Theorem 8.1.15(1) implies that B is well-ordered by $\in$. We now show that B is a transitive set. Let $\alpha \in \beta$ and $\beta \in B$. Then β is an ordinal, and Lemma 8.1.3(3) implies that α is an ordinal. So $\alpha \in B$ by (8.8). Hence, B is an ordinal, and thus, $B \in B$ by (8.8), which contradicts Lemma 8.1.3(2). $\quad\square$

Theorem 8.1.21. *Let $\psi(\xi)$ be any formula. If there is an ordinal ξ for which $\psi(\xi)$ holds, then there is a least ordinal α such that $\psi(\alpha)$.*

Proof. Assume that ξ is an ordinal so that $\psi(\xi)$ holds. Thus, the set of ordinals $A = \{\gamma \in \xi^+ : \psi(\gamma)\}$ is nonempty. By Theorem 8.1.15(1), A has a least element α. Hence, α is the least ordinal satisfying $\psi(\alpha)$. $\qquad\square$

Theorem 8.1.22 (Ordinal Induction). *Let $\psi(\xi)$ be a formula and suppose for every ordinal β we have that*

$$if\, (\forall \xi \in \beta)\psi(\xi), \text{ then } \psi(\beta). \tag{8.9}$$

Then we can conclude that $\psi(\beta)$ holds for all ordinals β.

Proof. Assume that (8.9) holds for each ordinal β. Suppose, for a contradiction, that there is an ordinal ξ for which $\neg\psi(\xi)$ holds. By Theorem 8.1.21, there is a least ordinal α that satisfies $\neg\psi(\alpha)$. Since α is the least such ordinal, it follows that $(\forall \xi \in \alpha)\psi(\xi)$. Thus, (8.9) implies that $\psi(\alpha)$, a contradiction. $\qquad\square$

So to prove that a statement $\psi(\beta)$ holds for every ordinal β, it is sufficient to show that (8.9) holds for each ordinal β. Such a proof is called *proof by transfinite induction*, and it, typically, has the following structure:

Let β be an ordinal.
    Assume that $(\forall \xi \in \beta)\psi(\xi)$. [induction hypothesis]
        Case 1: Assume $\beta = 0$.
            Prove $\psi(\beta)$.
        Case 2: Assume β is a successor ordinal.   (8.10)
            Prove $\psi(\beta)$.
        Case 3: Assume β is a limit ordinal.
            Prove $\psi(\beta)$.

We shall present several proofs by transfinite induction in the next section (e.g., see Theorem 8.2.9).

Exercises 8.1

*1. Suppose that $(A, \preccurlyeq)$, $(B, \preccurlyeq^*)$, and $(C, \preccurlyeq^\#)$ are all well-ordered structures. Prove the following:
   (a) $(A, \preccurlyeq) \cong (A, \preccurlyeq)$.
   (b) If $(A, \preccurlyeq) \cong (B, \preccurlyeq^*)$, then $(B, \preccurlyeq^*) \cong (A, \preccurlyeq)$.
   (c) If $(A, \preccurlyeq) \cong (B, \preccurlyeq^*)$ and $(B, \preccurlyeq^*) \cong (C, \preccurlyeq^\#)$, then $(A, \preccurlyeq) \cong (C, \preccurlyeq^\#)$.
   Conclude that two well-ordered structures have the same order type if and only if the structures are isomorphic.

*2. Let α be an ordinal, and so $(\alpha, \in)$ is a well-ordered structure. By the proof of Theorem 8.1.4, there is a function H with domain α satisfying

$$H(u) = \{H(x) : x \in u\} \text{ for all } u \in \alpha.$$

Show that $H(u) = u$ for all $u \in \alpha$ and conclude that $\text{ran}(H) = \alpha$.

3. Let α and β be ordinals. Prove that either $\alpha \subseteq \beta$ or $\beta \subseteq \alpha$.

4. Suppose $\omega \in \alpha$ where α is an ordinal. Show that α is an infinite set.

*5. Let γ be an ordinal and let $\beta \in \gamma$. Since γ is well-ordered by $\in$, the initial segment up to β is the set $s^\in(\beta) = \{x \in \gamma : x \in \beta\}$ (see (6.3) on page 143). Show that $s^\in(\beta) = \beta$.

6. Suppose $(A, \preceq)$ is a well-ordered structure, and let H and G be functions with domain A satisfying

$$H(u) = \{H(x) : x \prec u\}, \text{ for all } u \in A,$$

$$G(u) = \{G(x) : x \prec u\}, \text{ for all } u \in A.$$

Prove, without appealing to Theorem 6.2.5, that $H = G$.

*7. Let $(A, \preceq)$ and $(B, \preceq^*)$ be well-ordered structures with order type δ and δ^*, respectively. Suppose that $f: A \to B$ is a one-to-one function satisfying $x \prec y$ if and only if $f(x) \prec^* f(y)$, for all x and y in A. Let H and H^* be as in the proof of Theorem 8.1.10. Thus, $\delta = \operatorname{ran}(H)$ and $\delta^* = \operatorname{ran}(H^*)$.
  (a) Prove that $H(u) \in H^*(f(u))$ for all $u \in A$.
  (b) Prove that $\operatorname{ran}(H) \subseteq \operatorname{ran}(H^*)$.
  (c) Prove that $\delta \in \delta^*$.

*8. Let $(A, \preceq)$ be a well-ordered structure with order type α. Let γ be the order type of $(C, \preceq_C)$ where $C \subseteq A$ and $\preceq_C$ is the induced relation (see Definition 3.4.15). Using Exercise **7**, show that $\gamma \in \alpha$.

*9. Let α and β be ordinals.
  (a) Prove that there is no ordinal γ such that $\alpha \in \gamma \in \alpha^+$.
  (b) Suppose that $\alpha \in \beta$. Prove that $\alpha^+ \in \beta$.
  (c) Prove that if $\alpha \in \beta$, then $\alpha^+ \in \beta^+$.
  (d) Prove that if $\alpha^+ = \beta^+$, then $\alpha = \beta$.
  (e) Prove that if $\alpha \in \beta$, then $\alpha \cap \beta = \alpha$.

*10. Let α be a nonzero ordinal.
  (a) Prove that α is a limit ordinal iff $(\forall \lambda \in \alpha)(\exists \gamma)(\lambda \in \gamma \in \alpha)$.
  (b) Prove that α is a limit ordinal if and only if $\alpha = \sup \alpha$.

11. Suppose that $(A, \preceq)$ is a well-ordered structure and $a \in A$. Let $s(a) \subseteq A$ be defined by $s(a) = \{x \in A : x \prec a\}$. Thus, $(s(a), \preceq_{s(a)})$ is a well-ordered structure (see Exercise 5 on page 145). Let δ be the order type of $(A, \preceq)$ and let $H: A \to \delta$ be as in the proof of Theorem 8.1.4. Show that η is the order type of $(s(a), \preceq_{s(a)})$ if and only if $H(a) = \eta$.

12. Let $(A, \preceq)$ and $(C, \preceq^*)$ be well-ordered structures. Let α be the order type of $(A, \preceq)$ and let η be the order type of $(C, \preceq^*)$. Suppose that $\eta \in \alpha$. Prove that there exists an $a \in A$ such that $(s(a), \preceq_{s(a)}) \cong (C, \preceq^*)$, where $s(a) = \{x \in A : x \prec a\}$.

*13. Let $\delta = \sup C$ where C is a set of ordinals.
  (a) Show that if $\alpha \in C$, then $\alpha \subseteq \delta$.
  (b) Let γ be an ordinal. Show that if $\alpha \subseteq \gamma$ for all $\alpha \in C$, then $\delta \subseteq \gamma$. Conclude that δ is the least upper bound for C (see Definition 3.4.9).

*14. Let C be a nonempty set of ordinals and let γ be the $\subseteq$-least element in C. Prove that $\gamma = \inf C$.

15. Prove that there is no set that contains every successor ordinal.

Exercise Notes: For Exercise 1(b), assume f is an isomorphism from $(A, \preccurlyeq)$ onto $(B, \preccurlyeq^*)$. Show that f^{-1} is an isomorphism from $(B, \preccurlyeq^*)$ onto $(A, \preccurlyeq)$. For Exercise 1(c), let f be an isomorphism from $(A, \preccurlyeq)$ onto $(B, \preccurlyeq^*)$, and let g be an isomorphism from $(B, \preccurlyeq^*)$ onto $(C, \preccurlyeq^\#)$. Show that the composition $g \circ f$ is the desired isomorphism. For Exercise 7(a), the proof of Lemma 8.1.7(1) shows that $x \prec y$ if and only if $f(x) \prec^* f(y)$ for all x and y in A. Let $a \in A$ be the $\prec$-least such that $H^*(f(a)) \in H(a)$. Thus, $H^*(f(a)) = H(x)$ for some $x \prec a$. For Exercise 8, by Exercise 5 on page 145, $(C, \preccurlyeq_C)$ is a well-ordered structure. For Exercise 15, if B were a set containing every successor ordinal, show that $\bigcup B$ contains every ordinal.

8.2 Ordinal Recursion and Class Functions

The ordinal numbers form an extension of the natural numbers. Even though the collection of ordinals is not a set, it will be useful to have some notation for this class. We shall let On denote the class of ordinal numbers; that is, let $\text{On} = \{\alpha : \alpha \text{ is an ordinal}\}$.

Let γ be an ordinal. As γ is well-ordered by $\subseteq$, Theorem 6.2.5 and Exercise 5 on page 184 immediately imply the following version of transfinite recursion in which we explicitly reveal a parameter[1] p (see Remark 6.2.6).

Ordinal Recursion Theorem 8.2.1. *Let $\varphi(p, x, y)$ be a formula such that for all p and x, there exists a unique y such that $\varphi(p, x, y)$. If γ is an ordinal, then for each ordinal α there exists a unique function H_α with domain γ such that*

$$\varphi(\alpha, H_\alpha \upharpoonright \beta, H_\alpha(\beta)), \text{ for all } \beta \in \gamma.$$

If it is not necessary to reveal a particular parameter, then we have our next result that also follows from Theorem 6.2.5.

Theorem 8.2.2. *Suppose that $\varphi(x, y)$ is a functional formula. If γ is an ordinal, then there exists a unique function H with domain γ such that*

$\varphi(H \upharpoonright \beta, H(\beta))$, *for all $\beta \in \gamma$.*

An *ordinal class function* $\mathcal{F}$ is one for which there is a formula $\psi(x, y)$ such that $\mathcal{F} = \{\langle \beta, y \rangle : \beta \in \mathrm{On} \wedge \psi(\beta, y)\}$ is a "class function." So for all ordinals β,

$\mathcal{F}(\beta) = y$ if and only if $\psi(\beta, y)$.

Thus, an ordinal class function has On as its domain. Since the class of ordinals is not a set, it thereby follows that $\mathcal{F}$ is not a set. Hence, we should be somewhat cautious when working with $\mathcal{F}$; however, if we keep in mind that $\mathcal{F}$ is just an abbreviation for a "functional formula on the ordinals," then we can safely view $\mathcal{F}$ as if it were a set. For example, let β be an ordinal. Now let $\mathcal{F}[\beta]$ be the collection $\mathcal{F}[\beta] = \{\mathcal{F}(\eta) : \eta \in \beta\}$. Since β is a set, the replacement axiom implies that $\mathcal{F}[\beta]$ is a set. Hence, by the subset axiom, the restricted function $\mathcal{F} \upharpoonright \beta$ is also a set. So $\mathcal{F}$ can always be "approximated" by a set function. For these reasons, we shall informally work with $\mathcal{F}$ as though it were a set.

Now let us explore how one can define ordinal class functions by transfinite recursion. In Theorem 8.2.1, the function H_α has domain γ and depends on α; however, do the values of H_α depend on γ? In other words, suppose that γ' is such that $\gamma \in \gamma'$. Theorem 8.2.1 implies that there is a function H'_α with domain γ' that satisfies

$\varphi(\alpha, H'_\alpha \upharpoonright \beta, H'_\alpha(\beta))$, for all $\beta \in \gamma'$.

Does $H_\alpha(\lambda) = H'_\alpha(\lambda)$ for each $\lambda \in \gamma$? The proof of Theorem 6.2.5 shows that the answer is "yes." Thus, we can define an ordinal class function $\mathcal{H}_\alpha$ by

$\mathcal{H}_\alpha(\lambda) = y$ if and only if

$(\exists \gamma)(\exists H)[\lambda \in \gamma \wedge \vartheta(H, \gamma) \wedge (\forall \beta \in \gamma)\varphi(\alpha, H \upharpoonright \beta, H(\beta)) \wedge H(\lambda) = y]$

where $\vartheta(H, \gamma)$ abbreviates "H is a function with domain γ, and γ is an ordinal." It therefore follows that the ordinal class function $\mathcal{H}_\alpha$ satisfies

$\varphi(\alpha, \mathcal{H}_\alpha \upharpoonright \beta, \mathcal{H}_\alpha(\beta))$, for all $\beta \in \mathrm{On}$.

We now record the above conclusions in the following theorem.

Ordinal Class Recursion Theorem 8.2.3. *Suppose that $\varphi(p, x, y)$ is a formula such that for all p and x, there exists a unique y so that $\varphi(p, x, y)$. Then, for each ordinal α, one can define an ordinal class function $\mathcal{H}_\alpha$ such that*

$\varphi(\alpha, \mathcal{H}_\alpha \upharpoonright \beta, \mathcal{H}_\alpha(\beta))$, *for all $\beta \in \mathrm{On}$.*

Moreover, there is a formula $\theta(\alpha, \lambda, y)$ such that for all ordinals α and λ, we have

$$\mathcal{H}_\alpha(\lambda) = y \text{ iff } \theta(\alpha, \lambda, y).$$

Theorem 8.2.3 will be used in the next section to define the operations of addition and multiplication on the ordinal numbers. Theorem 8.2.3, by adding a "dummy" parameter, implies an ordinal class recursion theorem for formulas that may, or may not, contain parameters.

Theorem 8.2.4. *Suppose that the formula $\varphi(x, y)$ is functional. Then one can define an ordinal class function $\mathcal{H}$ such that*

$$\varphi(\mathcal{H} \upharpoonright \beta, \mathcal{H}(\beta)), \text{ for all } \beta \in \text{On}.$$

The following lemma can be useful in certain applications of Theorem 8.2.4.

Lemma 8.2.5. *Let A be a set. If $\mathcal{F}: \text{On} \to A$ is an ordinal class function, then $\mathcal{F}$ is not one-to-one.*

Proof. Let $\mathcal{F}: \text{On} \to A$ be an ordinal class function where A is a set. Since $\mathcal{F}$ is definable by a formula, the subset axiom implies that the range of $\mathcal{F}$

$$B = \{x \in A : (\exists \alpha \in \text{On})(\mathcal{F}(\alpha) = x)\}$$

is a set. We shall prove that $\mathcal{F}$ is not one-to-one. Suppose, to the contrary, that $\mathcal{F}$ is one-to-one. Let $\varphi(a, \alpha)$ be a formula that satisfies

$$\varphi(a, \alpha) \text{ iff } (a \in B \wedge \mathcal{F}(\alpha) = a) \vee (a \notin B \wedge \alpha = \varnothing)$$

for all a and α. Since $\mathcal{F}$ is one-to-one, it follows that $\forall a \exists! \alpha \varphi(a, \alpha)$. As B is a set, the replacement axiom implies that

$$S = \{\alpha : \exists a(a \in B \wedge \varphi(a, \alpha))\}$$

is also a set. Because $\mathcal{F}: \text{On} \to A$, we see that $(\forall \alpha \in \text{On})(\exists a \in B)\varphi(a, \alpha)$. Thus, S contains every ordinal, contradicting Theorem 8.1.20. $\square$

Theorem 8.2.4 and Lemma 8.2.5 can be used, together, to give new proofs of Zorn's Lemma and the well-ordering theorem (see Exercises 14 and 15). These proofs, of course, are different from those that were presented in Chapter 7.

Normal Class Functions

We shall now investigate certain ordinal class functions whose codomain is the class of ordinals. The results obtained in this section will be used in Section 8.3 where we define addition and multiplication on the ordinals.

Definition 8.2.6. A class function $\mathcal{F}: \text{On} \to \text{On}$ is **strictly increasing** when for all ordinals α and β, if $\alpha \in \beta$, then $\mathcal{F}(\alpha) \in \mathcal{F}(\beta)$.

The class function $\mathcal{F}: \text{On} \to \text{On}$ defined by $\mathcal{F}(\alpha) = \alpha^+$ is strictly increasing by Exercise 9(c) on page 185.

Definition 8.2.7. A class function $\mathcal{F}: \text{On} \to \text{On}$ is **continuous** provided that $\mathcal{F}(\beta) = \sup\{\mathcal{F}(\eta) : \eta \in \beta\}$ for all limit ordinals β.

Definition 8.2.8. A class function $\mathcal{F}: \text{On} \to \text{On}$ is said to be **normal** if it is strictly increasing and continuous.

Our next theorem provides a method for showing that a continuous class function $\mathcal{F}: \text{On} \to \text{On}$ is strictly increasing. The proof follows the structure given in (8.10) on page 184.

Theorem 8.2.9. *Let $\mathcal{F}: \text{On} \to \text{On}$ be continuous. If $\mathcal{F}(\gamma) \in \mathcal{F}(\gamma^+)$ for every ordinal γ, then $\mathcal{F}$ is strictly increasing.*

Proof. Let $\mathcal{F}: \text{On} \to \text{On}$ be continuous. Assume that ($\blacktriangle$) $\mathcal{F}(\gamma) \in \mathcal{F}(\gamma^+)$ for every ordinal γ. Let α be a fixed ordinal. We prove by transfinite induction that

$$\forall \beta(\text{if } \alpha \in \beta, \text{ then } \mathcal{F}(\alpha) \in \mathcal{F}(\beta)).$$

Let β be an ordinal. Assume the induction hypothesis

$$(\forall \xi \in \beta)(\text{if } \alpha \in \xi, \text{ then } \mathcal{F}(\alpha) \in \mathcal{F}(\xi)). \tag{IH}$$

We must prove that

$$\text{if } \alpha \in \beta, \text{ then } \mathcal{F}(\alpha) \in \mathcal{F}(\beta). \tag{8.11}$$

CASE 1: $\beta = 0$. Then the conditional statement (8.11) is vacuously true.

CASE 2: $\beta = \gamma^+$ for an ordinal γ. We shall prove (8.11) as follows:

$$\alpha \in \beta \Rightarrow \alpha \in \gamma^+ \qquad\qquad \text{because } \beta = \gamma^+$$
$$\Rightarrow \alpha \subseteq \gamma \qquad\qquad\qquad \text{by definition of } \gamma^+$$
$$\Rightarrow \mathcal{F}(\alpha) \subseteq \mathcal{F}(\gamma) \qquad\quad \text{by (IH)}$$
$$\Rightarrow \mathcal{F}(\alpha) \subseteq \mathcal{F}(\gamma) \in \mathcal{F}(\gamma^+) \quad \text{by } (\blacktriangle)$$
$$\Rightarrow \mathcal{F}(\alpha) \in \mathcal{F}(\beta) \qquad\qquad \text{as } \beta = \gamma^+.$$

CASE 3: β is a limit ordinal. So, $\mathcal{F}(\beta) = \sup\{\mathcal{F}(\xi) : \xi \in \beta\}$ because $\mathcal{F}$ is continuous. Thus, $(\blacklozenge)$ $\mathcal{F}(\xi) \subseteq \mathcal{F}(\beta)$ for all $\xi \in \beta$. We prove (8.11) as follows:

$$\alpha \in \beta \Rightarrow \alpha \in \alpha^+ \in \beta \qquad\quad \text{as } \beta \text{ is a limit ordinal}$$
$$\Rightarrow \mathcal{F}(\alpha) \in \mathcal{F}(\alpha^+) \subseteq \mathcal{F}(\beta) \quad \text{by } (\blacktriangle) \text{ and } (\blacklozenge)$$
$$\Rightarrow \mathcal{F}(\alpha) \in \mathcal{F}(\beta) \qquad\qquad\quad \text{by transitivity or equality.}$$

Therefore, (8.11) is true, and the proof is complete. $\qquad\qquad\qquad\square$

Definition 8.2.10. Let C be a nonempty set of ordinals. Then C is a **limit set** if for all $\alpha \in C$, there is a $\beta \in C$ such that $\alpha \in \beta$.

In particular, a limit ordinal is a limit set (see Exercise 10(a) on page 185).

Definition 8.2.11. Let D be a limit set. We say that C is **cofinal** in D if $C \subseteq D$ and for all $\alpha \in D$ there is a $\beta \in C$ such that $\alpha \in \beta$.

If C is a cofinal subset of a limit set D, then C is also a limit set. The proofs of the following three useful lemmas will be left as exercises (see Exercises 3–5).

Lemma 8.2.12. *Suppose that C is a limit set. Then*

(1) $\sup C$ *is a limit ordinal.*
(2) C *is cofinal in* $\sup C$.

Lemma 8.2.13. *Let $\mathcal{F}: \text{On} \to \text{On}$ be normal and let D be a limit set. If C is cofinal in D, then $\mathcal{F}[D]$ is a limit set and $\mathcal{F}[C]$ is cofinal in $\mathcal{F}[D]$.*

Lemma 8.2.14. *Let D be a limit set. If C is cofinal in D, then $\sup C = \sup D$.*

Lemma 8.2.15. *Suppose that $\mathcal{F}\colon \mathrm{On} \to \mathrm{On}$ is normal. If C is a limit set, then $\sup \mathcal{F}[C] = \mathcal{F}(\sup C)$.*

Proof. Let $\mathcal{F}\colon \mathrm{On} \to \mathrm{On}$ be normal, and let C be a limit set. Let $\gamma = \sup C$. Thus, γ is a limit ordinal and C is cofinal in γ, by Lemma 8.2.12. Therefore, Lemma 8.2.13 implies that $\mathcal{F}[C]$ is cofinal in $\mathcal{F}[\gamma]$. Consequently, we have that $\sup \mathcal{F}[C] = \sup \mathcal{F}[\gamma]$ by Lemma 8.2.14. Because $\mathcal{F}$ is continuous, we conclude that $\sup \mathcal{F}[\gamma] = \mathcal{F}(\gamma)$. Since $\gamma = \sup C$, we obtain

$$\sup \mathcal{F}[C] = \sup \mathcal{F}[\gamma] = \mathcal{F}(\gamma) = \mathcal{F}(\sup C).$$

Therefore, $\sup \mathcal{F}[C] = \mathcal{F}(\sup C)$. $\qquad\qquad\qquad\qquad\qquad\qquad\square$

Corollary 8.2.16. *Suppose that $\mathcal{F}\colon \mathrm{On} \to \mathrm{On}$ is normal. If γ is a limit ordinal, then $\mathcal{F}(\gamma)$ is also a limit ordinal.*

Let $\mathcal{F}$ be a normal class function $\mathcal{F}\colon \mathrm{On} \to \mathrm{On}$. Then Lemma 8.2.15 shows that $\sup \mathcal{F}[C] = \mathcal{F}(\sup C)$ whenever C is a limit set. Thus, $\mathcal{F}$ "commutes" with the supremum of a limit set. Our next theorem shows that $\mathcal{F}$ commutes with the supremum of any nonempty set of ordinals.

Theorem 8.2.17. *If $\mathcal{F}\colon \mathrm{On} \to \mathrm{On}$ is normal, then $\sup \mathcal{F}[C] = \mathcal{F}(\sup C)$, that is,*

$$\sup\{\mathcal{F}(\xi) : \xi \in C\} = \mathcal{F}(\sup C)$$

for every nonempty set C of ordinals.

Proof. Let $\mathcal{F}\colon \mathrm{On} \to \mathrm{On}$ be normal and let C be a nonempty set of ordinals. We shall prove that $\sup \mathcal{F}[C] = \mathcal{F}(\sup C)$. There are two cases to consider: either C is a limit set or it is not. If C is a limit set, then $\sup \mathcal{F}[C] = \mathcal{F}(\sup C)$ by Lemma 8.2.15.

Now suppose that C is not a limit set. Then C must have a largest element γ. Therefore, $\gamma = \sup C$, $\gamma \in C$, and $\xi \in \gamma$ for all $\xi \in C$. Since $\mathcal{F}$ is normal, we conclude that $\mathcal{F}(\xi) \in \mathcal{F}(\gamma)$ for all $\xi \in C$. Thus, $\sup\{\mathcal{F}(\xi) : \xi \in C\} = \mathcal{F}(\gamma)$. Hence, $\sup \mathcal{F}[C] = \mathcal{F}(\sup C)$. $\qquad\qquad\square$

Remark 8.2.18. Whenever X is a set, we shall write $f\colon X \to \mathrm{On}$ to mean that $f\colon X \to \rho$ for some ordinal ρ; that is, f is an *ordinal-valued set function*.

The mathematician Oswald Veblen was the first to investigate normal class functions, and our final theorem on normal class functions is due to him. Before we state and prove this theorem, we illustrate how one can apply Theorem 8.2.2 to define a function by ordinal recursion. Such a function will be used in the proof of Veblen's theorem.

Let $\mathcal{F}\colon \mathrm{On} \to \mathrm{On}$ be a class function and let γ be an ordinal. Now consider the following definition by cases:

$$y = \begin{cases} \gamma, & \text{if } f \text{ is the empty function;} \\ \mathcal{F}(f(k)), & \text{if } f\colon k^+ \to \mathrm{On} \text{ for some } k \in \omega; \\ \varnothing, & \text{if none of the above hold.} \end{cases} \tag{8.12}$$

Since $\mathcal{F}$ is definable by a formula, there exists a formula $\varphi(f, y)$ that expresses (8.12). Clearly, for each f, there is a unique y so that $\varphi(f, y)$. Theorem 8.2.2 implies that there is a set function $h\colon \omega \to \mathrm{On}$ such that $\varphi(h \restriction m, h(m))$, for all $m \in \omega$. So for each natural number m, we have that

$$h(m) = \begin{cases} \gamma, & \text{if } m = 0; \\ \mathcal{F}(h(n)), & \text{if } m = n^+ \text{ for some } n \in \omega. \end{cases} \tag{8.13}$$

Note that since $h \restriction 0$ is the empty function, the first condition in (8.12) produces the first condition in (8.13). Because $h \restriction k^+$ is a function with domain k^+, the second condition in (8.12) yields the corresponding condition in (8.13). Thus, h is a function with domain ω satisfying:

(1) $h(0) = \gamma$,
(2) $h(n^+) = \mathcal{F}(h(n))$, for all $n \in \omega$.

We will be using the function h in the following proof of Veblen's theorem.

Fixed-Point Theorem 8.2.19. *If $\mathcal{F}\colon \mathrm{On} \to \mathrm{On}$ is a normal class function, then for each ordinal γ, there exists an ordinal β such that $\gamma \subseteq \beta$ and $\mathcal{F}(\beta) = \beta$.*

Proof. Let $\mathcal{F}\colon \mathrm{On} \to \mathrm{On}$ be normal and let γ be an ordinal. If $\mathcal{F}(\gamma) = \gamma$, then we are done. If $\mathcal{F}(\gamma) \neq \gamma$, then $\gamma \in \mathcal{F}(\gamma)$ by Exercise 9. By Theorem 8.2.2, there is a function h from ω into the ordinals such that

(1) $h(0) = \gamma$,
(2) $h(n^+) = \mathcal{F}(h(n))$, for all $n \in \omega$.

Let $C = \{h(n) : n \in \omega\}$. As $\mathcal{F}$ is strictly increasing, we see that C is a limit set and $\mathcal{F}[C]$ is cofinal in C (see Exercise 11). Thus, (▲) $\sup \mathcal{F}[C] = \sup C$ by Lemma 8.2.14, and (◆) $\sup \mathcal{F}[C] = \mathcal{F}(\sup C)$ by Lemma 8.2.15. Hence, (▲) and (◆) imply that $\mathcal{F}(\sup C) = \sup C$. Letting $\beta = \sup C$, we conclude that $\mathcal{F}(\beta) = \beta$. Moreover, as $\gamma \in C$, Lemma 8.2.12(2) implies that $\gamma \in \beta$. $\square$

Exercises 8.2

1. Let C be a set of ordinals with largest element γ. Show that sup $C = \gamma$.

2. Let $\mathcal{F} \colon \mathrm{On} \to \mathrm{On}$ be strictly increasing. Prove that $\mathcal{F}$ is one-to-one.

*3. Prove Lemma 8.2.12.

*4. Prove Lemma 8.2.13.

*5. Prove Lemma 8.2.14.

6. Let C be a limit set. Using Theorem 4.2.1, define a one-to-one function $h \colon \omega \to C$. Conclude that C is infinite.

*7. Let $\mathcal{F} \colon \mathrm{On} \to \mathrm{On}$ be normal and let C be a limit set. Prove that $\mathcal{F}[C]$ is also a limit set.

8. Using Exercise 7, prove Corollary 8.2.16.

*9. Suppose that $\mathcal{F} \colon \mathrm{On} \to \mathrm{On}$ is strictly increasing. Prove that $\alpha \subseteq \mathcal{F}(\alpha)$ for all ordinals α.

*10. Suppose $\mathcal{F} \colon \mathrm{On} \to \mathrm{On}$ and $\mathcal{G} \colon \mathrm{On} \to \mathrm{On}$ are normal.
    (a) Prove that the composition $(\mathcal{F} \circ \mathcal{G}) \colon \mathrm{On} \to \mathrm{On}$ is normal.
    (b) Show that if $\mathcal{F}(\mathcal{G}(\alpha)) = \alpha$, then $\mathcal{G}(\alpha) = \alpha$ and $\mathcal{F}(\alpha) = \alpha$.

*11. Let $\mathcal{F}, C$, and h be as in the proof of Theorem 8.2.19.
    (a) Prove that $h(n) \in h(n^+)$ for all $n \in \omega$. Conclude that C is a limit set.
    (b) Prove that $\mathcal{F}[C]$ is cofinal in C.

12. Let $\mathcal{F} \colon \mathrm{On} \to \mathrm{On}$ be normal. Suppose $\mathcal{F}(\beta) = \beta$. Show that for all α, if $\alpha \in \beta$, then $\mathcal{F}(\alpha) \in \beta$.

13. Let $\mathcal{F} \colon \mathrm{On} \to \mathrm{On}$ be normal, and suppose that C is a limit set such that the set $D = \{\alpha \in C : \mathcal{F}(\alpha) = \alpha\}$ is cofinal in C. Let $\gamma = \sup C$. Prove that $\mathcal{F}(\gamma) = \gamma$.

*14. Let $(A, \preccurlyeq)$ be a poset in which every chain $C \subseteq A$ has an upper bound in A. Let $n_0 \notin A$. For each chain $C \subseteq A$, let C_p be the set of proper upper bounds for C. Let $\mathcal{D} = \{C_p : C_p \neq \varnothing \wedge C \subseteq A \text{ is a chain}\}$. By the axiom of choice, there is a function G such that $G(C_p) \in C_p$ for all $C_p \in \mathcal{D}$. Note that $G(C_p) \notin C$ whenever $C_p \in \mathcal{D}$. Let $\varphi(f, y)$ be a formula that asserts the following:

$$y = \begin{cases} G(f[\gamma]_p), & \text{if } f \colon \gamma \to A, \gamma \in \text{On}, f[\gamma] \subseteq A \text{ is a chain, and } f[\gamma]_p \in \mathcal{D}; \\ n_0, & \text{if the above does not hold.} \end{cases}$$

Hence, for all f, there exists a unique y such that $\varphi(f, y)$. Theorem 8.2.4 thus implies that there exists a class function $\mathcal{F} \colon \text{On} \to A \cup \{n_0\}$ such that $\varphi(\mathcal{F} \restriction \gamma, \mathcal{F}(\gamma))$, for all $\gamma \in \text{On}$. Hence, for each ordinal γ,

$$\mathcal{F}(\gamma) = \begin{cases} G(\mathcal{F}[\gamma]_p), & \text{if } \mathcal{F}[\gamma] \subseteq A \text{ is a chain and } \mathcal{F}[\gamma]_p \in \mathcal{D}; \\ n_0, & \text{if the above does not hold.} \end{cases}$$

(a) Using Lemma 8.2.5, show that $\mathcal{F}(\gamma) = n_0$ for some ordinal γ.
(b) Let α be the least ordinal such that $\mathcal{F}(\alpha) = n_0$. Show that $\mathcal{F}[\alpha] \subseteq A$ is a chain with no proper upper bound.
(c) By assumption, the chain $\mathcal{F}[\alpha]$ has an upper bound m. Show that m is a maximal element.

*15. (AC) Let A be nonempty set and let $n_0 \notin A$. Let $\mathcal{C} = \mathcal{P}(A) \setminus \{\varnothing\}$. Note that $\bigcup \mathcal{C} = A$. By Theorem 3.3.24, let $G \colon \mathcal{C} \to A$ be a choice function for $\mathcal{C}$. Let $\varphi(f, y)$ be a formula that asserts the following:

$$y = \begin{cases} G(A \setminus f[\gamma]), & \text{if } f \colon \gamma \to A, \gamma \in \text{On}, \text{ and } A \setminus f[\gamma] \neq \varnothing; \\ n_0, & \text{if the above does not hold.} \end{cases}$$

So for all f, there exists a unique y such that $\varphi(f, y)$. Theorem 8.2.4 implies that there exists a class function $\mathcal{F} \colon \text{On} \to A \cup \{n_0\}$ such that $\varphi(\mathcal{F} \restriction \gamma, \mathcal{F}(\gamma))$, for all $\gamma \in \text{On}$. Thus,

$$\mathcal{F}(\gamma) = \begin{cases} G(A \setminus \mathcal{F}[\gamma]), & \text{if } \mathcal{F}[\gamma] \subseteq A \text{ and } A \setminus \mathcal{F}[\gamma] \neq \varnothing; \\ n_0, & \text{if the above does not hold.} \end{cases}$$

(a) Using Lemma 8.2.5, show that $\mathcal{F}(\gamma) = n_0$ for some ordinal γ.
(b) Let γ be the least ordinal such that $\mathcal{F}(\gamma) = n_0$. Show that the set function $(\mathcal{F} \restriction \gamma) \colon \gamma \to A$ is a bijection.
(c) Show that A has a well-ordering.

Exercise Notes: For Exercise 10(a), use Exercise 7. For Exercise 10(b), apply Exercise 9. For Exercise 15(c), use (b) and Theorem 6.1.4. Exercise 14 outlines another proof of Zorn's Lemma 7.1.1. Exercise 15 offers an alternative proof of Theorem 7.3.1, the well-ordering theorem.

8.3 | Ordinal Arithmetic

In Chapter 4, we used the recursion theorem to define the arithmetic operations of addition and multiplication on the natural numbers. We shall now define such operations on the ordinals by transfinite recursion, that is, by applying Theorem 8.2.3. We first show how to define ordinal addition. Consider the following definition by cases:

$$y = \begin{cases} \alpha, & \text{if } f \text{ is the empty function;} \\ f(\gamma)^+, & \text{if } f: \gamma^+ \to \text{On for some ordinal } \gamma; \\ \sup f[\beta], & \text{if } f: \beta \to \text{On for some limit ordinal } \beta; \\ \emptyset, & \text{if none of the above hold.} \end{cases} \tag{8.14}$$

Let $\varphi(\alpha, f, y)$ be a formula that expresses (8.14). Hence, for all α and f, there exists a unique y so that $\varphi(\alpha, f, y)$. Theorem 8.2.3 implies that for each ordinal α, there is a class function $\mathcal{A}_\alpha: \text{On} \to \text{On}$ such that $\varphi(\alpha, \mathcal{A}_\alpha \restriction \beta, \mathcal{A}_\alpha(\beta))$, for all $\beta \in \text{On}$ (see Exercise 1). Thus,

$$\mathcal{A}_\alpha(\beta) = \begin{cases} \alpha, & \text{if } \beta = 0; \\ \mathcal{A}_\alpha(\gamma)^+, & \text{if } \beta = \gamma^+ \text{ for some ordinal } \gamma; \\ \sup \mathcal{A}_\alpha[\beta], & \text{if } \beta \text{ is a limit ordinal.} \end{cases} \tag{8.15}$$

Since $\mathcal{A}_\alpha \restriction 0$ is the empty function, the first condition in (8.15) corresponds to the first condition in (8.14). Moreover, for each ordinal β, we see that $\mathcal{A}_\alpha \restriction \beta$ is an ordinal-valued set function with domain β. Thus, the last two conditions of (8.15) correspond, respectively, to conditions two and three of (8.14). We can now define the operation of addition on the ordinal numbers.

Definition 8.3.1. For all ordinals α and β, define $\alpha + \beta$ to be the unique ordinal $\mathcal{A}_\alpha(\beta)$; that is, $\alpha + \beta = \mathcal{A}_\alpha(\beta)$.

So, using the above definition of ordinal addition, (8.15) yields the following rules for the addition of ordinals α, γ, and limit ordinal β:

$$\alpha + 0 = \alpha \tag{A1}$$

$$\alpha + \gamma^+ = (\alpha + \gamma)^+ \tag{A2}$$

$$\alpha + \beta = \sup\{\alpha + \eta : \eta \in \beta\}. \tag{A3}$$

The rules (A1)–(A3) extend those given in Theorem 4.3.3 for the addition of natural numbers. Some of the properties of addition on the natural numbers also extend to ordinal addition, whereas others do not. For example, addition on the natural numbers is commutative; however, addition on the ordinals is not commutative. To verify this, let us evaluate $\omega + 1$ and $1 + \omega$. We compute $\omega + 1$ as follows:

$$
\begin{aligned}
\omega + 1 &= \omega + 0^+ && \text{because } 1 = 0^+ \\
&= (\omega + 0)^+ && \text{by rule (A2)} \\
&= \omega^+ && \text{by rule (A1).}
\end{aligned}
$$

Now we determine the value of $1 + \omega$. First, note that the set $\{1 + n : n \in \omega\}$ is cofinal in ω. Now, since ω is a limit ordinal, we obtain the following:

$$
\begin{aligned}
1 + \omega &= \sup\{1 + n : n \in \omega\} && \text{by rule (A3)} \\
&= \sup \omega && \text{by Lemma 8.2.14} \\
&= \omega && \text{by Exercise 10(b) on page 185.}
\end{aligned}
$$

Because $\omega^+ \neq \omega$, we see that $\omega + 1 \neq 1 + \omega$. Thus, addition of ordinal numbers is not necessarily commutative. The following lemma will allow us to prove that ordinal addition is associative.

Lemma 8.3.2. *For each ordinal α, the class function $\mathcal{A}_\alpha : \mathrm{On} \to \mathrm{On}$ is normal.*

Proof. See Exercise 9. $\square$

Using the operation of addition on the ordinals, we can define multiplication using Theorem 8.2.3. The method used to define addition can be modified[2] to show that for every ordinal α, there exists a class function $\mathcal{M}_\alpha : \mathrm{On} \to \mathrm{On}$ satisfying the following:

$$
\mathcal{M}_\alpha(\beta) = \begin{cases}
0, & \text{if } \beta = 0; \\
\mathcal{M}_\alpha(\gamma) + \alpha, & \text{if } \beta = \gamma^+ \text{ for some ordinal } \gamma; \\
\sup \mathcal{M}_\alpha[\beta], & \text{if } \beta \text{ is a limit ordinal.}
\end{cases} \tag{8.16}
$$

The operation of multiplication on the ordinal numbers can now be defined.

Definition 8.3.3. Let α and β be ordinals. Define $\alpha \cdot \beta$ to be the unique ordinal $\mathcal{M}_\alpha(\beta)$; that is, $\alpha \cdot \beta = \mathcal{M}_\alpha(\beta)$.

Definition 8.3.3 and (8.16) yield the following rules for the multiplication of ordinals α, γ, and limit ordinal β:

$$\alpha \cdot 0 = 0, \tag{M1}$$

$$\alpha \cdot \gamma^+ = (\alpha \cdot \gamma) + \alpha, \tag{M2}$$

$$\alpha \cdot \beta = \sup\{\alpha \cdot \eta : \eta \in \beta\}. \tag{M3}$$

Lemma 8.3.4. *If α is a nonzero ordinal, then $\mathcal{M}_\alpha$: On $\to$ On is normal.*

Proof. See Exercise 10. □

Our next lemma will be used to prove that ordinal multiplication distributes over addition. The proof also illustrates how one can prove that arithmetic compositions result in a normal class function.

Lemma 8.3.5. *Let α and β be ordinals. Then*

(a) *The class function $\mathcal{F}$: On $\to$ On defined by $\mathcal{F}(\xi) = \alpha \cdot (\beta + \xi)$ is normal.*
(b) *The class function $\mathcal{G}$: On $\to$ On defined by $\mathcal{G}(\xi) = \alpha \cdot \beta + \alpha \cdot \xi$ is also normal.*

Proof. Let ξ be an ordinal. Then

$$\mathcal{F}(\xi) = \alpha \cdot (\beta + \xi) \quad \text{by definition of } \mathcal{F}$$

$$= \mathcal{M}_\alpha(\beta + \xi) \quad \text{by Definition 8.3.3}$$

$$= \mathcal{M}_\alpha(\mathcal{A}_\beta(\xi)) \quad \text{by Definition 8.3.1.}$$

Hence, $\mathcal{F} = \mathcal{M}_\alpha \circ \mathcal{A}_\beta$. Since $\mathcal{M}_\alpha$ and $\mathcal{A}_\beta$ are normal, Exercise 10 on page 193 implies that $\mathcal{F}$ is normal. A similar argument will show that $\mathcal{G}$ is normal. □

Some of the properties of ordinal addition and multiplication are familiar.

Theorem 8.3.6. *For all ordinals α, β, and γ,*

(1) $(\alpha + \beta) + \gamma = \alpha + (\beta + \gamma)$,
(2) $(\alpha \cdot \beta) \cdot \gamma = \alpha \cdot (\beta \cdot \gamma)$,
(3) $\alpha \cdot (\beta + \gamma) = \alpha \cdot \beta + \alpha \cdot \gamma$,
(4) $\alpha + 0 = 0 + \alpha = \alpha$,
(5) $\alpha \cdot 1 = 1 \cdot \alpha = \alpha$,
(6) $\alpha \cdot 0 = 0 \cdot \alpha = 0$.

Proof. We shall prove only (3) and leave the rest for the exercises. Thus, we assume (1). Let α and β be ordinals. We prove by transfinite induction that

$$\forall \gamma (\alpha \cdot (\beta + \gamma) = \alpha \cdot \beta + \alpha \cdot \gamma).$$

Let γ be an ordinal. Assume the induction hypothesis

$$(\forall \xi \in \gamma)(\alpha \cdot (\beta + \xi) = \alpha \cdot \beta + \alpha \cdot \xi). \tag{IH}$$

We must prove that

$$\alpha \cdot (\beta + \gamma) = \alpha \cdot \beta + \alpha \cdot \gamma. \tag{8.17}$$

CASE 1: $\gamma = 0$. Then $\alpha \cdot (\beta + 0) = \alpha \cdot \beta$ by rule (A1), and

$$\alpha \cdot \beta + \alpha \cdot 0 = \alpha \cdot \beta + 0 \quad \text{by rule (M1)}$$
$$= \alpha \cdot \beta \quad\quad\ \text{by rule (A1)}.$$

Therefore, $\alpha \cdot (\beta + 0) = \alpha \cdot \beta + \alpha \cdot 0$.

CASE 2: $\gamma = \eta^+$ for an ordinal η. We shall prove (8.17) as follows:

$$\alpha \cdot (\beta + \eta^+) = \alpha \cdot (\beta + \eta)^+ \quad\quad\ \text{by rule (A2)}$$
$$= \alpha \cdot (\beta + \eta) + \alpha \quad\quad \text{by rule (M2)}$$
$$= (\alpha \cdot \beta + \alpha \cdot \eta) + \alpha \quad \text{by (IH) as } \eta \in \gamma$$
$$= \alpha \cdot \beta + (\alpha \cdot \eta + \alpha) \quad \text{by item (1) of the theorem}$$
$$= \alpha \cdot \beta + \alpha \cdot \eta^+ \quad\quad\ \text{by rule (M2)}.$$

Therefore, $\alpha \cdot (\beta + \eta^+) = \alpha \cdot \beta + \alpha \cdot \eta^+$.

CASE 3: γ is a limit ordinal. We shall show that $\alpha \cdot (\beta + \gamma) = \alpha \cdot \beta + \alpha \cdot \gamma$ as follows:

$$\alpha \cdot (\beta + \gamma) = \sup\{\alpha \cdot (\beta + \xi) : \xi \in \gamma\} \quad \text{by Lemma 8.3.5(a)}$$
$$= \sup\{\alpha \cdot \beta + \alpha \cdot \xi : \xi \in \gamma\} \quad \text{by (IH)}$$
$$= \alpha \cdot \beta + \alpha \cdot \gamma \quad\quad\quad\quad\quad\ \text{by Lemma 8.3.5(b)}.$$

Therefore, (8.17) is true, and the proof is complete. $\qquad\qquad\square$

As $\mathcal{A}_\alpha$ and $\mathcal{M}_\alpha$ are normal class functions, the following order-preserving and *left* cancellation laws hold.

Theorem 8.3.7. *For all ordinals α, β, and γ,*

(1) $\beta \in \gamma$ iff $\alpha + \beta \in \alpha + \gamma$,
(2) *if* $1 \subseteq \alpha$, *then* $\beta \in \gamma$ iff $\alpha \cdot \beta \in \alpha \cdot \gamma$,
(3) *if* $\alpha + \beta = \alpha + \gamma$, *then* $\beta = \gamma$,
(4) *if* $1 \subseteq \alpha$, *then* $\alpha \cdot \beta = \alpha \cdot \gamma$ *implies* $\beta = \gamma$.

We now define exponentiation on the ordinals. By Theorem 8.2.3, for each ordinal α such that $2 \subseteq \alpha$, there is a class function[3] $\mathcal{E}_\alpha : \mathrm{On} \to \mathrm{On}$ satisfying

$$\mathcal{E}_\alpha(\beta) = \begin{cases} 1, & \text{if } \beta = 0; \\ \mathcal{E}_\alpha(\gamma) \cdot \alpha, & \text{if } \beta = \gamma^+ \text{ for some ordinal } \gamma; \\ \sup \mathcal{E}_\alpha[\beta], & \text{if } \beta \text{ is a limit ordinal.} \end{cases} \tag{8.18}$$

Definition 8.3.8. Let α and β be ordinals where $2 \subseteq \alpha$. Define α^β to be the unique ordinal $\mathcal{E}_\alpha(\beta)$; that is, $\alpha^\beta = \mathcal{E}_\alpha(\beta)$.

If $2 \subseteq \alpha$, the following exponent rules for ordinal γ and limit ordinal β hold:

$$\alpha^0 = 1, \tag{E1}$$

$$\alpha^{\gamma^+} = (\alpha^\gamma) \cdot \alpha, \tag{E2}$$

$$\alpha^\beta = \sup\{\alpha^\eta : \eta \in \beta\}. \tag{E3}$$

Lemma 8.3.9. *Let α be an ordinal. If $2 \subseteq \alpha$, then $\mathcal{E}_\alpha : \mathrm{On} \to \mathrm{On}$ is normal.*

Proof. See Exercise 11. $\qquad\qquad\qquad\qquad\qquad\qquad\qquad\qquad\qquad\qquad\square$

Theorem 8.3.10. *For all ordinals α, β, and γ, if $2 \subseteq \alpha$, then*

(1) $\alpha^{\beta+\gamma} = \alpha^\beta \cdot \alpha^\gamma$,
(2) $(\alpha^\beta)^\gamma = \alpha^{\beta\cdot\gamma}$,
(3) *if* $\alpha^\beta = \alpha^\gamma$, *then* $\beta = \gamma$,
(4) $\beta \in \gamma$ iff $\alpha^\beta \in \alpha^\gamma$.

Exercises 8.3

*1. Let α be an ordinal and let $\varphi(\alpha, f, y)$ be a formula that expresses (8.14). Let $\mathcal{A}$ be an ordinal class function satisfying $\varphi(\alpha, \mathcal{A} \upharpoonright \beta, \mathcal{A}(\beta))$, for all $\beta \in \mathrm{On}$. Prove, by transfinite induction, that $\mathcal{A}(\beta) \in \mathrm{On}$, for all $\beta \in \mathrm{On}$.

2. Show that $\omega \in \omega + 1$ and that $2 \cdot \omega \neq \omega \cdot 2$.

3. Show that $(\omega + 1) \cdot 2 \neq \omega \cdot 2 + 1 \cdot 2$ and $(2 \cdot 2)^\omega \neq 2^\omega \cdot 2^\omega$.

4. Show that $2 + \omega \notin 3 + \omega$, $2 \cdot \omega \notin 3 \cdot \omega$, and $2^\omega \notin 3^\omega$.

5. Show that $\omega^2 = \sup\{\omega \cdot n : n \in \omega\}$.

6. Let α be an ordinal such that $1 + \alpha = \alpha$.
   (a) Prove that $1 + \alpha \cdot (1 + n) = \alpha \cdot (1 + n)$ for all $n \in \omega$,
   (b) Show that $1 + \alpha \cdot \omega = \alpha \cdot \omega$.
   (c) Show that $1 + \omega^{n+1} = \omega^{n+1}$ for each $n \in \omega$.
   (d) Let $i \in \omega$ and $n \in \omega$. Show that $\omega^i + \omega^{i+(n+1)} = \omega^{i+(n+1)}$.

7. Prove that for all ordinals α, if $\omega \subseteq \alpha$, then $1 + \alpha = \alpha$.

8. Suppose $\omega \subseteq \alpha$. Prove for all ordinals β, if $\alpha^2 \subseteq \beta$, then $\alpha + \beta = \beta$.

*9. Prove Lemma 8.3.2.

*10. Prove Lemma 8.3.4.

*11. Prove Lemma 8.3.9.

12. For every ordinal α, prove that there exists an ordinal $\beta \neq 0$ such that $\alpha + \beta = \beta$.

13. Let α be a nonzero ordinal. Prove that $\beta + \alpha \neq \beta$, for all ordinals β.

14. Prove Lemma 8.3.5(b).

15. Prove Theorem 8.3.6(1).

16. Prove Theorem 8.3.6(2).

17. Prove Theorem 8.3.7(1).

18. Prove the following:
   (a) Let α and γ be ordinals. If $\gamma \subseteq \alpha \in \gamma + \omega$, then $\alpha = \gamma + n$ for some $n \in \omega$.
   (b) For all ordinals ζ, we have that $\zeta + \omega$ is a limit ordinal.
   (c) For all ordinals ζ, if $\zeta \neq 0$, then $\omega \cdot \zeta$ is a limit ordinal.
   (d) If α is a limit ordinal, then $\alpha = \omega \cdot \zeta$ for some nonzero ordinal ζ.
   (e) If α is a successor ordinal, then $\alpha = \omega \cdot \zeta + n$ for some ordinal ζ and $n \in \omega$.
   (f) There exists an ordinal α such that $\alpha = \omega \cdot \alpha$.

19. Prove Theorem 8.3.10(1).

20. Prove Theorem 8.3.10(2).

21. (AC) Let α be a countable ordinal. Prove that for all ordinals β, if β is countable, then $\alpha + \beta$, $\alpha \cdot \beta$, and α^{β} are countable.

22. Let α be an ordinal. For each ordinal β, let $W^{\beta} = (\{0\} \times \alpha) \cup (\{1\} \times \beta)$. Define the relation $\preccurlyeq^{\beta}$ on W^{β} as follows: For all $\langle i, \zeta \rangle$ and $\langle j, \rho \rangle$ in W^{β},

$$\langle i, \zeta \rangle \preccurlyeq^{\beta} \langle j, \rho \rangle \ \text{iff}\ i \in j \vee (i = j \wedge \zeta \subseteq \rho).$$

Prove the following:
(a) $(W^{\beta}, \preccurlyeq^{\beta})$ is a well-ordered structure, for each ordinal β.
(b) For all ordinals β, the order type of $(W^{\beta}, \preccurlyeq^{\beta})$ equals $\alpha + \beta$.

23. Let α be an ordinal. For each ordinal β, let $W^{\beta} = \alpha \times \beta$. Define the relation $\preccurlyeq^{\beta}$ on W^{β} as follows: For all $\langle \mu, \zeta \rangle$ and $\langle v, \rho \rangle$ in W^{β},

$$\langle \mu, \zeta \rangle \preccurlyeq^{\beta} \langle v, \rho \rangle \ \text{iff}\ \zeta \in \rho \vee (\zeta = \rho \wedge \mu \subseteq v).$$

Prove the following:
(a) $(W^{\beta}, \preccurlyeq^{\beta})$ is a well-ordered structure for every ordinal β.
(b) For all ordinals β, the order type of $(W^{\beta}, \preccurlyeq^{\beta})$ equals $\alpha \cdot \beta$.

Exercise Notes: For Exercises 9–11, apply Theorem 8.2.9. Use Lemma 8.3.2 to prove Lemma 8.3.4, and then apply Lemma 8.3.4 to prove Lemma 8.3.9. For Exercise 18(d)(e), let η be the least ordinal such that $\alpha \in \omega \cdot \eta$. For Exercise 21, use Corollary 5.2.10 together with transfinite induction. For Exercises 22–23, see Exercise 19 on page 76 and Exercise 6 on page 174; also, use transfinite induction for part (b). Exercises 22–23 offer another way of defining ordinal addition and multiplication.

8.4 The Cumulative Hierarchy

The cumulative hierarchy is a collection of sets that are indexed by the ordinal numbers. We will be defining the sets in this collection by ordinal recursion. The intent is to build up sets by starting with the empty set and then proceed to construct new sets using the power set operation in a step by step manner, where the ordinals act as the steps.

Let $\varphi(f, y)$ be a formula that asserts the following definition by cases:

$$y = \begin{cases} \varnothing, & \text{if } f \text{ is the empty function;} \\ \mathcal{P}(f(\gamma)), & \text{if } f \text{ is a function with domain } \gamma^{+} \text{ for some ordinal } \gamma; \\ \bigcup f[\alpha], & \text{if } f \text{ is a function with domain } \alpha, \text{ a limit ordinal;} \\ \varnothing, & \text{if none of the above hold.} \end{cases}$$

Hence, for all f, there exists a unique y so that $\varphi(f, y)$. Theorem 8.2.4 implies that there exists an ordinal class function $\mathcal{V}$ satisfying $\varphi(\mathcal{V} \upharpoonright \alpha, \mathcal{V}(\alpha))$, for all $\alpha \in$ On. For each ordinal α, we let $V_\alpha = \mathcal{V}(\alpha)$, and thus,

$$
V_\alpha = \begin{cases}
\varnothing, & \text{if } \alpha = 0; \\
\mathcal{P}(V_\gamma), & \text{if } \alpha = \gamma^+ \text{ for some ordinal } \gamma; \\
\bigcup\{V_\gamma : \gamma \in \alpha\} & \text{if } \alpha \text{ is a limit ordinal.}
\end{cases}
\tag{8.19}
$$

The replacement axiom implies that $\{V_\gamma : \gamma \in \alpha\}$ is a set, for each $\alpha \in$ On. The class $\{V_\gamma : \gamma \in \text{On}\}$ is called the **cumulative hierarchy** of sets. Clearly, (8.19) asserts that

- $V_0 = \varnothing$,
- $V_{\gamma^+} = \mathcal{P}(V_\gamma)$ for any ordinal γ,
- $V_\alpha = \bigcup_{\gamma \in \alpha} V_\gamma$ for any limit ordinal α.

Theorem 8.4.1. *For every ordinal α, we have that V_α is a transitive set and that $(\forall \eta \in \alpha)(V_\eta \subseteq V_\alpha)$.*

Proof. We apply proof by transfinite induction. Let α be an ordinal. Assume the induction hypothesis

$$(\forall \xi \in \alpha)(V_\xi \text{ is a transitive set and } (\forall \eta \in \xi)(V_\eta \subseteq V_\xi)). \tag{IH}$$

We must prove that V_α is a transitive set and that $(\forall \eta \in \alpha)(V_\eta \subseteq V_\alpha)$.

CASE 1: $\alpha = 0$. Then $V_0 = \varnothing$ is a transitive set and $(\forall \eta \in 0)(V_\eta \subseteq V_0)$.

CASE 2: $\alpha = \gamma^+$ for an ordinal γ. Let $x \in y$ and $y \in V_{\gamma^+}$. Since $V_{\gamma^+} = \mathcal{P}(V_\gamma)$, we see that $y \subseteq V_\gamma$. As $x \in y$, we conclude that $x \in V_\gamma$. From (IH), because $\gamma \in \alpha$, we see that V_γ is a transitive set. So $x \subseteq V_\gamma$. Hence, $x \in \mathcal{P}(V_\gamma)$; that is, $x \in V_{\gamma^+}$. Thus, V_α is a transitive set. As $V_\alpha = \mathcal{P}(V_\gamma)$, it follows that $V_\gamma \in V_\alpha$. Hence, $V_\gamma \subseteq V_\alpha$, because V_α is a transitive set. Since $\alpha = \gamma^+$, the induction hypothesis (IH) now implies that $(\forall \eta \in \alpha)(V_\eta \subseteq V_\alpha)$.

CASE 3: α is a limit ordinal. Thus, $V_\alpha = \bigcup\{V_\eta : \eta \in \alpha\}$ by (8.19). Therefore, $(\forall \eta \in \alpha)(V_\eta \subseteq V_\alpha)$. Let $x \in y$ and $y \in V_\alpha$. So $y \in V_\xi$ for some $\xi \in \alpha$. By (IH), V_ξ is a transitive set. Since $x \in y$ and $y \in V_\xi$, we infer that $x \in V_\xi$. So $x \in V_\alpha$, and thus, V_α is a transitive set. $\qquad \square$

Theorem 8.4.2. *For all ordinals α, we have that $\alpha \subseteq V_\alpha$.*

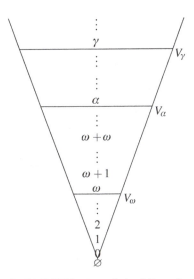

Figure 8.1. The cumulative hierarchy.

Proof. We will again apply proof by transfinite induction. Let α be an ordinal. Assume the induction hypothesis

$$(\forall \eta \in \alpha)(\eta \subseteq V_\eta). \tag{IH}$$

We must prove that $\alpha \subseteq V_\alpha$.

CASE 1: $\alpha = 0$. Clearly $0 \subseteq V_0$, as $0 = \varnothing$.

CASE 2: $\alpha = \gamma^+$ for an ordinal γ. Hence, $V_\alpha = \mathcal{P}(V_\gamma)$. So, ($\blacktriangle$) $\gamma \subseteq V_\gamma$ by (IH), since $\gamma \in \alpha$. Therefore,

(i) $\gamma \subseteq V_\alpha$ as $\gamma \subseteq V_\gamma \subseteq V_\alpha$, by ($\blacktriangle$) and Theorem 8.4.1;
(ii) $\gamma \in V_\alpha$ because $\gamma \subseteq V_\gamma$ and $V_\alpha = \mathcal{P}(V_\gamma)$.

Because $\alpha = \gamma \cup \{\gamma\}$, items (i) and (ii) imply that $\alpha \subseteq V_\alpha$.

CASE 3: α is a limit ordinal. Thus, $V_\alpha = \bigcup \{V_\eta : \eta \in \alpha\}$. Let $\eta \in \alpha$. By (IH), we have that $\eta \subseteq V_\eta$. Therefore, $\eta \in V_{\eta^+}$. Since α is a limit ordinal, it follows that $\eta^+ \in \alpha$. By Theorem 8.4.1, $V_{\eta^+} \subseteq V_\alpha$. Hence, $\eta \in V_\alpha$. Thus, $\alpha \subseteq V_\alpha$. $\square$

The cumulative hierarchy of sets and Theorems 8.4.1–8.4.2 are illustrated in Figure 8.1. We now discuss some consequences of the regularity axiom. One of which (Theorem 8.4.5) will allow us to determine the complexity of any set.

Regularity Axiom. Every nonempty set A contains an element that is disjoint from A.

The regularity axiom implies that $x \notin x$ for all x. In mathematics, there are no sets that are members of themselves. For example, the set of real numbers $\mathbb{R}$ is not a real number. So $\mathbb{R} \notin \mathbb{R}$. The regularity axiom, also known as the axiom of foundation, eliminates sets that are not relevant for standard mathematics. The next theorem identifies three basic consequences of regularity.

Theorem 8.4.3. *The axiom of regularity implies the following:*
(1) *No set is a member of itself.*
(2) *There are no sets x and y such that $x \in y$ and $y \in x$.*
(3) *No function f with domain ω satisfies $f(n^+) \in f(n)$ for all $n \in \omega$.*

Proof. We shall prove only item (3). Suppose, for a contradiction, that there exists a function f with domain ω satisfying ($\blacktriangle$) $f(n^+) \in f(n)$ for all $n \in \omega$. Let $A = \{f(n) : n \in \omega\}$. So A is nonempty. The regularity axiom implies that there is a set $a \in A$ such that $a \cap A = \varnothing$. Since $a \in A$, there is an $n \in \omega$ such that $a = f(n)$. Clearly, $f(n^+) \in A$. Moreover, because $a = f(n)$, we conclude from ($\blacktriangle$) that $f(n^+) \in a$. Hence, $f(n^+) \in a \cap A$ and therefore $a \cap A \neq \varnothing$, which is a contradiction. $\square$

Let V be the class $V = \bigcup_{\alpha \in \text{On}} V_\alpha$; that is, $x \in V$ if and only if $x \in V_\alpha$ for some ordinal α. More formally, $V = \{x : \exists \alpha(\alpha \in \text{On} \wedge x \in V_\alpha)\}$. The next lemma shows that if a set is a "subset" of V, then it is in V.

Lemma 8.4.4. *Let A be a set. If for all $x \in A$, there is an ordinal α such that $x \in V_\alpha$, then $A \in V_\beta$ for some ordinal β.*

Proof. Assume that for all $x \in A$, there is an ordinal α such that $x \in V_\alpha$. Let $\varphi(x, y)$ be a formula that expresses the following definition by cases:

$$y = \begin{cases} \alpha, & \text{if } \alpha \text{ is the least ordinal such that } x \in V_\alpha; \\ \varnothing, & \text{if the above fails to hold.} \end{cases} \quad (8.20)$$

Hence, for all x, there exists a unique y so that $\varphi(x, y)$. Thus, by our assumption, for all $x \in A$, there is a unique nonzero ordinal α such that $\varphi(x, \alpha)$. By the replacement axiom there is a set S such that $\alpha \in S$ iff $\varphi(x, \alpha)$ for some $x \in A$. Let $\gamma = \sup S$. To show that $A \subseteq V_\gamma$, let $x \in A$. Thus, $x \in V_\alpha$ for some $\alpha \in S$. As $\alpha \leq \gamma$, Theorem 8.4.1 implies that $x \in V_\gamma$. Hence, $A \subseteq V_\gamma$ and $A \in V_{\gamma^+}$. $\square$

Theorem 8.4.5. *Every set is an element of the class* V.

Proof. Let A be a set. By Theorem 6.2.8, let $\overline{A}$ be a transitive set so that $A \subseteq \overline{A}$. We will first show that

$$\text{for each } x \in \overline{A}, \text{ there is an ordinal } \beta \text{ such that } x \in V_\beta. \qquad (8.21)$$

Suppose, to the contrary, that this is not the case. Therefore, the set

$$B = \{x \in \overline{A} : x \notin V_\beta \text{ for all ordinals } \beta\}$$

is nonempty. By the regularity axiom, there is a set $b \in B$ such that $b \cap B = \varnothing$. Hence, (▲) $(\forall x \in b)(x \notin B)$. Since $b \in \overline{A}$ and $\overline{A}$ is a transitive set, we infer that $b \subseteq \overline{A}$. Thus, (▲) implies that for all $x \in b$, there is an ordinal α such that $x \in V_\alpha$. So by Lemma 8.4.4, we must have that $b \in V_\beta$ for some ordinal β. Therefore, $b \notin B$, a contradiction. So (8.21) holds and Lemma 8.4.4 implies that $\overline{A} \subseteq V_\gamma$ for some ordinal γ. Since $A \subseteq \overline{A}$, we have that $A \subseteq V_\gamma$. Hence, $A \in V_{\gamma+}$. $\qquad\square$

Theorem 8.4.5 allows us now identify V as the **universe of sets**. Since every set belongs to V, we can now measure the complexity of any set, also called the rank of a set. The proof of Theorem 8.4.5 shows that for each set A, there is an ordinal γ such that $A \subseteq V_\gamma$.

Definition 8.4.6. The **rank** of a set A, denoted by $\text{rank}(A)$, is the least ordinal γ such that $A \subseteq V_\gamma$.

Thus, if $\text{rank}(A) = \gamma$, then $A \in V_{\gamma+}$. The rank of a set allows us to specify the particular stage of the cumulative hierarchy at which the set is first constructed.

Theorem 8.4.7. *The following two properties of the rank operation hold:*

(1) *If* $a \in A$, *then* $\text{rank}(a) \in \text{rank}(A)$.
(2) $\text{rank}(A) = \sup\{\text{rank}(a)^+ : a \in A\}$.

Proof. First, we prove item (1). Let $\alpha = \text{rank}(A)$. So $A \subseteq V_\alpha$. Let $a \in A$. Hence, $a \in V_\alpha$. Thus, $a \in \mathcal{P}(V_\beta)$ for some $\beta \in \alpha$ (see Exercise 11). Therefore, $a \subseteq V_\beta$. Since $\beta \in \alpha = \text{rank}(A)$, we conclude that $\text{rank}(a) \in \text{rank}(A)$.

To prove (2), the replacement axiom implies that $\{\text{rank}(a)^+ : a \in A\}$ is a set, that is, a set of ordinals. Hence, by Theorem 8.1.15(2) and Definition 8.1.16, we conclude that $\beta = \sup\{\text{rank}(a)^+ : a \in A\}$ is an ordinal. We will now show that $\text{rank}(A) \in \beta$ and $\beta \subseteq \text{rank}(A)$. First, observe that $\text{rank}(a) \in \text{rank}(a)^+ \in \beta$, for each $a \in A$. So by Exercise 15, $a \in V_\beta$ for all $a \in A$. Thus, $A \subseteq V_\beta$. Hence, $\text{rank}(A) \in \beta$. Now we show that $\beta \subseteq \text{rank}(A)$. From item (1), we conclude that $\text{rank}(a) \in \text{rank}(A)$ for all $a \in A$. Hence, $\text{rank}(a)^+ \subseteq \text{rank}(A)$ for all $a \in A$ (see Exercise 9(b) on page 185). Therefore, $\beta \subseteq \text{rank}(A)$ by Exercise 13 on page 186. Since $\text{rank}(A) \in \beta$ and $\beta \subseteq \text{rank}(A)$, Corollary 8.1.14(2) implies the equality $\text{rank}(A) = \beta$. □

Exercises 8.4

1. Prove Theorem 8.4.3(1).

2. Prove Theorem 8.4.3(2).

3. Prove that if $a \in x$, then $a \neq x$.

4. Prove that if $x^+ = y^+$, then $x = y$.

5. Prove, by induction, that V_n is finite, for all $n \in \omega$.

6. Using Exercise 5 and Theorem 5.2.9, show that V_ω is countable.

7. Suppose that $A \in V_\omega$. Using Exercise 5, prove that A is finite.

8. Suppose that A is finite and $A \subseteq V_\omega$. Prove that $A \in V_k$ for some $k \in \omega$.

9. Let α be a limit ordinal. Prove that V_α is an inductive set.

10. Show that if $A \subseteq B$, then $\text{rank}(A) \subseteq \text{rank}(B)$.

*11. Suppose that $a \in V_\alpha$. Show that $a \in \mathcal{P}(V_\beta)$ for some $\beta \in \alpha$.

12. Let α be a limit ordinal. Let $x \in V_\alpha$ and $y \in V_\alpha$. Prove that $\{x, y\} \in V_\alpha$.

13. Let $X \in V_\alpha$ where α is a limit ordinal. Prove that $\bigcup X \in V_\alpha$ and prove that $\mathcal{P}(X) \in V_\alpha$.

14. Prove that for all ordinals α and γ, if $\gamma \in \alpha$, then $V_\gamma \in V_\alpha$.

*15. Let α be an ordinal. Prove that $a \in V_\alpha$ if and only if $\text{rank}(a) \in \alpha$.

16. Prove that if $\text{rank}(A) = \alpha$, then α is the least ordinal such that $A \in V_{\alpha^+}$.

17. Let A be a set. Prove that $\text{rank}(\{A\}) = (\text{rank}(A))^+$.

18. Let A be a set. Prove that $\mathrm{rank}(\mathcal{P}(A)) = (\mathrm{rank}(A))^+$.

19. Let A be a set. Prove that $\mathrm{rank}(\bigcup A) \in \mathrm{rank}(A)$.

20. Let x and y be sets and let $\alpha = \max(\mathrm{rank}(x), \mathrm{rank}(y))$.
    (a) Show that $\mathrm{rank}(x \cup y) = \alpha$.
    (b) Show that $\mathrm{rank}(\{x, y\}) = \alpha^+$.

21. Prove the following:
    (a) $\alpha = \sup\{\beta^+ : \beta \in \alpha\}$, whenever α is an ordinal.
    (b) For every ordinal α, the equality $\mathrm{rank}(\alpha) = \alpha$ holds.

22. Let $\overline{A}$ be the transitive closure of A. Prove that $\mathrm{rank}(A) = \mathrm{rank}(\overline{A})$.

23. Let A be a set and let $\varphi(x, y)$ be a formula. Suppose that

$(\forall x \in A)\exists y \varphi(x, y)$.

Prove that there exists a set B such that $(\forall x \in A)(\exists y \in B)\varphi(x, y)$.

Exercise Notes: For Exercise **1**, let a be a set, and then consider the set $\{a\}$. In Exercise **6**, V_ω is said to be the set of all *hereditarily finite sets*. For Exercise **8**, use Exercise 6 on page 116. For Exercise **22**, review both Definition 6.2.9 and Exercise 6 on page 155.

9

Cardinals

In Chapter 5, we showed that the natural numbers allow us to measure the size of every finite set; that is, for any finite A, we defined $|A| = n$ so that the natural number n identifies the number of elements in A. Theorem 5.1.7 showed that two finite sets have the same number of elements if and only if there exists a one-to-one correspondence between the two sets. On the other hand, we were not able to define the "number" $|A|$ when A is an infinite set; however, we did define the expression $|A| =_c |B|$ to mean that "A has the same cardinality as B." In other words, $|A| =_c |B|$ means that "there is a one-to-one function $f: A \to B$ that is onto B." This was done because we did not have "infinite numbers." Now that we have the ordinals, we can define $|A|$ to be a particular ordinal whenever A can be well-ordered. We will then be able to show that the ordinals $|A|$ and $|B|$ are equal if and only if there is a bijection $f: A \to B$.

9.1 Cardinal Numbers

His discovery of uncountable sets led Cantor to develop the cardinal numbers, which John von Neumann later refined via the following definition.

Definition 9.1.1. A **cardinal number**, or **cardinal**, is an ordinal κ such that for every $\beta \in \kappa$ there is no one-to-one function $f: \kappa \to \beta$.

Theorem 9.1.2. *Every natural number is a cardinal. The ordinal ω is a cardinal. Every infinite cardinal is a limit ordinal.*

Proof. Let $n \in \omega$. Corollary 5.1.9 states that for every $m \in n$, there does not exist a one-to-one function $f: n \to m$. Thus, every natural number is a cardinal. Theorem 5.1.10 implies that for every $n \in \omega$, there is no one-to-one function $f: \omega \to n$. Therefore, ω is also a cardinal. Let κ be an infinite cardinal.

Thus, $\omega \in \kappa$. We will now prove that κ is a limit ordinal. Suppose, for a contradiction, that κ is a successor ordinal. Hence, $\kappa = \alpha^+$ for some ordinal α. So $\alpha \in \kappa$. Consider the function $f : \kappa \to \alpha$ defined by

$$f(\xi) = \begin{cases} 0, & \text{if } \xi = \alpha; \\ \xi^+, & \text{if } \xi \in \omega; \\ \xi, & \text{if } \omega \subseteq \xi. \end{cases}$$

Clearly f is one-to-one, and thus, κ is not a cardinal. $\square$

Theorem 9.1.2 shows that ω^+ is not a cardinal. We now show that an ordinal α is a cardinal if and only if $|\alpha| \neq_c |\gamma|$, for all $\gamma \in \alpha$.

Theorem 9.1.3. *Let α be an ordinal number. Then α is a cardinal if and only if for every $\gamma \in \alpha$, there is no bijection $f : \alpha \to \gamma$.*

Proof. Let α be an ordinal. Assume that α is a cardinal. Let $\gamma \in \alpha$. As there is no one-to-one function $g : \alpha \to \gamma$, there is no bijection $f : \alpha \to \gamma$. Conversely, assume that ($\blacktriangle$) for every $\gamma \in \alpha$, there is no bijection $f : \alpha \to \gamma$. Suppose, for a contradiction, that α is not a cardinal. Hence, for some $\gamma \in \alpha$, there exists a function $f : \alpha \to \gamma$ that is one-to-one. Let $C = \text{ran}(f)$. Since $C \subseteq \gamma$, if follows that $(C, \in_C)$ is a well-ordered structure (see Exercise 5 on page 145). Let β be the order type of $(C, \in_C)$. Thus, by Exercise 8 on page 185, we have that $\beta \subseteq \gamma$. Hence, $\beta \in \alpha$. By Theorem 8.1.4, $H : C \to \beta$ is a bijection. Because $f : \alpha \to C$ is a bijection, $(H \circ f) : \alpha \to \beta$ is a bijection, which contradicts ($\blacktriangle$). $\square$

Theorem 9.1.4. *Let A be a set. Then the following are equivalent:*

(1) *A has a well-ordering.*
(2) *There is a bijection $f : \alpha \to A$ for some ordinal α.*
(3) *There is a one-to-one function $f : A \to \alpha$ for some ordinal α.*

Proof. We shall show that $(1) \Rightarrow (2) \Rightarrow (3) \Rightarrow (1)$.

$(1) \Rightarrow (2)$: Assume A has a well-ordering. Theorem 8.1.4 and its proof show that there is an ordinal δ and a bijection $H : A \to \delta$. Theorem 3.3.18 therefore implies that $H^{-1} : \delta \to A$ is a bijection.

$(2) \Rightarrow (3)$: Suppose that $f : \alpha \to A$ is a bijection for some ordinal α. Then the inverse function $f^{-1} : A \to \alpha$ is one-to-one by Theorem 3.3.18.

$(3) \Rightarrow (1)$: Assume that there is an injection $f: A \to \alpha$ for some ordinal α. Exercise 6 shows that A has a well-ordering. □

The next definition is a straightforward generalization of Definition 5.1.2.

Definition 9.1.5. If the set A has a well-ordering, then $|A|$ is the least ordinal κ for which there is a one-to-one function $f: A \to \kappa$.

Thus, if κ is a cardinal, then $|\kappa| = \kappa$ (see Exercise 2).

Lemma 9.1.6. *Suppose that A has a well-ordering, and let $\kappa = |A|$. Then κ is a cardinal, and there is a bijection $H: A \to \kappa$.*

Proof. Let A be a set with a well-ordering and let $\kappa = |A|$. Exercise 9 implies that κ is a cardinal. Since $\kappa = |A|$, there exists a function $f: A \to \kappa$ that is one-to-one. Define the relation $\preccurlyeq$ on A by $x \preccurlyeq y$ iff $f(x) \in f(y)$, for all x and y in A. Exercise 6 shows that $(A, \preccurlyeq)$ is a well-ordered structure. Let λ be the order type of $(A, \preccurlyeq)$. Exercise 7 on page 185 and Lemma 8.1.11 imply that $\lambda \in \kappa$. Theorem 8.1.10 implies that there is a bijection $H: A \to \lambda$. Because H is one-to-one, $\lambda \in \kappa$, and $|A| = \kappa$, it follows from Definition 9.1.5 that $\lambda = \kappa$. Therefore, $H: A \to \kappa$ is the desired bijection. □

Lemma 9.1.7. *Let A and B be well-ordered sets. Then*

(1) $|A| \leq_c |B|$ iff $|A| \subseteq |B|$,
(2) $|A| =_c |B|$ iff $|A| = |B|$,
(3) $|A| <_c |B|$ iff $|A| \in |B|$.

Proof. See Exercise 11. □

Let C be the collection $C = \{\kappa : \kappa \text{ is a cardinal}\}$. Is C a set, or is C a proper class? Friedrich Hartogs, in 1915, proved the following theorem from ZF alone (that is, without using the axiom of choice). This theorem implies that C is a proper class (see Exercise 4).

Hartogs' Theorem 9.1.8. *For every set X, there is a cardinal κ so that there is no one-to-one function $f: \kappa \to X$.*

Proof. We begin by defining the set W (via Theorem 2.1.3) to be

$$W = \{(A, \preceq) : A \subseteq X \text{ and } (A, \preceq) \text{ is a well-ordered structure}\}.$$

By Theorem 8.1.4, for each $(A, \preceq) \in W$, there exists a unique ordinal δ that is the order type of $(A, \preceq)$ (see Definition 8.1.9). Thus, the replacement axiom implies that the following collection is a set:

$$\kappa = \{\delta : \delta \text{ is the order type of some } (A, \preceq) \in W\}. \tag{9.1}$$

We will show that κ is the desired cardinal by means of the following three claims. The proof of the first claim shows that κ is an ordinal, that is, κ is a transitive set and it is well-ordered by the relation $\in$.

Claim 1. κ is an ordinal.

Proof. We apply Definition 8.1.1. To show that κ is a transitive set, let $\gamma \in \delta$ and $\delta \in \kappa$. As $\delta \in \kappa$, there is a well-ordered structure $(A, \preceq) \in W$ that has order type δ. By Theorem 8.1.10, there is a bijection $H : A \to \delta$. Let $B = H^{-1}[\gamma]$. Thus, $B \subseteq A \subseteq X$. Since H is a bijection, it follows that $(H \restriction B) : B \to \gamma$ is also a bijection. Exercise 6 implies that there is a well-ordering on B of order type γ. Hence, $\gamma \in \kappa$. Thus, κ is a transitive set. Theorem 8.1.15(1) implies that κ is well-ordered by $\in$, as κ is a set of ordinals. So κ is an ordinal.

(Claim 1) □

Claim 2. There is no one-to-one function $f : \kappa \to X$.

Proof. Suppose that $f : \kappa \to X$ is one-to-one. Let $A = \text{ran}(f)$. Thus, $A \subseteq X$ and $f : \kappa \to A$ is a bijection. So $f^{-1} : A \to \kappa$ is also a bijection. Exercise 6 shows that there is a well-ordering $\preceq$ on A such that κ is the order type of $(A, \preceq)$. Thus, $(A, \preceq) \in W$ and $\kappa \in \kappa$, contradicting Lemma 8.1.3(2).

(Claim 2) □

Claim 3. κ is a cardinal.

Proof. Assume, to the contrary, that there is an injection $g : \kappa \to \alpha$ for some $\alpha \in \kappa$. Since $\alpha \in \kappa$, (9.1) implies there is a well-ordered structure $(A, \preceq)$ where $(A, \preceq)$ has order type α and $A \subseteq X$. So there is a bijection $h : \alpha \to A$.[1] Hence, $(h \circ g) : \kappa \to X$ is one-to-one, which contradicts Claim 2. (Claim 3) □

Thus, κ is a cardinal, and there is no injection $f : \kappa \to X$. (Theorem) □

Corollary 9.1.9. *For every ordinal γ, there is a cardinal κ such that $\gamma \in \kappa$.*

Proof. Let γ be an ordinal. Theorem 9.1.8 implies that there exists a cardinal κ such that there is no one-to-one function $f: \kappa \to \gamma$. Thus, $\gamma \in \kappa$. $\qquad\square$

Corollary 9.1.9 implies that there is a cardinal κ such that $\omega \in \kappa$. It follows that κ is uncountable. To see this, observe that if κ were countable, then there would be a function $f: \kappa \to \omega$ that is one-to-one, and this would imply that κ is not a cardinal. Hence, κ is uncountable.

Theorem 9.1.10. *Let C be a set of cardinals. Then*

(1) sup C *is a cardinal.*
(2) *Let $\kappa = $ sup C. If C is a limit set, then $\delta \in \kappa$ for all $\delta \in C$.*

Proof. See Exercise 5. $\qquad\square$

The aleph number system is used to identify the infinite cardinal numbers. The system is defined by transfinite recursion. As ω is the first infinite cardinal, ω will be the starting value of this recursion. We also need to define a "next cardinal" function. By Corollary 9.1.9, we know that for each ordinal α, there is a cardinal that is "larger" than α. Let $\mathcal{S}: \mathrm{On} \to \mathrm{On}$ be defined by

$$\mathcal{S}(\alpha) = \text{the least cardinal } \kappa \text{ such that } \alpha \in \kappa.$$

Clearly, $\mathcal{S}$ is a class function. Consider the following definition by cases:

$$y = \begin{cases} \omega, & \text{if } f \text{ is the empty function;} \\ \mathcal{S}(f(\gamma)), & \text{if } f: \gamma^+ \to \mathrm{On} \text{ for some ordinal } \gamma^2 \\ \sup f[\beta], & \text{if } f: \beta \to \mathrm{On} \text{ for some limit ordinal } \beta; \\ \varnothing, & \text{if none of the above hold.} \end{cases} \tag{9.2}$$

Let $\varphi(f, y)$ be a formula that expresses (9.2). Hence, by Corollary 9.1.9, for every set f, there exists a unique y so that $\varphi(f, y)$. Thus, Theorem 8.2.4 implies that there exists a class function $\aleph: \mathrm{On} \to \mathrm{On}$ such that[3]

$$\aleph(\beta) = \begin{cases} \omega, & \text{if } \beta = 0; \\ \mathcal{S}(\aleph(\gamma)), & \text{if } \beta = \gamma^+ \text{ for some ordinal } \gamma; \\ \sup \aleph[\beta], & \text{if } \beta \text{ is a limit ordinal.} \end{cases} \tag{9.3}$$

Note that $\mathcal{S}(\aleph(\gamma)) = \aleph(\gamma + 1)$. Corollary 9.1.9 and Theorem 9.1.10 imply that $\aleph(\beta)$ is an infinite cardinal for every ordinal β. In addition, for each infinite cardinal κ, there is an ordinal α such that $\aleph(\alpha) = \kappa$ (see Theorem 9.1.11).

Exercise 15 and (9.3) imply that $\aleph: \mathrm{On} \to \mathrm{On}$ is normal. For each ordinal β, we now let $\aleph_\beta = \aleph(\beta)$. With this slight change in notation, (9.3) becomes

$$\aleph_\beta = \begin{cases} \aleph_0, & \text{if } \beta = 0; \\ \aleph_{\gamma+1}, & \text{if } \beta = \gamma^+ \text{ for some ordinal } \gamma; \\ \sup\{\aleph_\alpha : \alpha \in \beta\}, & \text{if } \beta \text{ is a limit ordinal.} \end{cases} \quad (9.4)$$

The recursive definition (9.4) allows us to use the ordinals to identify all of the infinite cardinal numbers, in increasing order, as follows:

$$\aleph_0 \in \aleph_1 \in \cdots \in \aleph_\omega \in \aleph_{\omega+1} \in \cdots \in \aleph_\alpha \in \aleph_{\alpha+1} \in \cdots.$$

It follows that $\aleph_0 = \omega$. So, $\aleph_0$ is the first infinite cardinal. $\aleph_0$ is often referred to as *aleph-naught*. Since $\omega \in \aleph_1$, we see that $\aleph_1$ (aleph-one) is the smallest uncountable cardinal.

Our next theorem verifies that the class function $\aleph: \mathrm{On} \to \mathrm{On}$ is increasing and "enumerates" all of the cardinals.

Theorem 9.1.11. *For all ordinals α and β, if $\alpha \in \beta$, then $\aleph_\alpha \in \aleph_\beta$. Moreover, for every infinite cardinal κ, there is an ordinal α such that $\aleph_\alpha = \kappa$.*

Proof. See Exercise 15 and Exercise 20. $\qquad\qquad\qquad\qquad\qquad\qquad\square$

Let γ be a limit ordinal. Recalling Definition 8.2.11, a set $C \subseteq \gamma$ is said to be **cofinal** in γ if for all $\alpha \in \gamma$, there is a $\beta \in C$ such that $\alpha \in \beta$.

Definition 9.1.12. Let γ be a limit ordinal. Then the **cofinality** of γ, denoted by $\mathrm{cf}(\gamma)$, is the least ordinal $\alpha \in \gamma$ for which there exists a function $f: \alpha \to \gamma$ such that $f[\alpha]$ is cofinal in γ.

Theorem 9.1.13. *Let γ be a limit ordinal. Then $\mathrm{cf}(\gamma)$ is a cardinal.*

Proof. See Exercise 13. $\qquad\qquad\qquad\qquad\qquad\qquad\qquad\qquad\qquad\quad\square$

Since every infinite cardinal κ is a limit ordinal, the cofinality of κ is defined.

Definition 9.1.14. An infinite cardinal κ is said to be **regular** if $\mathrm{cf}(\kappa) = \kappa$. If $\mathrm{cf}(\kappa) \in \kappa$, then κ is said to be **singular**.

Thus, an infinite cardinal κ is regular if and only if whenever $f: \alpha \to \kappa$ and $\alpha \in \kappa$, then there is $\gamma \in \kappa$ such that $f[\alpha] \subseteq \gamma$. The cardinal $\aleph_0$ is regular because for every $n \in \omega$ and each function $f: n \to \omega$, the set $f[n]$ is finite (see Exercise 5 on page 116). Hence, $f[n]$ cannot be cofinal in ω. Moreover, since ω is a limit ordinal, we see that $\aleph_\omega = \sup\{\aleph_n : n \in \omega\}$ by (9.4). So $\aleph_\omega$ is a singular cardinal as the function $f: \omega \to \aleph_\omega$ defined by $f(n) = \aleph_n$ is such that $f[\omega]$ is cofinal in $\aleph_\omega$.

Theorem 9.1.15. *Let γ be a limit ordinal. Then* $\mathrm{cf}(\gamma)$ *is a regular cardinal.*

Proof. See Exercise 14.   □

Theorem 9.1.16 (AC). *The cardinal $\aleph_1$ is regular.*

Proof. Suppose, to the contrary, that $\aleph_1$ is not regular. Thus, there is an ordinal $\alpha \in \aleph_1$ and a function $f: \alpha \to \aleph_1$ such that $f[\alpha]$ is cofinal in $\aleph_1$. Exercise 21 implies that α is countable and that $f(\eta)$ is countable for all $\eta \in \alpha$. Hence, $\bigcup\{f(\eta) : \eta \in \alpha\}$ is countable by Corollary 5.2.10. Furthermore, Exercise 22 implies that $\bigcup\{f(\eta) : \eta \in \alpha\} = \aleph_1$. Thus, $\aleph_1$ is countable, a contradiction.   □

In the next section, we will extend Theorem 9.1.16 by showing that $\aleph_{\alpha+1}$ is a regular cardinal, for each ordinal α.

For each ordinal γ, we know that the relation $\in$ yields a well-ordering on γ. In our next lemma, we will prove that there also exists a well-ordering on the set $\gamma \times \gamma$. We first define the relation $\preccurlyeq_\gamma$ on $\gamma \times \gamma$, which orders the ordinal pairs in $\gamma \times \gamma$, first by maximum, then by first component, and then by second component. Recall that $\max(\alpha, \beta)$ is the larger of the two ordinals α and β.

Definition 9.1.17. Let γ be an ordinal. The relation $\preccurlyeq_\gamma$ on $\gamma \times \gamma$ is defined by $\langle \alpha, \beta \rangle \preccurlyeq_\gamma \langle \delta, \gamma \rangle$ if and only if

$\max(\alpha, \beta) \in \max(\delta, \gamma)$

$\quad \vee\, [\max(\alpha, \beta) = \max(\delta, \gamma) \wedge \alpha \in \delta]$

$\quad \vee\, [\max(\alpha, \beta) = \max(\delta, \gamma) \wedge \alpha = \delta \wedge \beta \in \gamma].$

Lemma 9.1.18. *For each ordinal γ, the relation $\preccurlyeq_\gamma$ is a well-ordering on $\gamma \times \gamma$.*

Proof. Let γ be an ordinal. It is straightforward to show that $\preccurlyeq_\gamma$ is a total order on $\gamma \times \gamma$ (see Exercise 24). So we will show only that $\preccurlyeq_\gamma$ is a well-ordering. Let $A \subseteq \gamma \times \gamma$ be a nonempty set. Let $M = \{\max(\alpha, \beta) : \langle \alpha, \beta \rangle \in A\}$. Clearly, M is a nonempty set of ordinals, and so it has an $\in$-least element μ. Now let

$$K = \{\alpha \in \gamma : (\exists \beta \in \gamma)(\langle \alpha, \beta \rangle \in A \wedge \max(\alpha, \beta) = \mu)\}.$$

Thus, K is another nonempty set of ordinals with an $\in$-least element ν. Finally, let $S = \{\beta \in \gamma : \langle \nu, \beta \rangle \in A \wedge \max(\nu, \beta) = \mu\}$. Since S is also nonempty, it has an $\in$-least element η. One can now show that $\langle \nu, \eta \rangle \in A$ and $\langle \nu, \eta \rangle \preccurlyeq_\gamma \langle \alpha, \beta \rangle$ for all $\langle \alpha, \beta \rangle \in A$. Therefore, $\preccurlyeq_\gamma$ is a well-ordering on $\gamma \times \gamma$. $\square$

The following result is fairly easy to verify (see Exercise 25).

Proposition 9.1.19. *Let γ be an ordinal. For each $\alpha \in \gamma$,*

$$\alpha \times \alpha = \{\langle \xi, \eta \rangle \in \gamma \times \gamma : \langle \xi, \eta \rangle \prec_\gamma \langle 0, \alpha \rangle\}.$$

A set $C \subseteq \gamma \times \gamma$ is said to be *cofinal* in $\gamma \times \gamma$ if for all $\langle \alpha, \beta \rangle \in \gamma \times \gamma$, there is a $\langle \eta, \xi \rangle \in C$ such that $\langle \alpha, \beta \rangle \preccurlyeq_\gamma \langle \eta, \xi \rangle$.

Lemma 9.1.20. *Let γ be an ordinal and δ be the order type of $\preccurlyeq_\gamma$. Then $\gamma \subseteq \delta$, and there is a function $f : \gamma \to \gamma \times \gamma$ such that $f[\gamma]$ is cofinal in $\gamma \times \gamma$.*

Proof. Let γ be an ordinal and $f : \gamma \to \gamma \times \gamma$ be defined by $f(\alpha) = \langle \alpha, \alpha \rangle$. It thus follows that $\alpha \in \beta$ if and only if $f(\alpha) \prec_\gamma f(\beta)$, for all α and β in γ. Thus, by Lemma 8.1.11 and Exercise 7 on page 185, we have $\gamma \subseteq \delta$ where δ is the order type of $\preccurlyeq_\gamma$. Let $\langle \eta, \xi \rangle \in \gamma \times \gamma$ and let $\alpha = \max(\eta, \xi)$. Thus, $\langle \eta, \xi \rangle \preccurlyeq_\gamma \langle \alpha, \alpha \rangle$, and therefore, $f[\gamma]$ is cofinal in $\gamma \times \gamma$. $\square$

So for each ordinal γ, we now know that the set $\gamma \times \gamma$ has a well-ordering. Lemma 9.1.6 therefore implies that the cardinal $|\gamma \times \gamma|$ is defined. When γ is an infinite ordinal, one might think that the cardinal $|\gamma \times \gamma|$ would be larger than the cardinal $|\gamma|$. In fact, it is not, as we will now establish.

Theorem 9.1.21. *Let γ be an infinite ordinal. Then $|\gamma \times \gamma| = |\gamma|$.*

Proof. Suppose, for a contradiction, that there exists an infinite ordinal γ such that ($\blacktriangle$) $|\gamma \times \gamma| \neq |\gamma|$. By Theorem 8.1.21, we shall assume (without loss of generality) that γ is the least such ordinal. Thus,

$$(\forall \alpha \in \gamma)(\omega \subseteq \alpha \to |\alpha \times \alpha| = |\alpha|). \tag{9.5}$$

Exercise 9 on page 116 implies that $|\omega \times \omega| = |\omega|$. Hence, $\omega \in \gamma$. We now consider two cases: either γ is not a cardinal, or it is a cardinal.

CASE 1: γ is not a cardinal. So Theorem 9.1.3 implies that there is a bijection $f: \gamma \to \alpha$ for some ordinal $\alpha \in \gamma$. Since γ is infinite, α is also infinite. Thus, by (9.5), we have that $|\alpha \times \alpha| = |\alpha|$. Using f, one can easily define a bijection $g: \gamma \times \gamma \to \alpha \times \alpha$ and conclude (via Lemma 9.1.7(2)) that $|\gamma \times \gamma| = |\gamma|$, which contradicts (▲).

CASE 2: γ is a cardinal. Because γ is an ordinal, Lemma 9.1.18 implies that $(\gamma \times \gamma, \preccurlyeq_\gamma)$ is a well-ordered structure. Let δ be the order type of this structure, and let $H: \gamma \times \gamma \to \delta$ be the isomorphism from $(\gamma \times \gamma, \preccurlyeq_\gamma)$ onto $(\delta, \in)$ (see Remark 8.1.8). Lemma 9.1.20 states that $\gamma \in \delta$. As H is a bijection, (▲) implies that $\gamma \in \delta$. Hence, there exists $\eta \in \gamma$ and $\nu \in \gamma$ so that $H(\langle \eta, \nu \rangle) = \gamma$. Let $\alpha = \max(\eta, \nu) + 1$. So, as γ is a limit ordinal, $\alpha \in \gamma$. Since $\langle \eta, \nu \rangle \prec_\gamma \langle 0, \alpha \rangle$, Remark 8.1.8 and Proposition 9.1.19 imply that the restricted function $H^{-1} \upharpoonright \gamma$ satisfies

$$(H^{-1} \upharpoonright \gamma): \gamma \to \alpha \times \alpha.$$

Because $H^{-1} \upharpoonright \gamma$ is one-to-one and γ is infinite, it follows that α is infinite. So $|\alpha \times \alpha| = |\alpha|$ by (9.5). Since $H^{-1} \upharpoonright \gamma$ is one-to-one and $|\alpha \times \alpha| = |\alpha|$, there is an injection $f: \gamma \to \alpha$. As $\alpha \in \gamma$, we infer that γ is not a cardinal. □

Corollary 9.1.22. *If κ is an infinite cardinal, then $|\kappa \times \kappa| = \kappa$.*

Corollary 9.1.23. *Suppose that α and β are nonzero ordinals. If either α or β is infinite, then $|\alpha \times \beta| = \max(|\alpha|, |\beta|)$.*

Proof. Suppose that at least one of the nonzero ordinals α and β is infinite. We shall assume, without loss of generality, that $\alpha \in \beta$. Thus, β is infinite, and as $0 \in \alpha$, we have that

$$\{0\} \times \beta \subseteq \alpha \times \beta \subseteq \beta \times \beta.$$

Since $\beta \times \beta$ has a well-ordering, every subset of $\beta \times \beta$ also has a well-ordering. Clearly, $|\{0\} \times \beta| = |\beta|$ and $|\beta \times \beta| = |\beta|$. So $|\alpha \times \beta| = |\beta|$. □

Corollary 9.1.24. *Let κ and δ be nonzero cardinals. Then $|\kappa \times \delta| = \max(\kappa, \delta)$ if at least one of the cardinals is infinite.*

1. Let α be an ordinal. Show that $|\alpha| \subseteq \alpha$.

*2. Let κ be a cardinal. Prove that $|\kappa| = \kappa$.

3. Prove that $|\omega^+| = \omega$.

*4. Prove that the class $\{\kappa : \kappa \text{ is a cardinal}\}$ is not a set.

*5. Prove Theorem 9.1.10.

*6. Let A be a set and α be an ordinal. Suppose that $f : A \to \alpha$ is one-to-one. Define a relation $\preccurlyeq$ on A by $x \preccurlyeq y$ if and only if $f(x) \subseteq f(y)$, for all x and y in A. Prove that $\preccurlyeq$ is a well-ordering on A. If $f : A \to \alpha$ is a bijection, then prove that α is the order type of $(A, \preccurlyeq)$.

7. Let α be an ordinal. Suppose $g : \alpha \to X$ is a function. Let $A \subseteq \alpha$. Prove that $g[A]$ has a well-ordering and that $|g[A]| \subseteq |A|$.

*8. Assume that A has a well-ordering and let $B \subseteq A$. Prove that $|B| \subseteq |A|$ using Definition 9.1.5.

*9. Prove that if A has a well-ordering and $\kappa = |A|$, then κ is a cardinal.

10. Let κ be a cardinal. Prove that if $f : A \to \kappa$ is a bijection, then $\kappa = |A|$.

*11. Prove Lemma 9.1.7.

12. Show that the collection W in the proof of Hartogs' Theorem 9.1.8 is a set.

*13. Prove Theorem 9.1.13.

*14. Prove Theorem 9.1.15.

15. Prove that $\aleph : \mathrm{On} \to \mathrm{On}$ is strictly increasing.

16. Let α be a limit ordinal. Prove that $\mathrm{cf}(\aleph_\alpha) = \mathrm{cf}(\alpha)$.

17. Prove that for all ordinals β, we have that $\beta \subseteq \aleph_\beta$.

18. Prove for every ordinal α there is an ordinal β such that $\alpha \in \beta$ and $\aleph_\beta = \beta$.

19. Prove there is an ordinal ξ such that $\mathrm{cf}(\xi) = \omega$ and $\aleph_\xi = \xi$.

*20. Let κ be an infinite cardinal. Prove that $\aleph_\alpha = \kappa$ for some ordinal α.

*21. Let α be an ordinal so that $\aleph_0 \in \alpha \in \aleph_1$. Prove that α is countable.

*22. Let $\alpha \in \gamma$ where α is an ordinal and γ is a limit ordinal. Let $f : \alpha \to \gamma$ be such that $f[\alpha]$ is cofinal in γ. Prove that $\bigcup \{f(\xi) : \xi \in \alpha\} = \gamma$.

23. Assume that for any two sets X and Y, either $|X| \leq_c |Y|$ or $|Y| \leq_c |X|$. Now, using Theorem 9.1.8, prove that every set can be well-ordered.

*24. Let γ be an ordinal. Show that $\preccurlyeq_\gamma$ is a total order on $\gamma \times \gamma$.

*25. Prove Proposition 9.1.19.

Exercise Notes: For Exercise **5**, recall that $\sup C = \bigcup C$. Theorem 8.1.15(2) implies that $\bigcup C$ is an ordinal. For Exercise **6**, review Exercise 12 on page 76, Theorem 8.1.10, and Lemma 8.1.11. For Exercise **15**, apply Theorem 8.2.9. For Exercise **17**, use Theorem 8.1.22, and Exercise **15** in the limit case. For Exercise **19**, read the proof of Theorem 8.2.19. For Exercise **20**, by Exercise **17**, there is a least ordinal α such that $\kappa \in \aleph_\alpha$. For Exercise **21**, use Exercises **1** and Lemma 9.1.7(2) to show that $|\alpha| = \aleph_0$. For Exercise **23**, use Theorems 9.1.8 and 9.1.4. Exercise **23** shows that the Comparability Theorem 7.1.8 implies the axiom of choice.

9.2 Cardinal Arithmetic

In Section 8.3, we defined three arithmetic operations on the ordinals; that is, we defined addition, multiplication, and exponentiation on the ordinals. So if α and β are ordinals, then $\alpha + \beta$, $\alpha \cdot \beta$, and α^β (where $2 \in \alpha$) are also ordinals. For example, ω is an ordinal, and hence, $\omega + \omega$ and $\omega \cdot \omega$ are ordinals; however, $\omega + \omega$ and $\omega \cdot \omega$ are not cardinals.

In this section, we will define arithmetic operations on the cardinals in such a way that when we add or multiply two cardinals, the result will be a cardinal. Thus, ordinal arithmetic and cardinal arithmetic will be different operations. For example, cardinal addition is commutative (see Lemma 9.2.2), whereas ordinal addition is not commutative (see page 196).

For a well-ordered set A, recall that in Definition 9.1.5 we identified a unique cardinal number κ so that $|A| = \kappa$. In addition, Lemma 9.1.6 shows that there also exists a bijection $H : A \to \kappa$.

If δ and κ are cardinals, the proof of Corollary 9.1.23 allows us to conclude that the two sets $(\{0\} \times \delta) \cup (\{1\} \times \kappa)$ and $\delta \times \kappa$ have well-orderings. Thus, we can make the following definition.

Definition 9.2.1. Let δ and κ be cardinals. Then the **cardinal sum** $\delta + \kappa$ and the **cardinal product** $\delta \cdot \kappa$ are defined, respectively, to be

(1) $\delta + \kappa = |(\{0\} \times \delta) \cup (\{1\} \times \kappa)|$,
(2) $\delta \cdot \kappa = |\delta \times \kappa|$.

In this section, the Greek letters $\delta, \theta, \kappa, \lambda$ will used only to denote cardinals. When using these particular Greek letters, the operations $\delta + \kappa$ and $\delta \cdot \kappa$ are to be viewed as the cardinal operations given in Definition 9.2.1. The other Greek letters will be used to represent ordinals that may (or may not) be cardinals. In our next lemma, we show that cardinal addition and multiplication satisfy some very familiar properties.

Lemma 9.2.2. *Let $\delta, \theta,$ and κ be cardinals. Then*

(1) $\delta + \kappa = \kappa + \delta$,
(2) $\delta \cdot \kappa = \kappa \cdot \delta$,
(3) $\theta \cdot (\delta + \kappa) = \theta \cdot \delta + \theta \cdot \kappa$,
(4) $\theta + (\delta + \kappa) = (\theta + \delta) + \kappa$,
(5) $\theta \cdot (\delta \cdot \kappa) = (\theta \cdot \delta) \cdot \kappa$.

Proof. Let $\delta, \theta,$ and κ be cardinals. We first prove (1). Since one can easily define a bijection

$$f : (\{0\} \times \delta) \cup (\{1\} \times \kappa) \to (\{0\} \times \kappa) \cup (\{1\} \times \delta),$$

Definition 9.2.1 and Lemma 9.1.7(2) imply that $\delta + \kappa = \kappa + \delta$. To prove (2), Exercise 17 on page 139 shows that $|\delta \times \kappa| =_c |\kappa \times \delta|$. Thus, $\delta \cdot \kappa = \kappa \cdot \delta$ by Lemma 9.1.7(2). Items (3), (4), and (5) follow in a similar manner. $\square$

Remark 9.2.3. A natural number is also a cardinal, that is, it is a finite cardinal. The arithmetic operations defined in Definition 9.2.1, when applied to natural numbers, coincide with the operations of addition and multiplication presented in Chapter 4. For natural numbers m and n, let $m + n$ be the cardinal sum given in Definition 9.2.1, and let $m +_\omega n$ be the natural number resulting from the notion of addition identified in Theorem 4.3.3. Let $m \in \omega$. One can prove by mathematical induction that

$$|(\{0\} \times m) \cup (\{1\} \times n)| =_c |m +_\omega n|$$

for every natural number n (using Theorem 5.4.6(1) in the inductive step). Lemma 9.1.7 thus implies that

$$|(\{0\} \times m) \cup (\{1\} \times n)| = |m +_\omega n|.$$

Hence, $m + n = |m +_\omega n|$. Theorem 5.1.8 thus implies that $m + n = m +_\omega n$. In a similar fashion, one can prove that the cardinal product of two natural numbers also agrees with the notion of multiplication given in Chapter 4.

The following lemma confirms that the operations of cardinal addition and multiplication have the expected order-preserving properties.

Lemma 9.2.4. *Let $\delta, \theta, \kappa, \lambda$ be cardinals so that $\delta \in \theta$ and $\kappa \in \lambda$. Then*

(1) $\delta + \kappa \in \theta + \lambda$,
(2) $\delta \cdot \kappa \in \theta \cdot \lambda$.

Proof. Let $\delta, \theta, \kappa, \lambda$ be cardinals. Assume that $\delta \in \theta$ and $\kappa \in \lambda$. Thus, $\delta \subseteq \theta$ and $\kappa \subseteq \lambda$, as cardinals are also ordinals. Hence,

$$(\{0\} \times \delta) \cup (\{1\} \times \kappa) \subseteq (\{0\} \times \theta) \cup (\{1\} \times \lambda),$$

$$\delta \times \kappa \subseteq \theta \times \lambda.$$

Thus, by Exercise 8 on page 217, $\delta + \kappa \in \theta + \lambda$ and $\delta \cdot \kappa \in \theta \cdot \lambda$. □

Our next lemma, called the **absorption law of cardinal arithmetic**, shows that the addition and multiplication of two cardinals is quite easy to evaluate, whenever one of the cardinals is infinite.

Lemma 9.2.5. *Suppose that δ and κ are nonzero cardinals. If either δ or κ is infinite, then $\delta + \kappa = \delta \cdot \kappa = \max(\delta, \kappa)$.*

Proof. Suppose that at least one of the nonzero cardinals δ and κ is infinite. Corollary 9.1.24 thus implies that $\kappa \cdot \delta = \max(\kappa, \delta)$. We will now prove that $\kappa + \delta = \max(\kappa, \delta)$. Assume, without loss of generality, that $\delta \in \kappa$. Therefore, κ is infinite and, as $0 \in \kappa$ and $1 \in \kappa$, we have

$$\{1\} \times \kappa \subseteq (\{0\} \times \delta) \cup (\{1\} \times \kappa) \subseteq \kappa \times \kappa.$$

Clearly, $|\{1\} \times \kappa| = \kappa$ and $|\kappa \times \kappa| = \kappa$. Hence, $\delta + \kappa = \kappa$. □

As was noted in Remark 9.2.3, the sum and product of two finite cardinals agree with the sum and product of two natural numbers as defined in Chapter 4. Lemma 9.2.5 shows that the sum and product of two infinite cardinals equals the maximum of the two cardinals. Cardinal exponentiation, when applied to infinite cardinals, is not as easy to evaluate.

Cardinal Exponentiation

Given an infinite set A, recall that the cardinal $|A|$ is defined if and only if the set A has a well-ordering (see Theorem 9.1.4 and Definition 9.1.5). For this reason, we shall be assuming the axiom of choice whenever cardinal exponentiation is involved. The axiom of choice, via the Well-Ordering Theorem 7.3.1, implies that every set can be well-ordered. In particular, when δ and κ are cardinals, the set $^{\delta}\kappa$ can be well-ordered, where $^{\delta}\kappa$ is the set of all functions from δ to κ.

Definition 9.2.6 (AC). Let δ and κ be cardinals. Then the **cardinal power** κ^{δ} is defined to be the cardinal $\left|^{\delta}\kappa\right|$.

Note that $0^0 = 1$ and $0^{\kappa} = 0$ when $\kappa \neq 0$ (see Remark 3.3.7).

Lemma 9.2.7 (AC). *Let κ be a cardinal. Then $2^{\kappa} = |\mathcal{P}(\kappa)|$.*

Proof. Recall that $2 = \{0, 1\}$. Theorem 5.4.3 thus asserts that $|^{\kappa}2| =_c |\mathcal{P}(\kappa)|$. Lemma 9.1.7(2) now implies that $2^{\kappa} = |\mathcal{P}(\kappa)|$. ☐

Recall that Cantor's Theorem 5.4.13 asserts that $|A| <_c |\mathcal{P}(A)|$ for any set A. This theorem can now be restated in terms of cardinal exponentiation.

Theorem 9.2.8 (AC). *Let κ be a cardinal. Then $\kappa \in 2^{\kappa}$.*

Proof. Let κ be a cardinal, Theorem 5.4.13 states that $|\kappa| <_c |\mathcal{P}(\kappa)|$. Thus, $|\kappa| \in |\mathcal{P}(\kappa)|$ by Lemma 9.1.7(3). Now Lemma 9.2.7 implies that $\kappa \in 2^{\kappa}$. ☐

The next theorem identifies three basic facts about cardinal exponentiation.

Lemma 9.2.9 (AC). *Let δ, θ, and κ be cardinals. Then*

(1) $\kappa^{\delta+\theta} = \kappa^{\delta} \cdot \kappa^{\theta}$,
(2) $(\kappa \cdot \delta)^{\theta} = \kappa^{\theta} \cdot \delta^{\theta}$,
(3) $(\kappa^{\delta})^{\theta} = \kappa^{\delta \cdot \theta}$.

Proof. Let δ, θ, κ be cardinals. For (1), we first show that $\left|^{\delta+\theta}\kappa\right| =_c \left|^{\delta}\kappa \times {}^{\theta}\kappa\right|$. Let $A = \{0\} \times \delta$ and $B = \{1\} \times \theta$, which are clearly disjoint. Lemma 9.1.6 implies that

$$|A| =_c |\delta|,$$

$$|B| =_c |\theta|,$$

$$|A \cup B| =_c |\delta + \theta|,$$

$$\left|^{\delta}\kappa\right| =_c \left|\kappa^{\delta}\right|,$$

$$\left|^{\theta}\kappa\right| =_c \left|\kappa^{\theta}\right|.$$

Theorem 5.4.7(1) asserts that $\left|^{A \cup B}\kappa\right| =_c \left|^{A}\kappa \times {}^{B}\kappa\right|$. Therefore, Theorem 5.4.6 now implies that $\left|^{\delta+\theta}\kappa\right| =_c \left|^{\delta}\kappa \times {}^{\theta}\kappa\right| =_c \left|\kappa^{\delta} \times \kappa^{\theta}\right|$. Thus, $\kappa^{\delta+\theta} = \kappa^{\delta} \cdot \kappa^{\theta}$ by Lemma 9.1.7(2). A similar argument will establish items (2) and (3).   □

In our next lemma, it shall be shown that cardinal exponentiation possesses an order-preserving property that closely corresponds to a property identified in Theorem 5.4.16.

Lemma 9.2.10 (AC). *Let $\delta, \theta, \lambda,$ and κ be cardinals. If $\delta \subseteq \theta$ and $\lambda \subseteq \kappa$, then $\lambda^{\delta} \subseteq \kappa^{\theta}$ when $\delta \neq 0$.*

Proof. Let $\delta, \theta, \lambda, \kappa$ be cardinals and assume that $\delta \subseteq \theta$ and $\lambda \subseteq \kappa$. Clearly, $|\delta| \leq_c |\theta|$ and $|\lambda| \leq_c |\kappa|$. If $\delta \neq 0$, then $\left|^{\delta}\lambda\right| \leq_c \left|^{\theta}\kappa\right|$ by Theorem 5.4.16(3). Therefore, we have that $\lambda^{\delta} \subseteq \kappa^{\theta}$ by Lemma 9.1.7(1).   □

As an application of Lemmas 9.2.9 and 9.2.10, we have the following result.

Lemma 9.2.11 (AC). *Let κ be an infinite cardinal. Then $\kappa^{\kappa} = 2^{\kappa}$.*

Proof. Let κ be an infinite cardinal. Lemma 9.2.7 yields the inequality $\kappa \subseteq 2^{\kappa}$. Thus, $\kappa^{\kappa} \subseteq (2^{\kappa})^{\kappa}$ by Lemma 9.2.10. Therefore,

$$\kappa^{\kappa} \subseteq (2^{\kappa})^{\kappa} = 2^{\kappa \cdot \kappa} = 2^{\kappa} \subseteq \kappa^{\kappa},$$

where the last equality follows from Lemma 9.2.5. Hence, $\kappa^{\kappa} = 2^{\kappa}$.   □

Theorem 9.2.12 (AC). *Let $\mathcal{A}$ be a set and let κ be a cardinal. If $|X| \subseteq \kappa$ for all $X \in \mathcal{A}$, then $\left|\bigcup \mathcal{A}\right| \subseteq |\mathcal{A}| \cdot \kappa$.*

Proof. Let $\mathcal{A}$ be a set and κ be a cardinal. Suppose $|X| \in \kappa$, for each $X \in \mathcal{A}$. Let $\delta = |\mathcal{A}|$ and let $h\colon \delta \to \mathcal{A}$ be a bijection. Since $|h(\alpha)| \in \kappa$ for all $\alpha \in \delta$, the axiom of choice implies that there is a one-to-one function $f_\alpha \colon h(\alpha) \to \kappa$, for each $\alpha \in \delta$. Define the function $g\colon \bigcup\mathcal{A} \to \delta \times \kappa$ by

$$g(x) = \langle \alpha, f_\alpha(x) \rangle \text{ where } \alpha \text{ is the least } \alpha \in \delta \text{ such that } x \in h(\alpha)$$

for each $x \in \bigcup\mathcal{A}$. We shall prove that the above function g is one-to-one. Let x and y be in $\bigcup\mathcal{A}$ and assume that $g(x) = g(y)$. Let $\alpha \in \delta$ be the least such that $x \in h(\alpha)$, and let $\beta \in \delta$ be the least such that $y \in h(\beta)$. Since $g(x) = g(y)$, we have $\langle \alpha, f_\alpha(x) \rangle = \langle \beta, f_\beta(y) \rangle$. So $\alpha = \beta$ and $f_\alpha(x) = f_\beta(y)$. Therefore, $f_\alpha(x) = f_\alpha(y)$ and so $x = y$, since f_α is one-to-one. Hence, g is one-to-one. By Lemma 9.1.7(1), $|\bigcup\mathcal{A}| \in |\delta \times \kappa|$; that is, $|\bigcup\mathcal{A}| \in \delta \cdot \kappa$. So $|\bigcup\mathcal{A}| \in |\mathcal{A}| \cdot \kappa$. $\qquad\square$

Theorem 9.2.13 (AC). *For each ordinal α, the cardinal $\aleph_{\alpha+1}$ is regular.*

Proof. See Exercise 11. $\qquad\square$

Lemma 9.2.14 (AC). $|\mathbb{R}| = 2^{\aleph_0}$.

Proof. Theorem 5.4.20 asserts that $|\mathbb{R}| =_c |\mathcal{P}(\omega)|$. Lemma 9.1.7(2) therefore implies that $|\mathbb{R}| = |\mathcal{P}(\aleph_0)|$, as $\omega = \aleph_0$. Hence, $|\mathbb{R}| = 2^{\aleph_0}$ by Lemma 9.2.7. $\qquad\square$

We will now show that the continuum hypothesis has a different formulation when one assumes the axiom of choice. Recall that Cantor's hypothesis (see page 133) asserts that any set of real numbers A is either countable or has the same cardinality as the entire set $\mathbb{R}$. Also note that $\aleph_0 \in 2^{\aleph_0}$ by Theorem 9.2.8.

Continuum Hypothesis. There is no set $A \subseteq \mathbb{R}$ such that $|\omega| <_c |A| <_c |\mathbb{R}|$.

Lemma 9.2.15 (AC). *There exists a cardinal δ so that $\aleph_0 \in \delta \in 2^{\aleph_0}$ if and only if there exists an $A \subseteq \mathbb{R}$ such that $|\omega| <_c |A| <_c |\mathbb{R}|$.*

Proof. Let δ be a cardinal such that $\aleph_0 \in \delta \in 2^{\aleph_0}$. Hence, $\aleph_0 \in \delta \in |\mathbb{R}|$ by Lemma 9.2.14. Thus, there is a one-to-one function $f\colon \delta \to \mathbb{R}$. Let $A = f[\delta]$.

Hence, $|\omega| \in |A| \in |\mathbb{R}|$. Therefore, $|\omega| <_c |A| <_c |\mathbb{R}|$ by Lemma 9.1.7(3). The converse holds similarly. $\qquad\square$

Since $\aleph_1$ is the first cardinal number greater than $\aleph_0$, Lemma 9.2.15 implies that the continuum hypothesis is equivalent to the equality $2^{\aleph_0} = \aleph_1$.

Continuum Hypothesis (AC). $2^{\aleph_0} = \aleph_1$.

In a similar manner, assuming the axiom of choice, the GCH (see page 137) can also be formulated in terms of a cardinal identity.

Generalized Continuum Hypothesis (AC). $2^{\aleph_\alpha} = \aleph_{\alpha+1}$ for every ordinal α.

We will present a theorem due to König that can be viewed as an extension of Cantor's Theorem 9.2.8. First, we establish two lemmas.

Lemma 9.2.16 (AC). *Let θ be an infinite cardinal such that $\mathrm{cf}(\theta) \subseteq \kappa \in \theta$. If $|\mathcal{F}| = \theta$, then there is a set $\{A_\xi : \xi \in \kappa\}$ such that $\mathcal{F} = \bigcup_{\xi \in \kappa} A_\xi$ and $|A_\xi| \in \theta$ for all $\xi \in \kappa$.*

Proof. Let θ be an infinite cardinal so that $\mathrm{cf}(\theta) \subseteq \kappa \in \theta$. Assume $|\mathcal{F}| = \theta$. Let $h \colon \theta \to \mathcal{F}$ be a bijection and $f \colon \kappa \to \theta$ be such that $f[\kappa]$ is cofinal in θ. For each $\xi \in \kappa$, let $A_\xi = h[f(\xi)]$. Thus, $A_\xi \subseteq \mathcal{F}$ and $|A_\xi| = |f(\xi)| \in \theta$. Clearly, $\bigcup_{\xi \in \kappa} A_\xi \subseteq \mathcal{F}$. To show that $\mathcal{F} \subseteq \bigcup_{\xi \in \kappa} A_\xi$, let $x \in \mathcal{F}$. Since h is a bijection, there is an $\alpha \in \theta$ such that $x = h(\alpha)$. Because $f[\kappa]$ is cofinal in θ, there is an $\xi \in \kappa$ such that $\alpha \in f(\xi)$. Hence, $x \in A_\xi$ and so $x \in \bigcup_{\xi \in \kappa} A_\xi$. $\qquad\square$

The proof of our next lemma employs a diagonal argument.

Lemma 9.2.17 (AC). *Let κ be a cardinal and let S be a set. If $\{A_\xi : \xi \in \kappa\}$ is such that $A_\xi \subseteq {}^\kappa S$ and $\{g(\xi) : g \in A_\xi\} \neq S$ for all $\xi \in \kappa$, then ${}^\kappa S \neq \bigcup_{\xi \in \kappa} A_\xi$.*

Proof. Let κ be a cardinal and let S be a set. Suppose that $\{A_\xi : \xi \in \kappa\}$ is as stated in the lemma. Therefore, $S \setminus \{g(\xi) : g \in A_\xi\}$ is nonempty for all $\xi \in \kappa$. By the axiom of choice there is a function $f \colon \kappa \to S$ such that

$$f(\xi) \in S \setminus \{g(\xi) : g \in A_\xi\}, \quad \text{for each } \xi \in \kappa. \tag{9.6}$$

So $f \in {}^\kappa S$. We will show that $f \notin \bigcup_{\xi \in \kappa} A_\xi$. Suppose to the contrary that $f \in A_\xi$ for some $\xi \in \kappa$. Then $f(\xi) \in \{g(\xi) : g \in A_\xi\}$, contradicting (9.6). $\qquad\square$

König's Theorem 9.2.18 (AC). *Let κ be an infinite cardinal. Then $\kappa \in \mathrm{cf}(2^\kappa)$.*

Proof. Let κ be an infinite cardinal. Suppose, to the contrary, that $\mathrm{cf}(2^\kappa) \in \kappa$. Recall that $\kappa \in 2^\kappa$, by Theorem 9.2.8. Thus, $\mathrm{cf}(2^\kappa) \in \kappa \in 2^\kappa$. Let $S = 2^\kappa$. So $|S| = 2^\kappa$. Moreover, $|{}^\kappa S| = 2^\kappa$ because

$$|{}^\kappa S| = |{}^\kappa (2^\kappa)| = (2^\kappa)^\kappa = 2^{\kappa \cdot \kappa} = 2^\kappa.$$

Since $|{}^\kappa S| = 2^\kappa$ and $\mathrm{cf}(2^\kappa) \in \kappa \in 2^\kappa$, Lemma 9.2.16 implies that there exists an indexed set $\{A_\xi : \xi \in \kappa\}$ so that ${}^\kappa S = \bigcup_{\xi \in \kappa} A_\xi$ and $|A_\xi| \in 2^\kappa$ for all $\xi \in \kappa$. Let $\xi \in \kappa$. Clearly,

$$\left|\{g(\xi) : g \in A_\xi\}\right| \in |A_\xi| \in 2^\kappa.$$

As $|S| = 2^\kappa$, we have that $\{g(\xi) : g \in A_\xi\} \neq S$. Therefore, by Lemma 9.2.17 ${}^\kappa S \neq \bigcup_{\xi \in \kappa} A_\xi$, a contradiction. $\qquad\square$

Corollary 9.2.19 (AC). $2^{\aleph_0} \neq \aleph_\omega$.

Proof. By Theorem 9.2.18, $\aleph_0 \in \mathrm{cf}(2^{\aleph_0})$. So $2^{\aleph_0} \neq \aleph_\omega$, as $\mathrm{cf}(\aleph_\omega) = \aleph_0$. $\qquad\square$

Exercises 9.2

1. (AC) Let $\delta, \theta, \lambda, \kappa$ be infinite cardinals so that $\delta \in \theta \in \lambda \in \kappa$. Simplify the cardinals $(\kappa^{\delta+\theta})^\kappa$ and $\kappa^{\aleph_1 \cdot \aleph_2 + \aleph_1 \cdot \kappa}$.

2. Prove Lemma 9.2.2(3).

3. Prove Lemma 9.2.2(4).

4. Prove Lemma 9.2.2(5).

5. Prove Lemma 9.2.9(2).

6. Prove Lemma 9.2.9(3).

7. Let α and β be ordinals. Show that
   (a) $\aleph_\alpha + \aleph_\beta = \aleph_{\max(\alpha,\beta)}$,
   (b) $\aleph_\alpha \cdot \aleph_\beta = \aleph_{\max(\alpha,\beta)}$,
   (c) $\aleph_\alpha + \aleph_\alpha = \aleph_\alpha$,
   (d) $\aleph_\alpha \cdot \aleph_\alpha = \aleph_\alpha$.

8. (AC) Let κ and λ be cardinals. Assume that $2 \in \kappa \in 2^\lambda$ and λ is infinite. Prove that $\kappa^\lambda = 2^\lambda$.

9. (AC) Let α be an ordinal. Prove that $\aleph_{\alpha+1} \subseteq 2^{\aleph_\alpha}$.

10. Let α be an ordinal. Prove that if $\beta \in \aleph_{\alpha+1}$, then $|\beta| \subseteq \aleph_\alpha$.

*****11.** Prove Theorem 9.2.13.

12. (AC) Let $\mathcal{A}$ be a set and let κ be an infinite cardinal. Suppose that $|\mathcal{A}| \subseteq \kappa$ and $|X| \subseteq \kappa$ for all $X \in \mathcal{A}$. Prove that $\left| \bigcup \mathcal{A} \right| \subseteq \kappa$.

13. (AC) Let $\mathcal{A}$ be a set and let θ be an infinite regular cardinal. Suppose that $|\mathcal{A}| \in \theta$ and $|X| \in \theta$ for all $X \in \mathcal{A}$.
 (a) Prove that $\sup\{|X| : X \in \mathcal{A}\} \in \theta$.
 (b) Prove that $\left| \bigcup \mathcal{A} \right| \in \theta$.

*****14.** (AC) Suppose that $h \colon A \to B$ is a function. Prove that $|h[A]| \subseteq |A|$.

15. (AC) Show that $2^{\aleph_0} \neq \aleph_{\alpha+\omega}$, for any ordinal α.

16. (AC) Show that there is a class function $\beth \colon \mathrm{On} \to \mathrm{On}$ such that
 1. $\beth(0) = \aleph_0$,
 2. $\beth(\gamma^+) = 2^{\beth(\gamma)}$,
 3. $\beth(\beta) = \sup\{\beth(\alpha) : \alpha \in \beta\}$ when β is a limit ordinal.

17. Prove that $\beth \colon \mathrm{On} \to \mathrm{On}$ is strictly increasing, where $\beth$ is as in Exercise **16**.

18. Let $\beth \colon \mathrm{On} \to \mathrm{On}$ be as in Exercise **16**. Prove the following:
 (a) For each ordinal γ, there is an ordinal β such that $\gamma \subseteq \beta$ and $\beth(\beta) = \beta$.
 (b) If $\beth(\beta) = \beta$, then β is a limit ordinal and $2^{\beth(\alpha)} \in \beta$ for all $\alpha \in \beta$.

*****19.** (AC) Let $\{A_i : i \in I\}$ and $\{B_i : i \in I\}$ be two indexed sets. Suppose that $|A_i| \in |B_i|$, for each $i \in I$. By the axiom of choice, let $p_i \colon A_i \to B_i$ be an injection for each $i \in I$.
 (a) Show that $|p_i[A_i]| \in |B_i|$ and conclude that $B_i \setminus p_i[A_i] \neq \varnothing$ for all $i \in I$.
 (b) Get a function h so that $\mathrm{dom}(h) = I$ and $h(i) \in B_i \setminus p_i[A_i]$ for all $i \in I$.
 (c) For each $x \in \bigcup_{i \in I} A_i$, define $g_x \in \prod_{i \in I} B_i$ (see (7.1) on page 156) by

$$g_x(i) = \begin{cases} p_i(x), & \text{if } x \in A_i; \\ h(i), & \text{if } x \notin A_i. \end{cases}$$

Define $F \colon \bigcup_{i \in I} A_i \to \prod_{i \in I} B_i$ by $F(x) = g_x$. Prove that F is one-to-one.

(d) For every $j \in I$, define $\pi_j \colon \prod_{i \in I} B_i \to B_j$ by $\pi_j(g) = g(j)$. Suppose $H \colon \bigcup_{i \in I} A_i \to \prod_{i \in I} B_i$ is an injection. Show that $|\pi_i[H[A_i]]| \in |B_i|$, for each $i \in I$. Prove that H is not a bijection.

(e) Conclude that $\left| \bigcup_{i \in I} A_i \right| <_c \left| \prod_{i \in I} B_i \right|$ and thus, $\left| \bigcup_{i \in I} A_i \right| \in \left| \prod_{i \in I} B_i \right|$.

Exercise Notes: For Exercise **11**, use Theorem 9.2.12 and Exercise **10**. Also review the proof of Theorem 9.1.16. For Exercises **12** and **13**(b), one can apply Theorem 9.2.12. In our proof of Theorem 9.2.18, we implicitly applied Exercise **14**. Exercise **16** defines the *beth numbers*, $\beth_\lambda = \beth(\lambda)$, for each ordinal λ. The Hebrew letter $\beth$ is called beth. For Exercise **19**(d), note that Exercise **14** implies that $|\pi_i[H[A_i]]| \in |H[A_i]|$. Exercise **19** outlines a proof of a theorem due to König: *If* $|A_i| \in |B_i|$ *for all* $i \in I$, *then* $\left| \bigcup_{i \in I} A_i \right| \in \left| \prod_{i \in I} B_i \right|$.

9.3 Closed Unbounded and Stationary Sets

We now discuss two concepts that play a prominent role in modern set theory, namely, closed unbounded sets and stationary sets.

Definition 9.3.1. Let κ be a limit ordinal and let $C \subseteq \kappa$. Then we say that

(1) C is **unbounded** in κ when C is cofinal in κ (see page 213).
(2) C is **closed** in κ when for all $\gamma \in \kappa$, if $C \cap \gamma \neq \varnothing$, then $\sup(C \cap \gamma) \in C$.
(3) C is **club** in κ when C is closed and unbounded in κ.

Let κ be a limit ordinal, and let $C = \{\beta \in \kappa : \alpha \in \beta\}$ where $\alpha \in \kappa$. Clearly C is unbounded in κ. If $\gamma \in \kappa$ and $C \cap \gamma \neq \varnothing$, then $\alpha \in \sup(C \cap \gamma) \in \kappa$. Thus, $\sup(C \cap \gamma) \in C$ and C is club in κ. On the other hand, if $\omega \in \kappa$, then the set $D = \{\gamma \in \kappa : \gamma \text{ is a successor ordinal}\}$ is unbounded in κ; however, D is not closed in κ. To see this, recall that every nonzero natural number is a successor ordinal (Theorem 4.1.6). So $D \cap \omega = \{n \in \omega : 0 \in n\}$ and $\sup(D \cap \omega) = \omega$, which is not in D, because ω is a limit ordinal. Thus, D is not closed in κ.

Definition 9.3.2. Let κ be a limit ordinal and let $A \subseteq \kappa$. We shall say that X is a **bounded subset** of A if and only if $X \subseteq A \cap \gamma$ for some $\gamma \in \kappa$.

Lemma 9.3.3. *Let κ be a limit ordinal and let $C \subseteq \kappa$. Then C is closed in κ if and only if* $\sup X \in C$ *whenever X is a nonempty bounded subset of C.*

Proof. Let κ be a limit ordinal and let $C \subseteq \kappa$.

($\Rightarrow$). Assume C is closed in κ. Let $X \neq \emptyset$ be so that $X \subseteq C \cap \gamma$ for some $\gamma \in \kappa$.

CASE 1: X is a limit set. Thus, by Lemma 8.2.12, $\delta = \sup X$ is a limit ordinal and X is a cofinal in δ. Moreover, $\delta \in \gamma \in \kappa$. As $X \subseteq \delta$, we see that $X \subseteq C \cap \delta$. Therefore, $C \cap \delta$ is also a cofinal subset of δ. Hence, $\sup(C \cap \delta) = \sup \delta = \delta$, by Lemma 8.2.14 and Exercise 10 on page 185. Since C is closed in κ, we conclude that $\delta \in C$. Thus, $\sup X \in C$.

CASE 2: X is a not a limit set. Therefore, X must have a largest element δ. So $\delta = \sup X$ and $\delta \in X$. Since $X \subseteq C$, we have that $\sup X \in C$.

($\Leftarrow$). Assume that $\sup X \in C$ whenever X is a nonempty bounded subset of C. Let $\gamma \in \kappa$ be such that $C \cap \gamma$ is nonempty. As $C \cap \gamma$ is clearly a bounded subset of C, the assumption implies that $\sup(C \cap \gamma) \in C$. Hence, C is closed in κ. $\qquad\square$

Theorem 9.3.4. *Let κ be a limit ordinal. Let $\mathcal{F}$ be a nonempty set such that C is closed in κ, for all $C \in \mathcal{F}$. Then $\bigcap \mathcal{F}$ is also closed in κ.*

Proof. Let $X \subseteq \bigcap \mathcal{F}$ be nonempty and bounded. Because $X \subseteq \bigcap \mathcal{F}$, we see that $X \subseteq C$ for each $C \in \mathcal{F}$. Since every $C \in \mathcal{F}$ is closed in κ, Lemma 9.3.3 implies that $\sup X \in C$, for all $C \in \mathcal{F}$. Therefore, $\sup X \in \bigcap \mathcal{F}$. Thus, by Lemma 9.3.3, we conclude that $\bigcap \mathcal{F}$ is closed in κ. $\qquad\square$

Before continuing our discussion on club sets, we will focus our attention on indexed functions (see Section 3.3.3) whose domain is an ordinal.

Definition 9.3.5. Whenever γ is an ordinal, an indexed function $\langle a_\beta : \beta \in \gamma \rangle$ shall be called a γ-**sequence**, or a **sequence** when the ordinal γ is understood.

So if $H = \langle a_\beta : \beta \in \gamma \rangle$ is a γ-sequence, then H is a function with domain γ. Thus, if $\beta \in \gamma$, then $H(\beta) = a_\beta$, and $H \restriction \beta = \langle a_\alpha : \alpha \in \beta \rangle$. Moreover, if $\gamma = 0$, then H is the empty sequence $\langle \rangle = \emptyset$. We now discuss how one can define a sequence by ordinal recursion. Let γ be an ordinal and let $\varphi(x, y)$ be a formula that is functional (see page 150). Theorem 8.2.2 thereby implies that there is a γ-sequence $\langle a_\beta : \beta \in \gamma \rangle$ satisfying

$$\varphi(\langle a_\alpha : \alpha \in \beta \rangle, a_\beta), \quad \text{for all } \beta \in \gamma.$$

Such a sequence is said to be *defined by ordinal recursion*. Let us consider an example. Suppose that $\langle \xi_\beta : \beta \in \gamma \rangle$ is a γ-sequence of ordinals. We can define a new γ-sequence $\langle \lambda_\beta : \beta \in \gamma \rangle$ of ordinals by the ordinal recursion

$$\lambda_\beta = \text{the least ordinal } \lambda \text{ such that } \xi_\beta \in \lambda \text{ and } \sup\{\lambda_\alpha : \alpha \in \beta\} \in \lambda \qquad (9.7)$$

for all $\beta \in \gamma$. Thus, (9.7) implies that $\xi_\beta \in \lambda_\beta$ for all $\beta \in \gamma$, and it also implies that the sequence $\langle \lambda_\beta : \beta \in \gamma \rangle$ is *strictly increasing*; that is, $\lambda_\alpha \in \lambda_\beta$ whenever $\alpha \in \beta \in \gamma$.

In this section, we shall be defining sequences by recursion, and we will then use these sequences to prove theorems about club sets in a regular uncountable cardinal. We must first make a useful observation. Suppose that κ is a regular uncountable cardinal and that $\langle \alpha_\xi : \xi \in \gamma \rangle$ is a sequence such that $\gamma \in \kappa$ and $\alpha_\xi \in \kappa$ for all $\xi \in \gamma$. Recalling Definition 9.1.14, it follows that there is an $\eta \in \kappa$ such that $\{\alpha_\xi : \xi \in \gamma\} \subseteq \eta$. Thus, the set $\{\alpha_\xi : \xi \in \gamma\}$ is bounded in κ and $\sup\{\alpha_\xi : \xi \in \gamma\} \subseteq \eta$.

Lemma 9.3.6. *Let κ be a regular uncountable cardinal and let C be closed in κ. Suppose that $\langle \gamma_\alpha : \alpha \in \lambda \rangle$ is a sequence where $0 \in \lambda \in \kappa$ and $\gamma_\alpha \in C$ for all $\alpha \in \lambda$. Then $\sup\{\gamma_\alpha : \alpha \in \lambda\} \in C$.*

Proof. Let κ, C, and $\langle \gamma_\alpha : \alpha \in \lambda \rangle$ be as in the statement of the lemma. Hence, $\gamma_\alpha \in \kappa$ for each $\alpha \in \lambda$. Since κ is regular and $\lambda \in \kappa$, there is an ordinal $\eta \in \kappa$ such that $\{\gamma_\alpha : \alpha \in \lambda\} \subseteq \eta$. Therefore, $\{\gamma_\alpha : \alpha \in \lambda\}$ is a bounded subset of C. Lemma 9.3.3 implies that $\sup\{\gamma_\alpha : \alpha \in \lambda\} \in C$. $\qquad \square$

Let κ be a regular uncountable cardinal. In our next theorem, we will prove that the intersection of two club sets is also a club set. Later we will prove that the intersection of fewer than κ club sets is also a club set.

Theorem 9.3.7. *Suppose that κ is a regular uncountable cardinal. If A and B are club in κ, then $A \cap B$ is club in κ.*

Proof. Let κ be a regular uncountable cardinal. Suppose that A and B are club sets in κ. Theorem 9.3.4 implies that $A \cap B$ is closed in κ. To show that $A \cap B$ is unbounded in κ, let $\nu \in \kappa$. We will show that there is an ordinal $\lambda \in A \cap B$

such that $v \in \lambda$. Since B is unbounded in κ, there is an ordinal $\beta_0 \in B$ such that $v \in \beta_0$. Because A and B are both unbounded in κ, there exists a recursively defined sequence $\langle \gamma_n : n \in \omega \rangle$ of ordinals such that $\gamma_0 = \beta_0$ and for all $n \in \omega$,

$$\gamma_{n^+} = \begin{cases} \alpha, & \text{if } n \text{ is even and } \alpha \text{ is the least } \alpha \in A \text{ such that } \gamma_n \in \alpha; \\ \beta, & \text{if } n \text{ is odd and } \beta \text{ is the least } \beta \in B \text{ such that } \gamma_n \in \beta. \end{cases} \quad (9.8)$$

Thus, ($\blacktriangle$) $v \in \gamma_n \in \gamma_{n^+}$ for all $n \in \omega$. Let $D = \{\gamma_n : n \in \omega\}$. Since κ is a regular uncountable cardinal, we have that $v \in \sup D = \lambda \in \kappa$. Observe that

$$X = \{\gamma_{2k+1} : k \in \omega\} \subseteq A \text{ and } X' = \{\gamma_{2k} : k \in \omega\} \subseteq B. \quad (9.9)$$

From ($\blacktriangle$), it follows that D is a limit set in which X and X' are both cofinal. Lemma 8.2.14 implies that $\sup X = \sup X' = \lambda$. Since X and X' are bounded in κ, Lemma 9.3.3 and (9.9) imply that $\lambda \in A$ and $\lambda \in B$. So $v \in \lambda \in A \cap B$. $\square$

Before proving our next lemma, we make another useful observation. Let κ be a regular cardinal and let $\gamma \in \kappa$. If $\langle \xi_\alpha : \alpha \in \gamma \rangle$ is a sequence such that $\xi_\alpha \in \kappa$ for all $\alpha \in \gamma$, then for any C that is unbounded in κ,

there is a $\beta \in C$ such that $\sup\{\xi_\alpha : \alpha \in \gamma\} \in \beta$. $\quad (9.10)$

The property (9.10) will be implicitly used in the proof of the next lemma.

Definition 9.3.8. A γ-sequence $\langle C_\alpha : \alpha \in \gamma \rangle$ of sets is said to be **nested** if and only if $C_\beta \subseteq C_\alpha$ whenever $\alpha \in \beta \in \gamma$.

Thus, if $\langle C_\alpha : \alpha \in \gamma \rangle$ is a nested sequence of sets, we have that

$$C_0 \supseteq C_1 \supseteq C_2 \supseteq \cdots \supseteq C_\alpha \supseteq \cdots \supseteq C_\beta \supseteq \cdots$$

where $\alpha \in \beta \in \gamma$. Of course, $A \supseteq B$ means that $B \subseteq A$.

Lemma 9.3.9. *Let κ be a regular uncountable cardinal. If $\langle C_\alpha : \alpha \in \gamma \rangle$ is a nested sequence of club sets in κ and $0 \in \gamma \in \kappa$, then $\bigcap_{\alpha \in \gamma} C_\alpha$ is club in κ.*

Proof. Let κ be a regular uncountable cardinal. Assume that $\langle C_\alpha : \alpha \in \gamma \rangle$ is a nested sequence of club sets in κ where $0 \in \gamma \in \kappa$. Theorem 9.3.4 implies that $\bigcap_{\alpha \in \eta} C_\alpha$ is closed in κ, for all nonzero $\eta \leq \gamma$. We prove that $\bigcap_{\alpha \in \gamma} C_\alpha$ is unbounded in κ. Suppose, to the contrary, that $\bigcap_{\alpha \in \gamma} C_\alpha$ bounded in κ. Let

$$\rho \leq \gamma \text{ be the least such that } \rho \neq 0 \text{ and } \bigcap_{\alpha \in \rho} C_\alpha \text{ is bounded in } \kappa. \quad (9.11)$$

So $2 \in \rho$, by Theorem 9.3.7. Hence, ρ is either is a successor or a limit ordinal.

CASE 1: $\rho = \eta^+$ for an ordinal η. Hence, ($\blacktriangle$) $\bigcap_{\alpha \in \rho} C_\alpha = (\bigcap_{\alpha \in \eta} C_\alpha) \cap C_\eta$. Since $\eta \in \rho$, (9.11) implies that $\bigcap_{\alpha \in \eta} C_\alpha$ is unbounded in κ. Thus, ($\blacktriangle$) and Theorem 9.3.7 imply that $\bigcap_{\alpha \in \rho} C_\alpha$ is unbounded, which contradicts (9.11).

CASE 2: ρ is a limit ordinal. Since $\bigcap_{\alpha \in \rho} C_\alpha$ bounded in κ, there exists a $\nu \in \kappa$ such that ($\blacklozenge$) $\bigcap_{\alpha \in \rho} C_\alpha \subseteq \nu$. Theorem 8.2.2 implies that there is a sequence of ordinals $\langle \beta_\alpha : \alpha \in \rho \rangle$ such that for all $\alpha \in \rho$,

$$\beta_\alpha = \text{the least } \beta \in C_\alpha \text{ such that } \nu \in \beta \text{ and } \sup\{\beta_\pi : \pi \in \alpha\} \in \beta. \tag{9.12}$$

It follows from (9.12) that

(1) $\nu \in \beta_\pi \in \beta_\alpha$ whenever $\pi \in \alpha \in \rho$,
(2) $X_0 = \{\beta_\alpha : \alpha \in \rho\}$ is a limit set,
(3) $X_\xi = \{\beta_\alpha : \xi \subseteq \alpha \in \rho\}$ is cofinal in X_0, for all $\xi \in \rho$,
(4) $X_\xi \subseteq C_\xi$ for all $\xi \in \rho$, because $\langle C_\alpha : \alpha \in \rho \rangle$ is nested.

As $X_0 \subseteq C_0$ and $\rho \in \kappa$, Lemma 9.3.6 and (1) imply that $\nu \in \sup X_0 = \lambda \in C_0$. Items (2)–(3) and Lemma 8.2.14 now yield the equation $\sup X_\xi = \lambda$, for each $\xi \in \rho$. Since $\lambda \in \kappa$, we conclude from Lemma 9.3.3 and (4) that $\lambda \in C_\xi$, for all $\xi \in \rho$. Therefore, $\lambda \in \bigcap_{\alpha \in \rho} C_\alpha$. This contradicts ($\blacklozenge$), because $\nu \in \lambda$. □

We can now generalize Theorem 9.3.7.

Theorem 9.3.10. *Let κ be a regular uncountable cardinal and let $0 \in \gamma \in \kappa$. If $\langle D_\alpha : \alpha \in \gamma \rangle$ is a sequence of club sets in κ, then $\bigcap_{\alpha \in \gamma} D_\alpha$ is club in κ.*

Proof. Let γ, κ, and $\langle D_\alpha : \alpha \in \gamma \rangle$ be as stated. Theorem 9.3.4 thus implies that $\bigcap_{\alpha \in \zeta} D_\alpha$ is closed in κ for every nonzero $\zeta \subseteq \gamma$. Suppose, for a contradiction, that $\bigcap_{\alpha \in \gamma} D_\alpha$ is bounded in κ. Let $\zeta \subseteq \gamma$ be the least such that $\zeta \neq 0$ and ($\blacktriangle$) $\bigcap_{\alpha \in \zeta} D_\alpha$ is bounded in κ. Theorem 9.3.7 implies that ζ is a limit ordinal (see case 1 of the above proof). Hence, for all $\alpha \in \zeta$, the set $C_\alpha = \bigcap_{\xi \in \alpha^+} D_\xi$ is club in κ. So $\langle C_\alpha : \alpha \in \zeta \rangle$ is a nested sequence of club sets (see Exercise 5). Since $\bigcap_{\alpha \in \zeta} D_\alpha = \bigcap_{\alpha \in \zeta} C_\alpha$, Lemma 9.3.9 implies that $\bigcap_{\alpha \in \zeta} D_\alpha$ is unbounded in κ, contradicting ($\blacktriangle$). □

Theorem 9.3.10 may not hold for a κ-sequence of club sets in κ. To verify this, let $\langle C_\alpha : \alpha \in \kappa \rangle$ be such that $C_\alpha = \{\beta \in \kappa : \alpha \in \beta\}$ where κ is a limit ordinal. Every C_α is club in κ; however, $\bigcap_{\alpha \in \kappa} C_\alpha$ is not club in κ because $\bigcap_{\alpha \in \kappa} C_\alpha = \varnothing$, which is bounded in κ. On the other hand, when κ is a regular uncountable cardinal, we will soon show that the "diagonal intersection" of a

κ-sequence of club in κ sets is also a club set. First, in the interest of notational correctness, we define $\bigcap_{\xi \in 0} C_\xi = \varnothing$.

Definition 9.3.11. Let $\langle C_\alpha : \alpha \in \kappa \rangle$ be a sequence of subsets of a limit ordinal κ. Then $\{\alpha \in \kappa : \alpha \in \bigcap_{\xi \in \alpha} C_\xi\}$ is the **diagonal intersection** of the sequence.

Lemma 9.3.12. *Let κ be a limit ordinal. If $\langle C_\alpha : \alpha \in \kappa \rangle$ is a sequence of closed sets in κ, then the diagonal intersection $\{\alpha \in \kappa : \alpha \in \bigcap_{\xi \in \alpha} C_\xi\}$ is closed in κ.*

Proof. Let κ and $\langle C_\alpha : \alpha \in \kappa \rangle$ be as stated in the lemma. We will prove that the set $D = \{\alpha \in \kappa : \alpha \in \bigcap_{\xi \in \alpha} C_\xi\}$ is closed in κ. Let $X \neq \varnothing$ be a bounded subset of D. Thus, $(\blacktriangle)\, X \subseteq D \cap \gamma$ for some $\gamma \in \kappa$. If X is not a limit set, then $\sup X \in D$ (see case 2 in the proof of Lemma 9.3.3). If X is a limit set, then let $\delta = \sup X$. By $(\blacktriangle)$, we have $\delta \in \kappa$. Lemma 8.2.12 implies that δ is a limit ordinal and that X is cofinal in δ. We now show that $\delta \in D$, that is, we prove the following claim.

Claim. $\delta \in \bigcap_{\xi \in \delta} C_\xi$.

Proof. Let $\xi \in \delta$. We will prove that $\delta \in C_\xi$. Let $X_\xi = \{\alpha \in X : \xi \in \alpha \in \delta\}$. It follows that X_ξ is cofinal in X. Hence, $\sup X_\xi = \delta$ by Lemma 8.2.14. Because $\delta \in \kappa$, we see that X_ξ is bounded in κ. We now show that $X_\xi \subseteq C_\xi$. Let $\alpha \in X_\xi$. So $\xi \in \alpha$ by the definition of X_ξ. Since $X_\xi \subseteq D$, it follows that $\alpha \in \bigcap_{\xi' \in \alpha} C_{\xi'}$. Hence, $\alpha \in C_\xi$ and thus, $X_\xi \subseteq C_\xi$. As X_ξ is a nonempty bounded subset of C_ξ, Lemma 9.3.3 implies that $\delta \in C_\xi$. Therefore, $\delta \in \bigcap_{\xi \in \delta} C_\xi$.   (Claim) □

Since $\sup X = \delta \in D$, Lemma 9.3.3 now implies that D is closed in κ.   □

Lemma 9.3.13. *Let κ be a regular uncountable cardinal. If $\langle C_\alpha : \alpha \in \kappa \rangle$ is a nested sequence of club sets in κ, then $\{\alpha \in \kappa : \alpha \in \bigcap_{\xi \in \alpha} C_\xi\}$ is unbounded in κ.*

Proof. Let $\langle C_\alpha : \alpha \in \kappa \rangle$ be a nested sequence of club sets in κ, where κ is a regular uncountable cardinal. We shall prove that $D = \{\alpha \in \kappa : \alpha \in \bigcap_{\xi \in \alpha} C_\xi\}$ is unbounded in κ. Let $\nu \in \kappa$. Recursively define the sequence $\langle \beta_n : n \in \omega \rangle$ as follows: Let $\beta_0 \in \kappa$ be such that $\nu \in \beta_0$. Then for all $n \in \omega$,

$$\beta_{n+} = \text{the least } \beta \in C_{\beta_n} \text{ such that } \sup\{\beta_i : i \subseteq n\} \in \beta. \tag{9.13}$$

It follows from (9.13) that

(1) $v \in \beta_i \in \beta_j$ whenever $i \in j \in \omega$,
(2) $X = \{\beta_i : i \in \omega\}$ is a limit set,
(3) $X_n = \{\beta_i : n \in i \in \omega\}$ is cofinal in X, for all $n \in \omega$,
(4) $X_n \subseteq C_{\beta_n}$ for all $n \in \omega$, because $\langle C_\alpha : \alpha \in \kappa \rangle$ is nested.

Let $\delta = \sup X$. Lemma 8.2.12 and (1)–(2) imply that $v \in \delta$ and that δ is a limit ordinal. As $\omega \in \kappa$ and κ is regular, we have $\delta \in \kappa$. We will prove that $\delta \in D$. Items (2)–(3) and Lemma 8.2.14 imply that $\sup X_n = \delta$, for each $n \in \omega$. Since $\delta \in \kappa$, we conclude from Lemma 9.3.3 and (4) that $\delta \in C_{\beta_n}$, for each $n \in \omega$. Because $\{\beta_n : n \in \omega\}$ is cofinal in δ and $\langle C_\alpha : \alpha \in \delta \rangle$ is nested, we have that $\delta \in C_\xi$ for every $\xi \in \delta$. Therefore, $\delta \in \bigcap_{\xi \in \delta} C_\xi$. So $v \in \delta \in D$, and thus, D is unbounded. $\qquad\square$

Theorem 9.3.14. *Let κ be a regular uncountable cardinal. If $\langle D_\alpha : \alpha \in \kappa \rangle$ is a κ-sequence of club sets in κ, then $\{\alpha \in \kappa : \alpha \in \bigcap_{\xi \in \alpha} D_\xi\}$ is club in κ.*

Proof. Let $\langle D_\alpha : \alpha \in \kappa \rangle$ and κ be as stated in the theorem. Thus, Lemma 9.3.12 implies that the set $\{\alpha \in \kappa : \alpha \in \bigcap_{\xi \in \alpha} D_\xi\}$ is closed in κ. For each $\alpha \in \kappa$, let $C_\alpha = \bigcap_{\xi \in \alpha^+} D_\xi$, which is club in κ by Theorem 9.3.10. Hence, $\langle C_\alpha : \alpha \in \kappa \rangle$ is a nested sequence of club sets. Because (see Exercise 5)

$$\{\alpha \in \kappa : \alpha \in \bigcap_{\xi \in \alpha} D_\xi\} = \{\alpha \in \kappa : \alpha \in \bigcap_{\xi \in \alpha} C_\xi\},$$

Lemma 9.3.13 implies that $\{\alpha \in \kappa : \alpha \in \bigcap_{\xi \in \alpha} D_\xi\}$ is unbounded in κ. $\qquad\square$

The Club Filter

Since the intersection of two club sets is again a club set, the club sets generate a natural filter.

Definition 9.3.15. Let κ be regular uncountable cardinal. Define $\mathcal{F} \subseteq \mathcal{P}(\kappa)$ by

$$\mathcal{F} = \{X \subseteq \kappa : C \subseteq X \text{ for some } C \text{ that is club in } \kappa\}.$$

The set $\mathcal{F}$ is called the **club filter** on κ.

Using Theorem 9.3.7, one can easily verify that $\mathcal{F}$ is a filter (see Exercise 8).

Theorem 9.3.16 (AC). *Let $\mathcal{F}$ be the club filter on κ, a regular uncountable cardinal. Then $\mathcal{F}$ is κ-**complete**; that is, whenever $\langle X_\alpha : \alpha \in \gamma \rangle$ is a γ-sequence of sets in $\mathcal{F}$ such that $0 \in \gamma \in \kappa$, then $\bigcap_{\alpha \in \gamma} X_\alpha \in \mathcal{F}$.*

Proof. Let $\mathcal{F}$ be the club filter on κ, a regular uncountable cardinal. Suppose that $\langle X_\alpha : \alpha \in \gamma \rangle$ is a γ-sequence of sets in $\mathcal{F}$ where $0 \in \gamma \in \kappa$. Definition 9.3.15 and the axiom of choice imply that there is a γ-sequence $\langle C_\alpha : \alpha \in \gamma \rangle$ of club sets in κ such that $C_\alpha \subseteq X_\alpha$, for all $\alpha \in \gamma$. Theorem 9.3.10 asserts that $\bigcap_{\alpha \in \gamma} C_\alpha$ is club in κ. Since $\bigcap_{\alpha \in \gamma} C_\alpha \subseteq \bigcap_{\alpha \in \gamma} X_\alpha$, we thus conclude that $\bigcap_{\alpha \in \gamma} X_\alpha \in \mathcal{F}$ by Definition 9.3.15. □

The above proof can be modified so as to show that a club filter is also closed under diagonal intersection.

Theorem 9.3.17 (AC). *Let $\mathcal{F}$ be the club filter on κ, a regular uncountable cardinal. Then $\mathcal{F}$ is **normal**; that is, whenever $\langle X_\alpha : \alpha \in \kappa \rangle$ is a κ-sequence of sets in $\mathcal{F}$, then $\{\alpha \in \kappa : \alpha \in \bigcap_{\xi \in \alpha} X_\xi\} \in \mathcal{F}$.*

Proof. See Exercise 9. □

Stationary Sets

The definition of a stationary set is due to Gérard Bloch (1953). Stationary sets permeate combinatorial set theory.

Definition 9.3.18. Let κ be a regular uncountable cardinal and let $S \subseteq \kappa$. Then S is **stationary** in κ if and only if $S \cap C \neq \varnothing$ for every club in κ set C.

A subset S of κ is stationary if and only if S has elements in common with every closed and unbounded subset of κ. For any $\gamma \in \kappa$, the set $\kappa \setminus \gamma^+$ is club in κ. Hence, there is an $\alpha \in S$ such that $\gamma \in \alpha$ whenever S is stationary in κ. So a stationary set is unbounded in κ. Also note that if S is stationary in κ, $\alpha \in \kappa$, and $S \subseteq T \subseteq \kappa$, then $S \setminus \alpha$ and T are also stationary in κ (see Exercise 2).

Theorem 9.3.7 implies that every club set is stationary in κ; however, there are stationary sets that are not club in κ. For example, one can easily show that the set $S = \kappa \setminus \{\omega\}$ is stationary in κ. Yet S is not closed, because $\omega \subseteq S$ and $\sup \omega = \omega \notin S$. Thus, S is a stationary set that is not a club. One can prove (using AC) that there exist two stationary sets that are disjoint. It thus follows that the club filter is not an ultrafilter (see Exercise 18).

Our next lemma makes a simple observation concerning Definition 9.3.18.

Lemma 9.3.19. *Let κ be a regular uncountable cardinal. Let $S \subseteq \kappa$. Then S is not stationary in κ if and only if there is a club in κ set C such that $S \cap C = \varnothing$.*

The following fundamental theorem was proved by Géza Fodor in 1956. Let κ be a regular uncountable cardinal and let $S \subseteq \kappa$. A function $f : \kappa \to \kappa$ is said to be **regressive** on S if $f(\alpha) \in \alpha$ for all $\alpha \in S \setminus \{0\}$. Fodor proved that if a function $f : \kappa \to \kappa$ is regressive on a stationary subset of κ, then f is thereby constant on a stationary set.

Theorem 9.3.20 (AC). *Let κ be a regular uncountable cardinal. Suppose that a function $f : \kappa \to \kappa$ is regressive on a stationary set. Then there exists an $\xi \in \kappa$ such that $\{\alpha \in \kappa : f(\alpha) = \xi\}$ is stationary.*

Proof. Let κ be a regular uncountable cardinal and let $f : \kappa \to \kappa$. Suppose that f is regressive on a stationary set. Hence, $S = \{\alpha \in \kappa : f(\alpha) \in \alpha\}$ is stationary. Assume, for a contradiction, that for all $\xi \in \kappa$ the set $X_\xi = \{\alpha \in \kappa : f(\alpha) = \xi\}$ is not stationary. Lemma 9.3.19 and the axiom of choice imply that there is a κ-sequence $\langle C_\xi : \xi \in \kappa \rangle$ of club sets such that ($\blacktriangle$) $C_\xi \cap X_\xi = \varnothing$, for all $\xi \in \kappa$. By Theorem 9.3.14, the diagonal intersection $D = \{\alpha \in \kappa : \alpha \in \bigcap_{\xi \in \alpha} C_\xi\}$ is club in κ. Since S is stationary and D is a club, there exists an $\alpha \in \kappa$ such that $\alpha \in S \cap D$. Thus, (i) $f(\alpha) \in \alpha$ and (ii) $\alpha \in \bigcap_{\xi \in \alpha} C_\xi$. So ($\blacktriangle$) and (ii) imply that $f(\alpha) \neq \xi$ for all $\xi \in \alpha$, which contradicts (i). $\qquad\square$

Because every set that is stationary in κ is also unbounded in κ, we have our next corollary. This corollary will then be followed by a proof of its converse.

Corollary 9.3.21 (AC). *Let κ be a regular uncountable cardinal. Suppose that E is stationary in κ. Then for all $f : \kappa \to \kappa$, if f is regressive on E, then there exists an $\xi \in \kappa$ such that $\{\alpha \in \kappa : f(\alpha) = \xi\}$ is unbounded in κ.*

Theorem 9.3.22. *Let κ be a regular uncountable cardinal and $E \subseteq \kappa$. Suppose that for all $f : \kappa \to \kappa$, if f is regressive on E, then there exists an $\xi \in \kappa$ such that $\{\alpha \in \kappa : f(\alpha) = \xi\}$ is unbounded in κ. Then E is stationary in κ.*

Proof. Let κ and E be as stated in the theorem. Assume that for all $f : \kappa \to \kappa$, if f is regressive on E, then there exists an $\xi \in \kappa$ such that $\{\alpha \in \kappa : f(\alpha) = \xi\}$ is unbounded in κ. Suppose, for a contradiction, that E is not stationary. Thus, by Lemma 9.3.19, there is club in κ set C so that $E \cap C = \varnothing$. Define $g: \kappa \to \kappa$ by $g(\alpha) = \sup(C \cap \alpha)$. Clearly, $g(\alpha) \subseteq \alpha$ for all $\alpha \in \kappa$. Since C is closed and $E \cap C = \varnothing$, it follows that $g(\alpha) \in \alpha$ for all $\alpha \in E \setminus \{0\}$. Hence, g is regressive on E. By our assumption, there is an $\xi \in \kappa$ such that $S = \{\alpha \in \kappa : g(\alpha) = \xi\}$ is unbounded in κ. Since C is unbounded, let $\gamma \in C$ be such that $\xi \in \gamma$. Because S is also unbounded, there is a $\beta \in S$ such that $\gamma \in \beta$. So $\gamma \in C$ and $\xi \in \gamma \in \beta$. Thus, by the definition of g, we have that $\xi \in g(\beta)$. Since $\beta \in S$, we also have that $g(\beta) = \xi$. This contradiction completes the proof. $\square$

The following definition will be used in some of the exercises.

Definition 9.3.23. Let κ be a regular uncountable cardinal and let $f: \kappa \to \kappa$. Then f is said to be **normal** if the following two conditions hold:

(1) $f(\alpha) \in f(\beta)$ whenever $\alpha \in \beta \in \kappa$ (f is strictly increasing),
(2) $f(\beta) = \sup\{f(\eta) : \eta \in \beta\}$ for every limit ordinal $\beta \in \kappa$ (f is continuous).

Exercises 9.3

1. Let C and D be closed sets in κ, a limit ordinal. Suppose that $X \subseteq C$ and $\sup X \in C$. Let $Y \subseteq X$ be nonempty. Show that if $Y \subseteq D$, then $\sup Y \in D$.

*2. Let κ be a regular uncountable cardinal and let $\alpha \in \kappa$. Show that if C is club in κ, then $C \setminus \alpha$ is club in κ. Show that if S is stationary in κ, then $S \setminus \alpha$ is stationary in κ.

3. Let $C = \{\alpha \in \kappa : \alpha \text{ is a limit ordinal}\}$ where κ is a regular uncountable cardinal. Show that C is club in κ.

4. Let $D = \{\alpha \in C : \alpha \text{ is a limit ordinal}\}$ where C is a club set in κ, a regular uncountable cardinal. Show that D is club in κ.

*5. Let ζ be a limit ordinal and let $\langle D_\alpha : \alpha \in \zeta \rangle$ be any sequence of sets. For each $\alpha \in \zeta$, let $C_\alpha = \bigcap_{\xi \in \alpha^+} D_\xi$.
(a) Prove that $\langle C_\alpha : \alpha \in \zeta \rangle$ is a nested sequence of sets.
(b) Prove that $\bigcap_{\xi \in \alpha} D_\xi = \bigcap_{\xi \in \alpha} C_\xi$, for all $\alpha \subseteq \zeta$.
(c) Prove that $\{\alpha \in \zeta : \alpha \in \bigcap_{\xi \in \alpha} D_\xi\} = \{\alpha \in \zeta : \alpha \in \bigcap_{\xi \in \alpha} C_\xi\}$.

6. Let $\langle C_\alpha : \alpha \in \kappa \rangle$ be such that $C_\alpha = \{\beta \in \kappa : \alpha \in \beta\}$ where κ is a limit ordinal. Evaluate $\{\alpha \in \kappa : \alpha \in \bigcap_{\xi \in \alpha} C_\xi\}$.

7. Let ζ be a limit ordinal and let $\langle C_\alpha : \alpha \in \zeta \rangle$ be a sequence of subsets of ζ. Prove that $\bigcap_{\alpha \in \zeta} C_\alpha \subseteq \{\alpha \in \zeta : \alpha \in \bigcap_{\xi \in \alpha} C_\xi\}$.

*8. Let κ be regular uncountable cardinal and let $\mathcal{F} \subseteq \mathcal{P}(\kappa)$ be the club filter on κ. Prove the following items:
   (a) $\varnothing \notin \mathcal{F}$ and $\kappa \in \mathcal{F}$.
   (b) If $Y \in \mathcal{F}$ and $Z \in \mathcal{F}$, then $Y \cap Z \in \mathcal{F}$.
   (c) If $Y \in \mathcal{F}$ and $Y \subseteq Z \subseteq \kappa$, then $Z \in \mathcal{F}$.

*9. Prove Theorem 9.3.17.

10. Let $\mathcal{F}$ be the club filter on κ, a regular uncountable cardinal. Suppose that $\langle X_\alpha : \alpha \in \gamma \rangle$ is a γ-sequence of sets in $\mathcal{F}$ where $\gamma \subseteq \kappa$. Explicitly explain how the axiom of choice (page 66) implies that there exists a γ-sequence $\langle C_\alpha : \alpha \in \gamma \rangle$ of club sets in κ such that $C_\alpha \subseteq X_\alpha$, for all $\alpha \in \gamma$.

11. Let κ be a regular uncountable cardinal and let $f \colon \kappa \to \kappa$. Prove that if f is normal, then the set $\{\alpha \in \kappa : f(\alpha) = \alpha\}$ is club in κ.

12. Let $f \colon \kappa \to \kappa$ be normal, where κ is regular uncountable cardinal. Prove that $f[\kappa] = \{f(\alpha) : \alpha \in \kappa\}$ is club in κ.

13. Let C be club in κ, a regular uncountable cardinal. Let $f \colon \kappa \to C$ be defined by the following recursion (see Theorem 8.2.2): For all $\alpha \in \kappa$,

$$f(\alpha) = \text{the least } \beta \in \kappa \text{ such that } \beta \in C \setminus f[\alpha].$$

Prove that f is normal, and that $f[\kappa] = C$.

14. Let κ be a regular uncountable cardinal and let $f \colon \kappa \to \kappa$. Prove that the set $\{\alpha \in \kappa : f[\alpha] \subseteq \alpha\}$ is club in κ.

15. Let κ be a regular uncountable cardinal. Suppose that S is stationary in κ and C is club in κ. Prove that $S \cap C$ is stationary.

16. (AC) Suppose that κ is a regular uncountable cardinal and that $\gamma \in \kappa$.
   (a) Let $\langle S_\alpha : \alpha \in \gamma \rangle$ be a γ-sequence of subsets of κ. Suppose that the set $\bigcup_{\alpha \in \gamma} S_\alpha$ is stationary in κ. Prove that S_α is stationary, for some $\alpha \in \gamma$.
   (b) Let $f \colon \kappa \to \gamma$. Prove that f is constant on a stationary set; that is, show that there is an $\alpha \in \gamma$ such that $\{\xi \in \kappa : f(\xi) = \alpha\}$ is stationary in κ.

17. (AC) Let κ be a regular uncountable cardinal. Suppose that $\langle S_\alpha : \alpha \in \kappa \rangle$ is a κ-sequence of subsets of κ. Prove that if $\{\alpha \in \kappa : \alpha \in \bigcup_{\xi \in \alpha} S_\xi\}$ is stationary, then S_α is stationary, for some $\alpha \in \kappa$.

*__18.__ Let $\mathcal{F}$ be the club filter on κ, a regular uncountable cardinal. Suppose that S and $S \setminus \kappa$ are both stationary in κ. Prove that $\mathcal{F}$ is not an ultrafilter.

19. (AC) Let $\mathcal{F}$ be the club filter on κ, a regular uncountable cardinal. Suppose that S is stationary in κ. Prove that there exists an ultrafilter $\mathcal{U}$ on κ such that $\mathcal{F} \subseteq \mathcal{U}$ and $S \in \mathcal{U}$.

NOTES

Preface
1. In naive set theory, a set is any well-defined collection of objects.

Chapter 1
1. In mathematics, "or" always means the inclusive "or," that is, we allow for both conditions to hold (see page 7). Thus, if $x \in A$ and $x \in B$, then $x \in A \cup B$.
2. In axiom 3 (8), the variable S is chosen so that it does not appear in the formula $\varphi\,(\psi)$.

Chapter 2
1. When proving $X = Y$, we shall say that X is the first set and Y is the second set.
2. Recall that the elements of a set are also sets.
3. The arrow $\Rightarrow$ shall be used to abbreviate the word "implies."

Chapter 3
1. For posets, we shall write $(A, \preccurlyeq)$ for the ordered pair $\langle A, \preccurlyeq \rangle$.

Chapter 4
1. The label (IH) will often be used to identify the induction hypothesis.
2. The material covered in this subsection will not be used elsewhere in the book.

Chapter 8
1. A *parameter* is an unassigned set that may occur in a formula (see page 22).
2. In (8.14), replace α with 0 and replace $f(\gamma)^+$ with $f(\gamma) + \alpha$.
3. In (8.14), replace α with 1 and replace $f(\gamma)^+$ with $f(\gamma) \cdot \alpha$.

Chapter 9
1. See Remark 8.1.8 and Definition 8.1.9 on page 179.
2. See Remark 8.2.18 on page 191.
3. $\aleph$ (aleph) is the first letter in the Hebrew alphabet.

REFERENCES

[1] Cunningham, Daniel W. 2012. *A Logical Introduction to Proof.* New York: Springer.

[2] Dauben, J.W. 1990. *Georg Cantor: His Mathematics and Philosophy of the Infinite.* History of science. Princeton: Princeton University Press.

[3] Dedekind, Richard. 2012. *Was sind und was sollen die Zahlen?* Cambridge: Cambridge University Press. Reprint of the 1893 edition.

[4] Ebbinghaus, Heinz-Dieter. 2007. *Ernst Zermelo. An Approach to His Life and Work.* Berlin: Springer. In cooperation with Volker Peckhaus.

[5] Enderton, Herbert B. 1977. *Elements of Set Theory.* New York: Academic Press.

[6] Enderton, Herbert B. 2001. *A Mathematical Introduction to Logic.* 2nd edn. Burlington, MA: Harcourt/Academic Press.

[7] Halmos, Paul R. 1974. *Naive Set Theory.* New York: Springer. Reprint of the 1960 edition published by Van Nostrand.

[8] Kanamori, Akihiro. 2012. In praise of replacement. *Bulletin of Symbolic Logic,* **18**(1), 46–90.

[9] Kunen, Kenneth. 2009. *The Foundations of Mathematics.* Studies in Logic, vol. 19. London: College Publications.

[10] Levy, Azriel. 2002. *Basic Set Theory.* Mineola, NY: Dover Publications. Reprint of the 1979 original published by Springer.

[11] Moore, Gregory H. 2012. *Zermelo's Axiom of Choice: Its Origins, Development, and Influence.* Mineola, NY: Dover Publications. Reprint of the 1982 original published by Springer.

[12] Moschovakis, Yiannis. 2006. *Notes on Set Theory.* 2nd edn. Undergraduate Texts in Mathematics. New York: Springer.

[13] Peano, Giuseppe. 1889. *Arithmetices Principia: Nova Methodo Exposita.* Turin: Fratres Bocca.

[14] Potter, Michael. 2004. *Set Theory and Its Philosophy.* New York: Oxford University Press.

INDEX OF SPECIAL SYMBOLS

INDEX